International Conference on Environmental and Sustainability Context

IOP Conference Series: Earth and Environmental Science
Volume 894

Online
23 August 2021

ISBN: 978-1-7138-4876-9
ISSN: 1755-1307

Printed from e-media with permission by:

Curran Associates, Inc.
57 Morehouse Lane
Red Hook, NY 12571

Some format issues inherent in the e-media version may also appear in this print version.

Institute of Physics
Dirac House, Temple Back
Bristol BS1 6BE UK

Phone: 44 1 17 929 7481
Fax: 44 1 17 920 0979

techtracking@iop.org

Additional copies of this publication are available from:

Curran Associates, Inc.
57 Morehouse Lane
Red Hook, NY 12571 USA
Phone: 845-758-0400
Fax: 845-758-2633
Email: curran@proceedings.com
Web: www.proceedings.com

TABLE OF CONTENTS

IOP Conf. Series: Earth and Environmental Science **894** (2021) 011001 doi:10.1088/1755-1315/894/1/011001

Preface

The International Conference on Environmental and Sustainability Context (ICoESCo) 2021 is the first international conference hosted by Environmental Sustainability Research Group (EnSi-RG) and co-hosted by Environmental Research Center, Lembaga Penelitian Universitas Trisakti, Environmental Engineering Department, Faculty of Engineering, Universitas Diponegoro, and Kitakyushu Indonesian Students' Union. This conference was successfully held on August 24, 2021 which brings the theme of "Emerging Challenges and Opportunities for Achieving Environmental Sustainability" since the theme of "*sustainability*" has become the most crucial issue of the world by now. This conference was attended by 81 participants that came from 34 universities across the world which include as follows.

1. Indonesian Universities and Institutions
    1) PT Sampoerna Agro Tbk
    2) Center for Advanced Material Science and Technology, Serpong,
    3) Universitas Diponegoro,
    4) Universitas Andalas,
    5) Universitas Islam Negeri Sunan Ampel Surabaya,
    6) Institut Pertanian Bogor,
    7) Universitas Sumatera Utara,
    8) Universitas Pasundan,
    9) Universitas Negeri Jakarta,
    10) Universitas Agung Podomoro,
    11) Universitas Trisakti,
    12) Universitas Gajah Mada,
    13) Universitas Gunadarma,
    14) Universitas Islam Negeri Sultan Syarif Kasim Riau,
    15) Institut Teknologi Sebelas Nopember,
    16) Universitas Jambi,
    17) Universitas Negeri Malang,
    18) Universitas Brawijaya,
    19) Universitas Bina Nusantara,
    20) Bogor Agricultural University
    21) Health Polytechnic, Ministry of Health, Indonesia
    22) Institut Teknologi Sepuluh Nopember,
    23) Institut Teknologi Nasional, and
    24) Institut Teknologi Pembangunan Surabaya.
2. Overseas universities
    1) Ndejje University – Uganda,
    2) International University of East Africa – Uganda,
    3) RWTH Aachen University – Germany,
    4) Mardoch University – Australia,
    5) Hochschule Ruhr West – Germany,

6) The University of Haripur – Pakistan,
7) Hydro-Quebec Institute of Research - Canada,
8) Universiti Tun Hussein Onn Malaysia - Malaysia
9) Helmholtz Centre for Environmental Research – Germany, and
10) The University of Kitakyushu – Japan.

The discussions cover some upcoming topics such as solid waste management and treatment, water and wastewater recycling, climate change and global warming, environmental management and economic, renewable energy, environmental health and risk, environmental education, nanotechnology and artificial intelligence and big data for environment.

Since the number of COVID-19 cases in Indonesia is still high in July – August 2021, the committee decided to hold the conference in a virtual setting. The committee used the Zoom Application to ensure the delivery of the research material could be done smoothly. The meeting is part of the annual meeting of the scientific consortium initiated by the Environmental Sustainability Research Group. Because of the importance of this meeting, the committee should conduct this conference even in a virtual situation. The invited keynote speakers all are as follows: Dr. M.V. Reddy (Hydro-Quebec Institute of Research, Canada); Dr. Astri Rinanti Nugroho, M.T. (Universitas Trisakti, Indonesia); Dr. Kamran Azam (The University of Haripur, Pakistan) and Prof. Hiroyuki Miyake (The University of Kitakyushu, Japan) who is divided into 2 plenary sessions. The speakers talked for 25 minutes and continued with 15 minutes of discussion. The committee used the breakout room function of the Zoom Application for facilitating an actual conference situation. The detailed of plenary, parallel, discussion, question, and answers sessiom are posted in the website https://www.icoesco.site/. There are 4 breakout rooms where 11-12 presenters were delivering their research results. The presenter has 10 minutes to present and 5 minutes to ask some questions from the audiences. All of the participants were allowed to join the rooms they wanted to. Even in a pandemic situation and done online, this conference is expected to give an interactive forum for sharing and exchanging knowledge on the latest and potential research collaboration in environmental science, engineering, education, and health. There are several scientific manuscripts produced through this conference which most of them are sent to the IOP Conference Series: Earth and Environmental Science for further process. The paper consists of the broad topic of environmental management, engineering, and science. All of the manuscripts are peer-reviewed by the expertise coming from the different scientific research fields.

The committee wants to thank all of the participants coming to this conference. The committee also wants to express the deepest gratitude to the keynote speakers, reviewers, and moderators who contribute to this conference's success. The committee also wants to acknowledge IOPP for publishing our conference proceedings. It is hoped that the readers can feel and extract the knowledge and get some memorable experience and information during the exploration of our scientific manuscript. The committee also wants to apologize for all of the mistakes found in the conference and the published papers.

Scientific Committee of ICoESCo 2021

IOP Conf. Series: Earth and Environmental Science **894** (2021) 011001 doi:10.1088/1755-1315/894/1/011001

ICOESCO 2021 Committees

Scientific Committee

Prof. Hiroyuki Miyake (The University of Kitakyushu, Japan)
Prof. Chafid Fandeli (Insitut Teknologi Yogakarta, Indonesia)
Prof. Syafrudin (EnSi-RG, Universitas Diponegoro, Indonesia)
Dr. M.V. Reddy (Hydro-Quebec Institute of Research, Canada)
Dr. Kamran Azam (The University of Haripur, Pakistan)
Dr. Nitish Pathak (BPIT, GGSIPU, India)
Dr. Astri Rinanti Nugroho, M.T.(Universitas Trisakti, Indonesia)
Dr. Yonik Meilawati Kustiani (Universitas Pasundan, Indonesia)
Ira Rumiris Hutagalung, M.Sc. (Universitas Islam Syekh Yusuf, Indonesia) Mochamad Arief Budihardjo, Ph.D. (EnSi-RG, Universitas Diponegoro, Indonesia) Dr. Badrus Zaman, M.T. (EnSi-RG, Universitas Diponegoro, Indonesia)

Organizing Committee

Dr. Yustina Metanoia Pusparizkita (EnSi-RG, Universitas Diponegoro, Indonesia) Bimastyaji Surya Ramadan, S.T., M.T. (EnSi-RG, Universitas Diponegoro, Indonesia) Indriyani Rachman, Ph.D. (The University of Kitakyushu, Japan)
Dr. Rini Setiati, M.T. (Universitas Trisakti, Indonesia)
Yudha Gusti Wibowo, S.Si., M.Si. (EnSi-RG, Universitas Diponegoro, Indonesia) Dendy Primanandi, S.T., M.T. (EnSi-RG, Universitas Diponegoro, Indonesia) Alya Karmilia, S.T. (EnSi-RG, Universitas Diponegoro, Indonesia)
Agung Santoso Wibowo, S.T. (Universitas Diponegoro, Indonesia)
Mellinia Ramadhani, S.T. (Universitas Diponegoro, Indonesia)
Reni Artati, S.T. (Universitas Diponegoro, Indonesia)

IOP Conf. Series: Earth and Environmental Science **894** (2021) 011002 doi:10.1088/1755-1315/894/1/011002

Photographs

Figure 1 Plenary Session

Figure 2 Closing Session

IOP Conf. Series: Earth and Environmental Science **894** (2021) 011002 doi:10.1088/1755-1315/894/1/011002

Figure 3 Closing Ceremony

IOP Conf. Series: Earth and Environmental Science **894** (2021) 011003 doi:10.1088/1755-1315/894/1/011003

Peer review declaration

All papers published in this volume of IOP Conference series: Earth and Environmental Science have been peer reviewed through processes administrated by the Editors. Reviews were conducted by expert referees to the professional and scientific standards expected of a proceedings journal published by IOP Publishing.

- **Type of peer review:** Single-blind
- **Conference submission management system:** Google Form
- **Number of submissions received:** 74
- **Number of submissions sent for review:** 59
- **Number of submissions accepted:** 46
- **Acceptance Rate (Number of Submissions Accepted / Number of Submissions Received X 100):** 62 %
- **Average number of reviews per paper:** 2
- **Total number of reviewers involved:** 10
- **Any additional info on review process:**
  Review Timeline
  1. The manuscripts were send to the guest editors for initial checking, including format, similarity, and language quality
  2. After manuscript acceptance by guest editor, the manuscripts were send to the reviewer for substantial evaluation.
  3. Among 59 manuscripts that were reviewed, 46 manuscripts were accepted for submitting to IOP EES.
  4. The manuscripts were sent back to the authors
  5. The authors sent the manuscripts for confirming the revisions to the guest editor
  6. The guest editor checked again the revised manuscript and asked the authors for signing the transfer agreement and answering the editorial comment (if any)
  7. The guest editor performed layout, language, and format checked
  8. The manuscripts were transferred to IOP EES

1st round =	
- Manuscript submission	= Jun 20, 2021 – Jul 7, 2021
- Editor check	= Jul 7, 2021 – Jul 9, 2021
- Reviewers handling	= Jul 9, 2021 – Aug 3, 2021
- Manuscript sent back to authors	= Aug 3, 2021
- Manuscript revised by the authors	= Aug 3, 2021 – Aug 20, 2021
2nd round =	
- Manuscript submission	= Jul 10, 2021 – Aug 5, 2021
- Editor check	= Aug 5, 2021 – Aug 7, 2021
- Reviewers handling	= Aug 7, 2021 – Aug 13, 2021

IOP Conf. Series: Earth and Environmental Science **894** (2021) 011003 doi:10.1088/1755-1315/894/1/011003

- Manuscript sent back to authors = Aug 13, 2021
- Manuscript revised by the authors = Aug 13, 2021 – Aug 30, 2021

Editor handling after revision = Aug 30, 2021 – Sep 18, 2021
Manuscripts sent back to authors = Sep 18, 2021 – Sep 26, 2021
Editor handling = Sep 26, 2021 – Oct 1, 2021
The manuscripts were sent to IOP = Oct 7, 2021

Notes:
Single-blind review means the manuscripts were reviewed by one peer-reviewer. The proofreader and layout editor are excluded for content review.

- **Contact person for queries:**
  Mochamad Arief Budihardjo (m.budihardjo@ft.undip.ac.id)
  Environmental Sustainability Research Group (EnSi-RG), Department of Environmental Engineering, Faculty of Engineering, Universitas Diponegoro

IOP Conf. Series: Earth and Environmental Science **894** (2021) 012001 doi:10.1088/1755-1315/894/1/012001

Effect of tannery wastewater exposure on chromium detected in the meat of tilapia (*Oreochromis niloticus L*)

T Ihsan[1]*, T Edwin[1] and R Kartika[1]

[1]Department of Environmental Engineering, Faculty of Engineering, Universitas Andalas, Kampus Unand Limau Manis, Padang, West Sumatra 25163, Indonesia

taufiqihsan@eng.unand.ac.id

Abstract. This study aimed to analyze the bioaccumulation of chromium (Cr) in the meat of tilapia (*Oreochromis niloticus L*) due to exposure to tannery wastewater. The wastewater concentration variations were 1.85% and 3.69% of the LC50 96-h. The study was carried out in triples, seven days of acclimatization, and 30 days of observation (exposed to wastewater). The measurement of Cr using Atomic Absorption Spectrophotometry (AAS) and calculation of BCF states the level of bioaccumulation in fish meat. The result showed the BCF value at a concentration of 1.85%, and 3.69% ranged 2.667-40.984, and 2.667-85.651 consecutively. It concluded that tilapia has a low accumulation rate.

1. Introduction

Chromium (Cr) was once considered a hazardous heavy metal; nonetheless, it is now used as a tanner in the leather tanning industry. Cr was difficult to depreciate naturally, and it was potentially toxic to fish and other species. [1,2]. It has a long half-life, especially in the body of aquatic biota [3]. Heavy metal exposure was physically and physiologically harmful to the fish, and it bioaccumulated in the fish's body as it grew older and heavier. [4]. Chemical bioaccumulation in water was a critical criterion for assessing an environment's ecology and pollution level. [5]. The ability of fish to collect heavy metals was measured by the bioconcentration factor (BCF). The BCF can be used to determine the biota's ability to absorb and retain contaminants. [6]. Toxicants absorbed by the gills or skin, then spread throughout the body through the circulation, resulting in the accumulation of heavy metals in the meat, are the source of metal accumulation in fish meat. In addition, fish meat is a good source of protein for humans. Fish meat can be harmed by dirty water, and the food cycle might be disrupted. [7].

Despite treatment, one of the largest tannery industries in West Sumatra, Indonesia, has discharged wastewater that exceeds the quality threshold. The effluent was released into the river by this industry. Tilapia (*Oreochromis niloticus L*), one of the river's most important species, was fed tannery wastewater directly. Tilapia has monetary worth since it is consumed by the community. This fish can be found in a variety of freshwater environments. It was not only widely distributed (because of its euryhaline nature) but also temperature and Dissolved Oxygen (DO) tolerant, unaffected by diseases, and able to eat a low-protein diet. The researchers were able to keep tilapia in the lab [8,9]. As a result, the goal of this study was to look into Cr bioaccumulation and compute the Cr bioconcentration factor in tilapia meat.

2. Methodology

The experimental treatment of tilapia is the same as the research of Edwin et al. [10]. We obtained tilapia from fish nurseries in the Bangek River, Padang City, West Sumatra, and acclimatized them seven days before the wastewater exposed fish for experiments. We put the tilapia into the aquarium (equipped with an aerator), then fed the fish three times a day (pellets). We also collected DO, pH, and temperature data for that seven days.

The aquarium consists of three different concentrations of tannery wastewater. We use three aquariums to observe the biological concentration of Cr in fish under each exposure change. The concentration used refers to the LC50 value of 25.85% [11] tanning wastewater in the previous study, which is 1/14 LC50-96 hours (1.85%) and 1/7 LC50-96 hours (3.69%) [12]. We conducted control experiments also, and no wastewater was exposed to the aquarium. The initial Cr concentration was 1 mg/l, the biochemical oxygen demand (BOD) was 53.27 mg/l, the chemical oxygen demand (COD) was 113.2 mg/l, the total suspended solids (TSS) was 64 mg/l, fat and oil It was 11.44 mg/l, ammonia was 1.94 mg/l, sulfide was 1.23 mg/l, and the pH was 9.4. Each aquarium contained five tilapias of 10 – 13 cm [13]. In the main experiment, wastewater exposed fish for 30 days. [14]. We dissected tilapia samples to harvest meat on the 0th, 10th, 20th, and 30th days of observation. Randomly catch fish from the aquarium, conduct a destruction test, and measure the Cr metal concentration on the fish meat using atomic absorption spectroscopy (AAS). After that, the BCF value is calculated as Equation 1 [6]. One-way analysis of variance (ANOVA) was also used to calculate the significance of wastewater concentration to Cr in fish.

$$BCF = C\ organism\ /\ C\ water \qquad (1)$$

3. Results and discussion

Before the bioaccumulation test, the test animals were acclimatized, including measuring dissolved oxygen (DO), pH, and temperature. Based on USEPA data, we discovered that these environmental conditions support tilapia life (2002). The findings reveal that the environmental quality matches the needs of tilapia in order for them to live. The DO level ranged from 6.9 to 8.4 mg/L for each concentration shift during the bioaccumulation test, with no significant change. The pH value under each concentration change is between 7.1 and 8.3, and there is no noticeable change in the bioaccumulation test. The temperature range is between 25.4-27.9°C. Observational data show that the DO, pH, and water temperature during the experiment are still within the recommended range of fish environment.

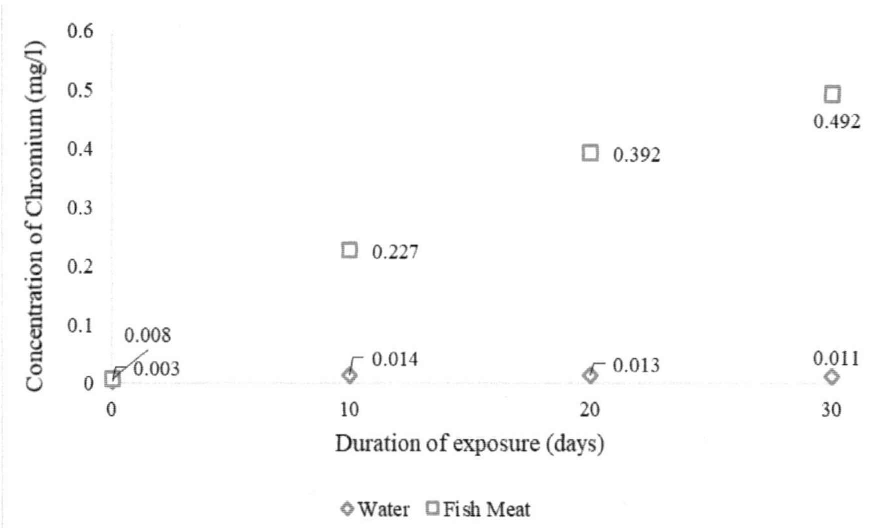

Figure 1. The concentration of chromium in water and fish meat (1,85% exposure).

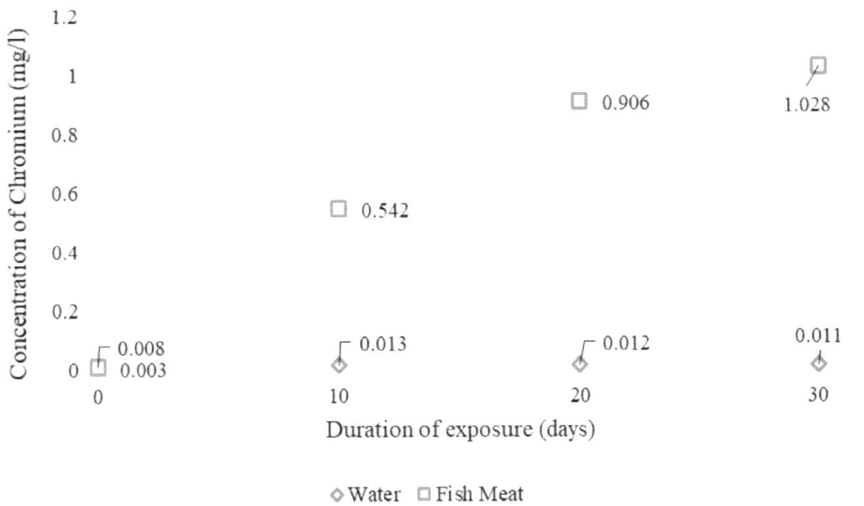

Figure 2. The concentration of chromium in water and fish meat (3.69% exposure).

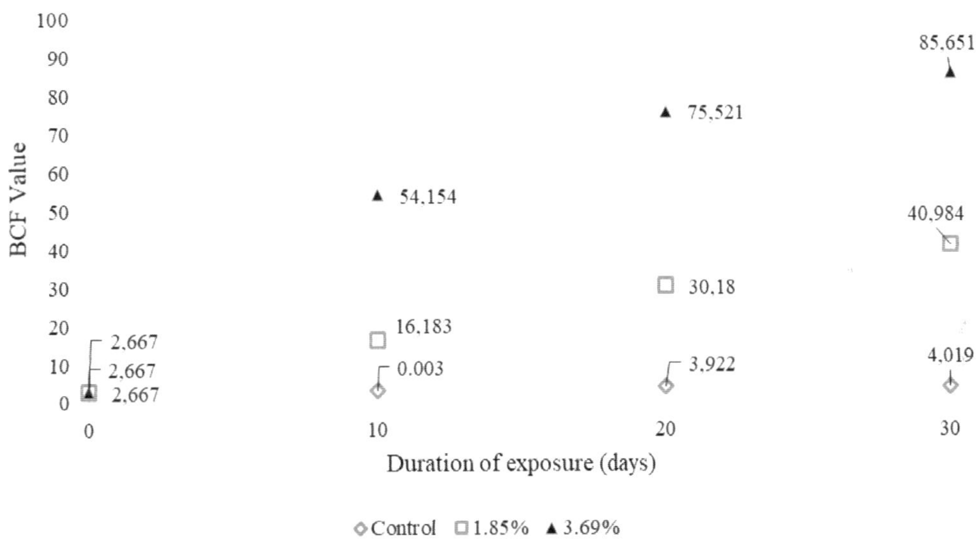

Figure 3. The experiment yielded a bioaccumulation factor.

Figures 1 and 2 show the measurement of Cr level in tilapia meat. In addition, Figure 3 illustrates the BCF calculation results for tilapia. The Cr concentration in fish organs increased with exposure time, while the Cr content in the control experiment was almost zero. The significance test of Cr in meat vs test water, which yielded a result of 0.05, indicates that the test water contains tannery effluent, which contributes to the Cr discovered in the meat experiment. The buildup of aquatic organisms is responsible for the high concentration of Cr in fish. Xenobiotics (e.g., heavy metals, pesticides) are obtained by aquatic organisms from water bodies or sediments and are concentrated 100-1000 times higher than the environment. [15–17].

The Cr concentration in tilapia meat at the beginning of the study (day 0) was 0.008 ± 0.000 mg/kg in three variations, while the concentration of Cr in tilapia meat after the study (30th day) of 0.013 ± 0.003 mg/kg (control), 0.492 ± 0.010 mg/kg (1.85%) and 1.028 ± 0.017 mg/kg (3.69%). In meat with a concentration of 3.69% wastewater on the 30th day, we observed the highest Cr concentration. Based on the Food and Agriculture Organization/FAO (1983), the quality standard for Cr in fish meat is 1 mg/kg. Compared with Cr concentration in control and 1.85%, Cr concentration in meat is still below the specified quality standard. In comparison, at a concentration of 3.69% on the 30th day of exposure, Cr concentration has exceeded the specified quality standard.

The BCF value in tilapia meat at the beginning of the experiment (day 0) was 2.667 ± 0.000 for all variations, while the BCF value in tilapia meat after the 30th day was 4.019±0.870 (control); 40.984±0.798 (1.85%) and 85.651±1.381 (3.69%). The bioconcentration factor value was at <100. It means the capability of chromium bioaccumulation by meat tilapia (throughout sublethal of tannery wastewater exposure to 30 days) is in the low accumulative category.

Although it is low, Cr in tilapia still has to be watched out for because of the cumulative nature of the heavy metal itself. If consumed continuously for an extended period, it will cause chronic poisoning. Cr's chronic poisoning caused by Cr causes damage to the fish body's physiological systems such as the kidneys, lungs, heart, and blood circulation system [10,18].

4. Conclusion

The Cr concentration in tilapia (*Oreochromis niloticus L*) meat increased as exposure time and tannery effluent Cr concentrations increased. The obtained bioconcentration factor value ranges from 4.019 to 85.651, indicating that fish can only accumulate Cr.

References

[1] Ihsan T, Edwin T and Vitri RY 2014 *Jur Dampak* **14(2)** 98-103
[2] Ihsan T, Edwin T, Zupit JR, Ananda SF and Nofrita 2021 *IOP Conf Ser Mater Sci Eng* **1041**, 1
[3] Ali H, Khan E and Ilahi I 2019 *J Chem* **2019** 1-14
[4] Yousif R, Choudhary M I, Ahmed S and Ahmed Q 2021 *Nusant Biosci* **13(1)** 73-84
[5] Ivanciuc T, Ivanciuc O and Klein D J 2006 *Mol Divers* **10(2)** 133-45
[6] Milošković A, Dojcinovic B, Đuretanović SK, Kojadinović NR, Radenković M, Milošević D and Simic VM 2016 *Environ Sci Pollut Res* **23(10)** 9918-33
[7] Bosch A, O'Neill B, Sigge G and Kerwath S 2015 *J Sci Food Agric* **96** 32-48
[8] Agah H, Leermakers M, Elskens M, Fatemi S M R and Baeyens W 2008 *Environ Monit Assess* **157** 499-514
[9] Charo HK 2006 *Selection for Growth of Nile tilapia (Oreochromis niloticus L.) in Low Input Environments* PhD Thesis
[10] Edwin T, Ihsan T and Tamsin H T 2019 *IOP Conf Ser Mater Sci Eng* **602** 1
[11] Edwin T, Ihsan T, Putra M A, and Guspariani G 2018 *Int J Adv Res* **6(5)** 742-8
[12] Halappa R and David M 2009 *Turkish J Fish Aquat Sci* **9(2)** 233-8
[13] Vinodhini R and Narayanan M 2008 *Int J Environ Sci Technol* **5(2)** 179-82
[14] Meena R, Chandrakala R and Indra N 2017 *Int J Appl Res* **3(1)** 468-70
[15] Ihsan T, Edwin T, Paramita D and Frimeli N 2021 *IOP Conf Ser Earth Environ Sci* **623** 1
[16] Ihsan T, Edwin T and Yanti RD, *Nat Environ Pollut Technol* **18**(4) 1399-403
[17] Kaoud H A 2015 *Eur J Acad* **2(9)** 23-6
[18] Kortei N K, Heymann M E, Essuman E K, Kpodo F, Lokpo S Y and Owusu N 2020 *Toxicol Reports* **7** 360-9

IOP Conf. Series: Earth and Environmental Science **894** (2021) 012002 doi:10.1088/1755-1315/894/1/012002

Implementation of criminal law enforcement concept of environmental sustainability (illegal logging in Indonesia)

I Rosyadi[1], M R Habibi[1], N Syam[2]

[1] Law Science, Faculty of Syariah and Law, Islamic State University Sunan Ampel, Surabaya, East Java, Indonesia
[2] Sociology, Faculty of Dakwah and Communication, Islamic State University Sunan Ampel, Surabaya, East Java, Indonesia

afithor@gmail.com

Abstract. Everyone in Indonesia has the right to a good and healthy environment as a form of human rights. Every generation has obligations and responsibilities in preserving the background to ensure the welfare and quality of life between ages. Sustainable development is an effort to guarantee these rights by managing natural resources wisely and rationally and simultaneously considering economic, social, and environmental aspects. Currently, humans tend to be greedy for existing natural resources to get the maximum benefit. Illegal logging is an activity to place forest resources on a large scale without paying attention to the needs of future generations. Efforts made by the government to enforce sanctions against perpetrators of illegal logging activities are the enactment of Law Number 32 of 2009 concerning Environmental Protection and Management and Law Number 18 of 2013 concerning Prevention and Eradication of Forest Destruction.

1. Introduction

The birth of the concept of "sustainable development" originated from the rise of pollution or environmental damage, which is fear to impact the current generation and future generations [1]. So that the impact is expected to be prevented, it is necessary to develop wisely and rationally that can utilise natural resources to fulfil their needs without harming the needs of future generations [2]. "Pay a attention to" and "considering" the current environmental capability in implementing sustainable development is very important to support human life [3].

The concept of sustainable development started in 1972 in Conference Stockholm or the UN Conference on the Human Environment that is expected to be carried out with attention and does not rule out environmental factors, because if the construction of the capability of the environment (eco-development) then it will be done well. The Stockholm Conference also discussed environmental problems and their solutions. With the holding of the Stockholm Conference as an effort from the United Nations to prevent and repair environmental damage [3].

Along with the times, modern humans have very high consumptive power and are very capitalistic, so that in meeting their needs, humans tend to be greedy for existing natural resources and ignore the needs of future generations can cause environmental damage [3]. Therefore, human safety and security can threat due to the lack of prudence or human intent in treating the environment to obtain the most significant possible benefit by ignoring the environment's carrying capacity. It is a severe problem, so it is necessary to protect the environment to reduce environmental abuse due to human activities [4].

In Indonesia, in utilising and protecting the environment, this has progressed with Law Number 4 of 1982 concerning Basic Provisions for Environmental Management or UULH, which made it the first step taken by the government to reduce environmental pollution damage. The UULH was replaced by Law Number 23 of 1997 concerning Environmental Management or UUPLH and then replaced by Law Number 32 of 2009 concerning Environmental Protection and Management or UUPPLH.

Illegal logging is an activity that causes environmental damage or, in this case, forest damage so that it can cause flooding, pollution, acid rain, and global warming. According to data from the Ministry of Forestry in 2006, 59.6 million hectares of the 120.35 million hectares of forest area in Indonesia were damaged. They could not carry out their functions optimally, the leading cause being massive deforestation in the last five years reaching 2.83 million hectares per year [6].

For sustainable development, it is necessary to enforce criminal law for those who commit Illegal logging crimes. The method used by the author in this study is a normative research method (library study) using primary legal materials collected from laws and regulations and secondary legal materials from books, articles, journals. To obtain the title of "Sustainable development: enforcement of environmental criminal law practice against illegal logging in Indonesia."

2. Result and discussion

2.1. The concept of sustainable development (sustainable development)

The requirement for every country to apply the concept of sustainable development is stated in the World Commission on Environment and Development (WCED) Report in 1987 or the "Brundtland Report," whose title is "Our Common Future." Together).[7] Not only that, the necessity for applying the concept in order to ensure the quality of people's lives and improve the welfare of the people mentioned in Article 33, paragraph 3 of the 1945 Constitution " with earth and water and natural resources contained in it are controlled by the state and used for the greatest prosperity of the people."

Study design in the form of a law-based environmental study, with the subject of environmental law conditions in Indonesia, located in the urban area of Surabaya, data collection with literature review.
Data analysis using parables in a limited discussion study with government officials in June 2021 guarantees its citizens their rights to enjoy natural resources whose management needs to optimally as possible by the state so that the community gets the maximum benefit sustainably.

The definition of the concept of sustainable development is stated, namely: "Sustainable development is a conscious and planned effort that integrates environmental, social and economic aspects into a development strategy to ensure the integrity of the environment as well as the safety, capabilities, welfare, and quality of life of present and future generations."

Based on this understanding, sustainable development is a process that is carried out wisely and rationally that can simultaneously manage natural resources by combining social, economic, and environmental aspects to fulfil their needs without harming future generations. Future, to ensure the welfare and quality of life together (sustainable living), which means it can be responsible for themselves and the surrounding community towards the environment, for that the concept of sustainable development will be achieved if it is accompanied by environmental insight [8].

Humans and other living creatures have the right to a healthy living environment. Therefore, to realise it, sustainable development with an environmental perspective needs to burden humans to carry out their obligations in meeting the interests of current and future generations [9].

Sustainable Development is carried out to be able to fulfil the needs of the present generation and future generations, so that efforts to manage natural resources are prohibited from being excessive, but must be carried out with due regard to the carrying capacity of the environment itself so that sustainable development goals can be carried out correctly. Thus making humans the centre of attention, this places humans as subjects and objects in realising sustainable development. As a subject because humans are the implementers of development. As an object because in terms of realising sustainable development of living things, especially in this case, humans are the goals of development results to meet their needs [10].

Sustainable development with an environmental perspective is a form of utilising natural resources to guarantee the quality of life for humans or other creatures in the current generation without neglecting the next generation [11]. Sustainable Development must contain economic, social, and environmental aspects, all of which must work together because they have their respective relationships in realising sustainable development. In development, one must be able to pay attention to resources and also the order of the environment, use natural resources should not be excessive, and must also be able to pay attention to the carrying capacity of the environment because, after all, the environment must be preserved. Development of a sector must be able to pay attention to the impact of another sector that is possible to account for as a result of development [12].

Furthermore, sustainable development must pay attention to the use of resources (SDA and HR), namely paying attention to the possible impacts arising from the development of a sector to humans as social beings with their environment [13]. Natural resources play a role in state revenues; the government has an essential role in natural resources that expect to manage them well [14].

2.2. The characteristics of sustainable development:

According to RM Gatot Soemartono, sustainable development that is environmentally sound has the following characteristics: [15]

- Strive to be able to conserve biodiversity, which is the main thing so that the ecosystem is balanced so that natural resources can always be available for current and future generations;
- In managing natural resources by paying attention to the carrying capacity of the environment because after all the environment must be preserved;
- Paying attention to the economic, social, and environmental aspects of development at the same time because all these aspects cannot be separated, which means that the three must be able to support each other in order to carry out sustainable development;
- Can conserve and increase biodiversity continuously so that natural resources are always available and can protect them;
- Using ways to preserve the environment that can pay attention to current generations without harming future generations.

In terms of utilisation of natural resources, however, humans can affect nature by giving positive or negative impacts (in this case, destructive). Therefore, development is required to pay attention to the possible impacts that will arise from these activities. The positive impact of this development is that it is beneficial for the management of human life, namely:

a. Natural resources play a role as state income, and if they can be appropriately used, they will certainly get good results for economic growth in current and future generations;
b. Technology will develop rapidly;
c. Humans can guarantee their lives with the opportunity to work equally and make their lives prosperous [16].

Next is the negative impact, if natural resources are not misused and only by paying attention to the needs of the current generation but not paying attention to the needs of future generations as well, this will be detrimental to humans as a result of these activities, namely:

a. Reducing natural resources, so that it affects future generations;
b. Can damage the ecosystem, causing the extinction of protected animals and plants;
c. Environmental damage or pollution; and so forth [17].

2.3. Strategy in realizing the concept of sustainable development

Askar Jaya stated that in a sustainable development strategy, the following must be considered: [18]

- Development by Providing Guarantees for Equity and Social Justice. The concept of sustainable development must be carried out fairly and equitably to the entire community [23]. Equity is a concept that cannot be measured (relatively). Realising it must be based on economic equality achieved through development, which is an effort to prosper the people. That justice, in this case, is to be able to carry out development without compromising the needs of the next generation but still being able to fulfil the needs of the current generation.

- Development that Values Diversity. For future generations can be fulfilled, biodiversity is needed, which is the main thing to balance the ecosystem. Therefore, maintaining diversity is a critical element in development so that natural resources are always available for each generation.
- Development Using an Integrative Approach. The integrative approach means that in sustainable development, there is a relationship between humans and nature. Humans can influence nature through positive or negative (in this case, destructive) impacts. Therefore, the relationship between humans and nature is carried out well by positively impacting the environment. Therefore, development with an integrative approach is a concept that is possible to implement and at the same time is a challenge for the government.
- Development with a Long-Term Perspective. Sustainable development implies that giving attention to the present and future generations makes the significant challenges in sustainable development for more people more critical, but the present and the future are equally important. So, the long-term perspective is a sustainable development perspective.

Based on this explanation, it can be concluded that sustainable development is based on 3 (three) things, namely: natural resources, environmental quality factors, and population. So if in implementing sustainable development, one of the aspects is not fulfilled, then the development will not be appropriately realised. Sustainable development must pay attention to the environment's carrying capacity so that existing resources can be used for the benefit of the current generation without ignoring the next generation's interests [5].

In realising sustainable development, some things must be considered. Emil Salim said that there are 5 (five) critical basic things that need to be considered, namely: [8]

a) Have a sense of awareness and attitude of cooperation on the importance of environmental, economic, and social aspects carried out together because all three need for each other in order to realise sustainable development that is beneficial for present and future generations;
b) Along with the times, the needs of humans or other living creatures continue to grow. Therefore, so that the current generation and future generations can still meet their needs, it is necessary to manage rationally and wisely natural resources while still paying attention to the carrying capacity of the environment;
c) Developing human resources with it is expected to be able to take action to deal with future environmental constraints or problems to minimise environmental distress;
d) Have a sense of awareness to take action in protecting or preserving environmental ecosystems among the community;
e) Strive for establishing institutions that aim to develop human resources to participate in managing or preserving the environment.

2.4. Principles of intergenerational justice

The implementation of sustainable development in UUPPLH is carried out wisely and rationally by managing current Natural Resources (SDA), which aims to fulfil the interests of the current generation without neglecting the interests of future generations. In the concept of sustainable development, there is the principle of justice, which justice is in the form of the principle of one generation justice (Intragenerational Equity Principle) and the principle of intergenerational equity (Intergenerational Equity Principle) [7].

- The principle of one generation justice, that one must be fair to the responsibility of an environmental problem in one generation. If there are environmental problems, they must be appropriately resolved not to impact future generations [5].
- The principle of intergenerational justice, human beings in the world have the right to a good and healthy environment in every generation, not to get a dire environmental situation or

condition as a result of lack of care or human intent in treating the environment in order to obtain the greatest possible benefit. Which also ignores the carrying capacity of the environment itself.

According to Edith Brown, there are 3 (three) human actions in the current generation that can harm future generations: [10]

a. In meeting their needs for consumptive power, modern humans tend to be very high on existing resources to impact future generations, namely in meeting their needs. They are required to pay more than the current generation even though they enjoy the same resources;
b. The activities of the current generation, namely in terms of resource utilisation, significantly impact future generations, so that if the utilisation is not carried out efficiently, it will cause future generations to pay dearly for the inefficiency of resource utilisation by the current generation;
c. Therefore, the massacre of natural resources claimed to fulfil the interests of the current generation can affect the next generation. It has an impact on the diversity of natural resources for future generations.

In the principle of intergenerational justice, there are 3 (three) basic things [20]:

a. So that the next generation can enjoy a large diversity of resources to meet their needs, as does the current generation, it is required for each generation to maintain and protect the diversity of existing environmental resources;
b. It is required for each generation to seek various ways in order to protect and maintain the environment, none other than because future generations also have the right to enjoy a healthy environment as it is in the current generation;
c. For various natural wealth inheritances, each generation must provide guarantees to get the same access rights because future generations are also entitled to the right to gain access to these natural wealth inheritances.

In the context of sustainable development, the principle of intergenerational and one-generation justice is indispensable to protect, manage, and preserve the environment. Article 2 letter b of the UUPPLH concerning the Principle of Sustainability and Sustainability, namely that each generation has obligations and responsibilities to future generations and one generation to preserve the environment.

A good and healthy environment is one of the fundamental rights for every Indonesian citizen [10]. Humans become the most critical part of sustainable development based on these rights, namely subjects and objects. Humans as subjects because they are those who carry out development. As an object because in terms of realising sustainable development of living things, especially in this case, humans are the goal of the development results to meet their needs [11]. Thus, the population's quality in sustainable development allows it to utilise natural resources optimally, wisely, and rationally to maintain environmental sustainability. Therefore, it is hoped that there will be a balance between the population and the environment's carrying capacity [12].

In his book Development with Environmental Insight, Emil Salim reveals that sustainable development must be able to manage as well as possible the existing natural resources so that the concept of sustainable development is implemented by being able to preserve or develop the environment (eco-development) [25]. The environment, which is the most critical element in realising the concept of sustainable development, aims to fulfil the interests of the current generation by taking into account the interests of future generations to ensure the welfare and quality of life between generations.

2.5. Impact of Illegal logging against development

Illegal logging in Indonesia has snowballed since the emergence of the timber industry in Indonesia in the 1970s to develop Indonesian wood production to benefit future generations. However, in reality, these activities tend to prioritise economic interests rather than the interests of the environment itself, which is not by its original purpose, namely for the benefit of future generations. These industries carry out activities against the law, namely Illegal logging, resulting in the loss of forest land by carrying out large-scale forest logging (deforestation). Indonesia began to lose its forest and even forest area until 2013 slowly began to disappear up to 72 per cent [5]. The widespread practice of Illegal logging impacts all sectors, namely the environment, economy, and society. Even though the environment has a critical role in the economy, the economic system will not function as expected without it, which indirectly cannot guarantee the welfare of the intergenerational community.

Illegal logging is an act of exploiting forest resources, not wisely or rationally. In other words, that is done excessively without paying attention to the needs of future generations. These activities include forestry activities against the law, starting from harvesting, managing, and trading timber [10].

Illegal logging causes forest damage, where forest damage is environmental damage. Environmental destruction is an activity that results in direct/indirect changes to the environmental ecosystem, which makes it unable to function as a supporter of sustainable development be stated in article 1 number 16 UUPPLH. Natural disasters that often hit various regions in Indonesia, such as floods and landslides due to Illegal logging activities, the extinction of several rare species of flora and fauna is also a form of environmental damage from illegal logging, as explained by experts, environmentalists and forestry [16].

Illegal logging activity exploits natural resources on a large scale to obtain the most significant possible profit by ignoring the environment's carrying capacity. Such activity would only harm the environment that is causing pollution and environmental damage, thus causing declining environmental quality, which makes the environment unable to perform its function like a system of life support (life support system) with him would harm living beings, especially humans [23]. As a result, sustainable, environmentally sound development to ensure the welfare and quality of life between generations cannot be implemented. Below are some of the consequences of Illegal logging practices : [25]

- Some areas in Indonesia, when it often rains, experienced floods or landslides due to the loss of trees in Indonesia
- Reduced sources of water, so that the interests of the community are disturbed by this;
- The extinction of flora and fauna;
- - Natural disasters that may occur as a result of Illegal logging practices
- In meeting their needs, in this case, future generations are required to pay more than the current generation, even though they consume the same resources;
- The reduction of trees or forests as a result of Illegal logging activities triggers global warming (global warming) due to the depletion of the ozone layer, as it is already known that this will result in decreased air quality, melting of polar ice caps, smog, making it a serious international issue.

Development that prioritises economic interests rather than the environment will impact environmental damage, not guarantee that sustainable development can be realised. Sustainable Development must be ecologically, socially, and economically sustainable, which means that sustainable development is required to be ecologically (environmentally), socially, and economically sustainable [22]. These aspects must be carried out simultaneously because they have their respective relationships, meaning that the economic, social, and environmental aspects must support each other to carry out sustainable development. The environment plays a role in the economic sector, so that if the management of natural resources is not carried out in a balanced manner and ignores the carrying capacity of the environment, the country's economy cannot run well. The economic sector must still be considered because if the economy is neglected, it will cause problems with the social aspect, where the economic aspect can affect the social aspect [20]. In utilising natural resources, it is necessary to pay attention to an activity of the surrounding community so that the development can have a positive impact [22] on the community's economy, so it is less likely that environmental damage will occur. These activities have a positive impact on the current generation and have an impact on future generations.

Based on the description above can be seen that the practice of Illegal logging (logging) do not reflect the goals of sustainable development, which is to manage natural resources wisely and rationally so that today's generation can be fulfilled interests without ignoring the interests of the next generation, to guarantee the well-being and quality of life intergenerational. Illegal logging is the act of collecting, cutting or receiving, buying or transporting, possessing timber forest products from forest areas that are suspected of being illegally carried out without a Certificate of Legal Forest Products (SKSHH). Illegal logging tends to be more concerned with economic interests than the environment itself in managing natural resources, even though the environment has a critical role in the economy, without which the economic system will not function as expected. Development that only pays attention to the economic system indirectly does not reflect aspects of justice and sustainability for intergenerational society [24].

2.6. Criminal law enforcement against illegal logging practices in Indonesia

The crime of Illegal logging is categorised as a great crime because these activities can harm the stability of the country's economy and people's lives, so that if the use of natural resources is not wisely and rationally and not If you pay attention to the carrying capacity of the environment, the country's economy cannot run well. However, the economic sector must still be considered because if the economy is neglected, it will cause problems with the social aspect, where the economic aspect can affect the social aspect [24].

However, considering that there is no official definition of Illegal logging itself, there are differences of opinion regarding its elements. They are:[24]

a) there is an activity
b) felling or transporting or processing or selling or buying wood illegally
c) which can destroy the forest
d) there is a law that prohibits
e) contrary to applicable law

From the explanation above, there is the most striking element, namely "which can damage the forest," forest damage is damage to the environment because it has a wide enough impact on economic, social, and environmental (ecological) aspects so that it cannot support the concept of sustainable development that is regulated. in Law 32/2009 on Environmental Protection and Management.

Along with the times, the quality of the environment gradually decreases due to rampant illegal logging activities, resulting in the reduction of trees or forests, which then triggers global warming due to the depleting ozone layer, as is well known that global warming has an impact on the environment. With an impact on the survival of living things, regulations, protection, and environment management are needed seriously [24].

Both the government, the state, and every stakeholder strive to protect and preserve the environment. So, on this basis, it is hoped that the government will try to improve legal products, in this case at the technical level, so that environmentally sound and sustainable development is a written concept but is expected to be implemented well [12].

In realising the concept of sustainable development, it is necessary to prevent and enforce the law as an effort to protect and secure forests:

- Preventive Efforts. They were intended to prevent Illegal logging activities by a. counselling on the dangers of forest destruction to create awareness among the community that forests are very useful in ensuring the quality of the environment; b. Forest patrols are carried out by the community and law enforcement officers to secure or protect the forest from irresponsible elements and other prevention programs.
- Repressive Effort. In criminal law against environmental crime (forestry), in this case, Illegal logging, the government has issued Presidential Instruction No. 3 of 2005 on the Eradication of logging illegally in the forest area is expected d apart reduce the number of illegal logging crimes (Illegal logging). Furthermore, following up on the perpetrators of Illegal logging, namely with Law 18/2013 on Forest Prevention and Eradication, namely: imprisonment and fines. In addition, without reducing criminal sanctions, it is obligatory for those responsible for violations without a permit and for permit holders to be subject to "administrative sanctions." For violations without a permit, sanctions are imposed to pay compensation by the level of damage or consequences caused to the state for rehabilitation costs, restoration of forest conditions, and other necessary actions. Meanwhile, permit holders are subject to sanctions in the form of fines, cessation of activities, reduction of area, or revocation of permits [12].

3. Conclusion

In the implementation of development, it will have a positive or negative impact depending on the humans themselves in managing or preserving, or protecting natural resources, if done wisely and rationally, which means that it is not excessive but can still meet the interests of the current generation and does not ignore the interests of future generations. Therefore will have a positive impact on environmental, economic, and social aspects. When done by slaughtering the natural resources on a large scale so that the needs of the present generation are met, but by ignoring the interests of future generations, it can have adverse effects that pollution da n damage to the environment, so that the welfare and quality of life between generations hard to guarantee.

The crime of illegal logging (illegal logging) is an extraordinary crime because these activities can harm the stability of the country's economy and people's lives, and of course forest damage, forest damage is also damage to the environment because it has a reasonably broad impact on the economic, social, and environment (ecology). In the enforcement of the criminal law, there are articles 82-106 of Law 18/2013 concerning the Prevention and Eradication of Environmental Destruction with the threat of imprisonment, fines, and in article 18 "administrative sanctions." The practice of illegal logging (logging) does not reflect the purpose of constructing air- the continuation of the public welfare by managing natural resources wisely and rationally so that the present generation and future generations can enjoy the results as a form of "justice."

Law 18/2013 concerning Prevention and Eradication of Environmental Destruction as a form of preventive and repressive efforts to the environment will not play many roles because sustainable development is. Therefore, it is hoped that there will be awareness in protecting and preserving environmental ecosystems among the community. So that the realisation of sustainable development that is environmentally sound to ensure the welfare and quality of life between generations.

Acknowledgement

This research received no specific grant from any funding agency in the public, commercial, or not-for-profit sectors.

References

[1] Keraf A S 2001 *Sustainable Development or Sustainable Ecology?* (Postgraduate Program, Faculty of Law, University of Indonesia) 1
[2] Sugandhy A and Rustam H 2007 *Basic Principles of Environmentally Friendly Sustainable Development Policy* (Jakarta: Bumi Aksara) 23
[3] Mas'ari A 2017 *STIKI Journal* 717
[4] Article 28 H of the 1945 Constitution
[5] Askar J 2004 *Concept of Sustainable Development* (Bogor: Bogor Agricultural Institute) 5-6
[6] Bambang Y 2013 *Building Citizens Awareness in Environmental Preservation* (Yogyakarta: Deepublish) 70
[7] Burhanuddin 2016 *EduTech Journal* **2(1)** 14
[8] Deasy S 2016 *Unsrat Law Journal* **21(5)** 1
[9] Edith Br w W 1997 Richard L.Revesz *Foundations of Environmental Law and Policy* (Oxford: Oxford University Press) p 309-312
[10] Edith B 1991 *Am J Int Law* **84** 201-202
[11] Emil S 1993 *Environmentally Friendly Development* (Jakarta: LP3ES) 184-185
[12] Evi P W 2018 *Environmental Law Development* **3(1)** 121
[13] Utama I M A 2007 *Environmental Law: Environmental Permit Law System for Sustainable Development* (Bandung: Pustaka Sutra) p 65
[14] Lisa A 2013 *Lex Administratum* **1(1)** p 5-6
[15] Luaxs-berjaya.blogspot.com, 2011, *Illegal Logging Crime (Law of the Republic of Indonesia Number 41 of 1999 concerning Forestry)*, in http://luaxs-

berjaya.blogspot.com/2011/10/tindak- criminal-illegal-logging law. HTML. Retrieved 11 July 2015

[16] Marhaeni R S 2017 *Environmental Law and Implementation of Sustainable Development in Indonesia* (Jakarta: PT Gramedia Pustaka Utama) 63

[17] Maryanto 2018 *Meta-Juridical Journal* **1(1)** p 45-46

[18] Poerwanti H P 2008 *Journal of Dimensia* **2(1)** 107

[19] Sumartono G P 1999 *Understanding Indonesian Environmental Law* (Jakarta: Sinar Graphic) 145-146

[20] Samsul W 2014 *Legal Dimensions of Environmental Protection and Management,* (Yogyakarta: Learning Library) 22

[21] Shaula F S 2018 *Sustainable Development in Law Number 32 of 2009 concerning Protection of Environmental Management from Political Legal Perspective in the Review of Islamic Law* (Thesis—Indonesian Islamic University, Yogyakarta) p 25-26

[22] Suardi *The Problems of Implementing Sustainable Development Principles in Environmental Management and Its Implications for the Fulfillment of Human Rights* (Palu: Tadulako University) p 97

[23] Surya P 2013 *Sustainable Development Process* (Yogyakarta) 10

[24] Winarno 2013*Yustisia* **2(2)** p 95

Reduction of synthetic fertilizer application and heavy metal absorption on oil palm nursery related to the application of humic substances

H Agusta[1,2*], D Asmono[3], M Fajri[1]

[1]Department of Agronomy and Horticulture, IPB-University, Bogor 16680, Indonesia.
[2]Surfactant and Bioenergy Research Center, IPB University, Bogor 16144, Indonesia
[3]PT Sampoerna Agro Tbk, Sampoerna Strategic Square, North Tower, Jakarta 12930, Indonesia

agusta@apps.ipb.ac.id

Abstract. Humic acid with amorphous and colloidal characteristics can improve soil chemical properties, especially tropical soil with lower fertility. To contribute to an environmentally friendly farming system, the capacity of humic acid to reduce the consumption of synthetic fertilizer on nursery oil palm was tested in this experiment. The experiment's purpose was to confirm the minimizing fertilizer application and reduce the heavy metal absorption in oil palm nurseries without any deterioration effects on the growth performance. Varied composition types of humic substance products were applied, i.e., in combination with fish meal, seaweed, and chicken manure.The concentration of heavy metals Ba, Cd, Ba and the absorption of macroelements N, P, K, Ca, and Mg in the growing plants was also observed. It showed that humic acid in the dosage of 40 g/plant in the main nursery was capable of reducing the application of mineral fertilizer by 25% base on recommended standard application and at the same time reducing heavy metal Ba and Cd absorption by the plant.

1. Introduction

Oil palm seedlings require additional fertilizer for their growth, where fertilizers and their application cost can reach about majorly at the processing cost in the plantation [1]. Excessive use of fertilizers can waste the expenditure and increase greenhouse gas emissions [2], sourced from applied inputs and emissions occurring in the field. Oil palm nurseries with germinated seeds can be grown through pre-nursery and then the main nursery before being planted at the field.

Humic acid application is proposed to reduce fertilizer consumption. The humic fraction is soluble in bases, colloidal dispersed, amorphous, yellow to blackish brown, and relatively high molecular weight. Humic acid is relatively resistant to degradation and contains a negative charge affected by pH. Humic acid can bind to toxic metal ions so that it becomes insoluble. Humic acid also plays a direct role in plant growth, which can stimulate the growth of roots and the upper part of plants. Related to this, humic acid can be used as a stimulant for plant growth and support plant growth and productivity. Fertilizer is a source of nutrients for plants and microflora. Humic acid helps move the micronutrients from soil to plant.

Cd, Hg, and Pb are not essential elements of plants. However, other heavy metals Fe, Cu, Zn, and Mn, are required in small quantities, where a shortage of essential heavy metals can be defective for living plants. Humic compounds as metal chelators in the soil react with metal ions, acting as a metal rectification in the plants.

Due to routine operations and maintenance of production lines in oil palm companies, the attention on environmental problems, weighty metal absorption issues by the plants related to intensive synthetic fertilizers application was not explored sufficiently. There is a challenging method to overcome this sustainable concern on oil plantations, whereas abundant humic substances resources in nature can be explored. The humic substances have advantages potency to be combined with the synthetic fertilizer, which will be expected to reduce mineral fertilizer application and as a consequence, it will support the greenhouse gas emission by reducing the use of synthetic fertilizer.

The experiment's purpose was to confirm the minimizing fertilizer application and reduce the heavy metal absorption in oil palm nurseries without any deterioration effects on the growth performance. The novelty of this research was to reveal the affectivity rate of various humic acid applications and its application rates on the reduction of consumption rate of standard fertilizer dosage at the main nursery stage of oil palm plantation, especially on its interaction with heavy metal Pb, Cd and Ba absorption by the plant without reducing the plant growth performance.

2. Methodology

This research was conducted in the Nursery Plantation of Hikmah Dua - PT. Sampoerna Agro Tbk, Ogan Komering Ilir Regency, South Sumatra. Soil analysis and humic acid content determination were conducted in Soil Science Laboratory, Univerity of Sriwijaya Palembang, and Laboratory of Nusa Pusaka Kencana- PT. Asian Agri Inc. Plant tissue analysis was conducted in Soil Science and Land Resources Laboratory, IPB, and IPB Integrated Laboratory, Bogor. This research was carried out in April 2008 – January 2009. The humic acid used in this study was the product of PT Green Planet Indonesia, whose main characteristics are presented in Table 1. PT Sampoerna Agro Tbk provided the oil palm cultivar SJ2 with female number 930/42 and male number 321/204 from a pre-nursery plantation at the exact location. Pesticides Deltamethrin 25 g/ℓ and Mancozeb 75% were applied for plant protection purposes. The planting media used was the subsoil of the plantation site. A relatively large polybag size 40 cm x 50 cm was prepared for the leading nursery transplanting until the plants were ready to be transplanted to the plantation.

Tabel 1. Description of applied humic acid product at oil palm main nursery.

Type	Description	Color	Odor level	Formulation	Hardness level
B	Fish meal (20 %) + Humic acid + Chicken manure	Brownish black	***	Pellet/ granular	###
C	Seaweed + Fish meal (20 %) + Humic acid + Chicken manure	Brownish black	***	Pellet/ granular	###
E	Fish meal (28 %) + Humic acid + Chicken manure	black	*****	Pellet/ granular	#
F	Seaweed + Fish meal (28 %) + Humic acid + Chicken manure	black	*****	Pellet/ granular	###
G	Humic acid + Chicken manure + Seaweed	Light brown+	**	Pellet/ granular	###

The design used in the main nursery was a split-split plot design with three doses of fertilizer as the main plot, three doses of humic acid as a subplot, and five types of humic acid as a sub-sub plot. A unit experiment consisted of 16 plants to be observed during the experiment time; there were 2160 plants to be observed. Like the main plot, the Fertilizer dose consisted of 3 levels of fertilizer dosage, i.e., 50%, 75%, and 100% of the company recommendation dosage. This recommendation dosage was set per plant during the main nursery growing time in 14 g N, 80 g P_2O_5, 60 g K_2O, and 88 g CaO. The dose of humic acid as a subplot consisted of 3 levels, i.e., 0, 20, and 40 g plant^{-1}. The sub-sub plot consisted of the humic acid types B, C, E, F, and G.

3. Result and discussion

The application of synthetic fertilizer as industrial inorganic fertilizer on the main nursery installment in the dosage of 50%, 75%, and 100% was an effort to reduce fertilizer consumption for the plant. This effort was also combined with the idea to activate the ecological activities with bioactive substances, which was expected to increase soil microorganic activity. The bioactive substances comprised humic acid mixed with other varieties of organic materials such as fish meal, seaweed, chicken manure. The application of these biomaterials should support environmental-friendly agriculture[3,4].

The reduction of fertilizer dosage for the plant at 50% and 75% level of a typical application and the application of the humic acid product in the quantity of 20 and 40 g plant^{-1} had no significant impact on the possible improvement of soil fertility parameters. Hence, soil quality was still in a range of environmental variability, and no significant improvement was found. The average value from all treatments of experimental soil parameters was as the following, pH H_20 4.01±1.19, N-total 0.20±0.06%, Soil C-organic content 2.27±0.69%, Phosphate Bray-1 192.0±148.2 ppm, CEC pH7 11.31±3.32 me 100 g^{-1} soil, K 1.43±0.66 me 100 g^{-1}, Ca 1.28±0.66 me 100 g^{-1}, Mg 3.17±1.71 me 100 g^{-1}, and Al 1.06±0.47 me 100 g^{-1}. The soil quality was categorized as low, with acidic pH value, low in mineral contents, and Al status should be considered for planting management.

The growth of the oil palm in the main nursery generally was not influenced by the increase of the application of humic acid dosage at 0, 20, 40 g plant^{-1} nor rate of recommended fertilizer dosage at 50%, 75%, and 100%. Only minor differentiation at the experimental growth parameters affected by the type of humic acid applications products was noticed. There was no significant difference in plant height (63.14±0.88 cm), fresh stem weight (268.82±16.81 g plant^{-1}), leaf number per plant (13.21±0.28), stem diameter (37.29±0.59 cm), number of secondary roots in 0-10 cm soil depth (751.32±67.54 plant^{-1}) and number of tertiary root in 20-30 cm (10327±1355) plant^{-1}. After the destruction of the samples, the total dry weight of biomass was an average of 214.70±63.66 g plant^{-1}.

However, there was an unsynchronous effect of the treatment observed, especially on dry root matter and plant leaf dry matter parameters. The application of 75% fertilizer dosage in combination with the application of 75% recommendation dosage of the fertilizer showed the highest plant root dry matter value. The lowest plant leaf dry matter was found under the application of 100% fertilizer dosage recommendation in combination with humic acid products in the amount of 20 g plant^{-1}. In this case, the 100% fertilizer dosage treatment would be safe only if no humic acid was applied (Table 2). On arable land with neutral soil pH value, additional fertilizing of oil palm nurseries is not reuired[5].

Regarding the plant leaves biomass production, the combination of humic acid products in the amount of 40 g plant^{-1} in combination with 75% fertilizer recommendation dosage exposed the best performance of the parameters and plant root development. It makes sense to combine both treatments in the recommended dose to reduce fertilizer application in the oil palm main nursery. The effect of the recommended fertilizer dosage level in combination with the varied humic acid products is presented in Table 2. The reduction of fertilizer application would save the environment. On acid soil, the fertilizer application in combination with extra organic carbon to promote microbial activity improved the performance of oil palm nursery plants [6].

The absorbed macro minerals from the soil at the nursery showed no significant difference due to the fertilizer recommendation dosage level nor the rate of humic acid application dosage. The total absorption of nitrogen, phosphor, potassium, calcium, and manganese exposed 1321±395 mg plant^{-1}, 291±86 mg plant^{-1}, 1112±329 mg plant^{-1}, 2726±803 mg plant^{-1}, and 216±66 mg plant^{-1}, respectively.

Table 2. Dry matter of root and Leaf oil palm main nursery related to fertilizer and humic acid and application.

Treatment	Dry matter weight		
	Humic acid dosage		
	0 g plant^{-1}	20 g plant^{-1}	40 g plant^{-1}
Fertilizer dosage	----------plant root (g plant^{-1}) ---------		
50%	416.70abcde	400.00bcde	333.30cde
75%	416.70abcde	400.00bcde	700.00a
100%	666.70ab	300.00e	366.70cde
Fertilizer dosage	-----------Plant leaf (g plant^{-1})---------		
50%	62.08b	61.45b	62.45b
75%	60.99b	64.81ab	68.73ab
100%	69.46ab	73.39a	60.40b

Note: Unequal indexed letters (a,b) at the same observed parameter show a significant difference level at p=0.05

The heavy metals Cd, Pb, and Ba showed no different values in the plant leaf and stem. The heavy metal concentration value in the leaf was observed at 0.05±0.11 ppm Cd, 0.94±1.04 ppm Pb, and 11.18±9.75 Ba, and in the stem were 0.05±0.09 ppm Cd, 1.13±1.12 ppm Pb, and 3.81±2.00. The affectivity of humic acid products to control Cd, Pb, and Ba in the soil was confirmed, that heavy metal contaminant in the soil was effectively removed achieved using humic acids [7], even more effective than the traditional chelating agent EDTA [8]. The generated metal-humic compound was more stable in the soil [9].

The absorption of Ba in the root was associate with the application of fertilizer. The higher fertilizer application increased the Pb concentration both in root and rachis. On the other side, the application of humic acid products reduces the absorption of heavy metals. Hence, the concentration of Pb and Cd at the application of 40 g humic product plant^{-1} exposed a lower value than without the application of humic acid (Table 3). The concentration of heavy metal Cd in oil palm plants demonstrated more variability in value ranging 12.7-76.2 ppm, where the Cd concentration in the soil ranged only 0.27-4.26 ppm[10].

Table 3. Plant heavy metal concentration in rachis and root of the main nursery related to fertilizer and humic acid application.

Treatment	Heavy metal concentration (ppm)					
	Rachis			Root		
	Cd	Pb	Ba	Cd	Pb	Ba
Fertilizer dosage						
50%	0.28	0.42	1.64	0.07	5.16	5.51a
75%	0.11	0.80	1.07	0.11	3.75	7.94ab
100%	0.33	0.74	2.91	0.09	2.24	12.41b
Humic acid						
0 g plant^{-1}	0.20	1.09a	1.93	0.14a	4.42a	8.03
20 g plant^{-1}	0.35	0.42b	3.12	0.06b	3.75ab	9.31
40 g plant^{-1}	0.36	0.26b	1.57	0.07b	2.98b	8.74

Note: Unequal indexed letters (a,b) at the same column and the same main plot or subplot show significant difference level at p=0.05

Total heavy metal absorption also showed an inherent pattern with the concentration in the leaf and root of the plant. The fertilizer application in a higher dosage from 50% to 100% recommendation dosage increased the absorbed Ba more than doubled. The increase of humic acid application reduced the Ba and Pb absorption of the plant. Any humification levels in the soil are commonly suitable for increasing the availability of soil fertilizer and play a role in improving contaminated soils by metals and organic pollutants [11]. The interactions that affect nitrogen fertilizing and humic acids application were described, that the increase of nitrate would reduce the humic acid formation, and the microbes were stimulated accordingly, especially those that can promote the humic acid degradation [12].

Table 4. Plant heavy metal absorption at the main nursery due to fertilizer humic acid and application.

Treatment	Heavy metal absorption		
	Cd	**Pb**	**Ba**
Fertilizer dosage	----------------g plant^{-1}--------------		
50%	0.10	1.55	3.62a
75%	0.05	1.37	4.75ab
100%	0.15	1.19	8.45b
Humic acid dosage			
0 g plant-1	0.09	1.69a	6.39a
20 g plant-1	0.12	1.33ab	5.97ab
40 g plant-1	0.11	1.09b	4.45b

Note: Unequal indexed letters (a,b) at the same column and the same main plot or subplot show a significant difference level at p=0.05

4. Conclusion

The application of fertilizer in a higher dosage from 50% to 100% recommendation dosage without the application of humic acid substances increased the absorbed Ba in the plant more than doubled. Humic acid application in the dosage of 40 g/plant in the main nursery was capable of reducing the application of mineral fertilizer by 25% base on recommended standard application and at the same time also reducing heavy metal Cd, Pb, and Ba absorption by the plant.

Acknowledgment

Fund and facilitation is supported by PT Sampoerna Agro Tbk 2008-2009.

References

[1] Foong S Z Y, Goh C K M, Supramaniam C V and Ng D K S 2021 *Sustain Prod Consum* (in Press)
[2] Kusin F M, Akhir N I M, Muhammat-Y F and Awang M 2015 *Atmósfera* **28** 243-250
[3] Rhebergen R, Fairhurst T, Giller K E and Zingore S 2019 *Agric Water Manag* **221** 377–387
[4] Wu J, Zhao Y, Qi H, Zhao X, Yang T, Du Y, Zhang H and Wei Z *Biores Technol* **244** 1193–1196
[5] Akpo E, Stomph TJ, Kossou DK, Omore AO and Struik PC 2014 *For Ecol Manag* **324** 28–36
[6] Murugan P, Ong SY, Hashim R, Kosugi A, Arai T and Sudesh K 2020 *Biocatal Agric Biotechnol* 28 101710
[7] Shi W, Lü C, He J, En H, Gao M, Zhao B, Zhou B, Zhou H, Liu H and Zhang Y 2018 *Ecotoxicol Environ* **154** 59–68
[8] Zingaretti D, Lombardi F and Baciocchi R 2018 *Sci Total Environ* **619–620** 1366–1374
[9] Rashid I, Murtaza G, Dar AA, and Wang Z 2020 *Ecotoxicol Environ Saf* **205** 111347
[10] Olafisoye OB, Fatoki OS, Oguntibejub and Osibote 2020 *Toxicol Rep* **7** 324–334
[11] Zhao Y, Wei Y, Zhang Y, Xin, Wen X, Xi B, Zhao X, Zhang X, Wei Z 2017 *Ecol Indic* **72** 473–480
[12] Shi M, Wei Z, Wang L, Wu J, Zhang D, Wei d, Tang Y and Zhao Y 2018 *Biores Technol* **258** 390–394

IOP Conf. Series: Earth and Environmental Science **894** (2021) 012004 doi:10.1088/1755-1315/894/1/012004

Environmental impact assessment of tofu production process: case study in SME Sugihmanik, Grobogan

S Hartini[1*], B S Ramadan[2], R Purwaningsih[1], S Sumiyati[2], M A A Kesuma[1]

[1]Department of Industrial Engineering, Universitas Diponegoro, Indonesia
[2]Department of Environmental Engineering, Universitas Diponegoro, Indonesia

srihartini@lecturer.undip.ac.id

Abstract. Tofu contains various substances that are very good when consumed to improve people's nutrition. In addition, tofu also has good taste. The problem is that the tofu production process produces products and non-product outputs in the form of waste that is very dangerous if directly disposed of in the environment. The BOD5 content of tofu small and medium-sized enterprises (SMEs) in Sugihmanik Village ranged from 3,667 – 4,933 mg/L and COD 7,668 – 9,736 mg/L. At the same time, the TSS values ranged from 701-1,189 mg/L. The BOD5 value in the river water content is 367 mg/L. It greatly exceeds the set Threshold Value. This study aims to measure the environmental impact using Life Cycle Assessment (LCA). LCA can identify the impact of each activity based on the impact category to identify the processes that contribute significantly to damaging the environment. This study found that the cooking and frying process had the highest impact, where the climate change category was the largest. Wastewater treatment plants, biogas from the biodigester as a substitute for electricity for water pumps, rice husks, and corn cobs are expected to reduce environmental impacts. The first section in your paper

1. Introduction

Tofu is a highly nutritious food from China that is produced from soybean extract [1]. According to the Directorate General of Public Health in 2018, 100 grams of tofu contains 80 kcal of energy, 10.90 grams of protein, 4.70 grams of fat, 0.80 grams of carbohydrates, 223 mg of calcium, 183 mg of phosphorus, 0.1 mg of vitamin B3. and 82.20 grams of water. Tofu is an essential and valuable nutrient because of its relatively large protein content, minerals, vitamins, lipids, and isoflavones [2]. The various substances contained in tofu are excellent when consumed to improve people's nutrition. The public has realized it as seen from the increase in the average consumption of tofu in Indonesia in 2017 by 0.157 kg/capita/week and 2018 to 0.158 kg/capita/week [3]. The increasing consumption of tofu is followed by the emergence of small businesses that produce tofu in Indonesia, so that it has a positive effect on the economic aspect. However, from the environmental aspect, many negative impacts due to the development of tofu SMEs are not followed by proper waste management.

Tofu production produces emissions, liquid and solid waste. Solid waste has been widely used as "*tempe gembus*," animal feed, and processed into crackers, shredded, and dry bread [4]. Meanwhile, solid waste from burning husks and firewood can be used as organic fertilizer [4]. Tofu liquid waste contains BOD (Biochemical Oxygen Demand), COD (Chemical Oxygen Demand), TSS (Total Suspended Solid), and high pH (acidity) [5]. The content of these organic compounds can produce

IOP Conf. Series: Earth and Environmental Science **894** (2021) 012004 doi:10.1088/1755-1315/894/1/012004

biogas when processed through an aerobic process [4]. However, tofu waste that is disposed of without being processed will cause negative impacts such as unpleasant odors, water pollution, sources of disease, and lowering the aesthetics of the surrounding environment. Some of the limitations that SMEs have to treat wastewater, for example, are limited funds, technology, and land ownership. To minimize these limitations, clean production practices should reduce the pollutant load to a certain extent [6].

Sugihmanik Village, Tanggungharjo, Grobogan has more than 30 SMEs that produce traditional tofu. The average production capacity is around 150 kg of soybeans per day for one SME. It means that the daily soybean requirement for Sugihmanik village is equivalent to 4 tons. Solid waste from tofu production has been used as animal feed and material for making "*tempe gembus*," while liquid waste is directly discharged into the river through pipes. This study has tested wastewater in some of these SMEs and river water. The BOD5 content of tofu SMEs ranged from 3,667 – 4,933 mg/L, COD 7,668 – 9,736 mg/L, and TSS values in the range 701-1,189 mg/L. While the BOD of river water is 367 mg/L and COD is 738 mg/L. This value exceeds the threshold value stipulated by Central Java Provincial Regulation No. 5 of 2012. The high organic load content creates anaerobic conditions that produce decomposition products in ammonia, carbon dioxide, acetic acid, and methane. These compounds cause a pungent, black river odor, damage the quality of healthy water around the river, and become one of the factors in decreasing agricultural productivity. This condition reflects the cottage industry as an industry with low energy efficiency and a high level of environmental pollution due to pollution [7].

This study aims to measure environmental impacts and provide recommendations for improvements to the tofu production process. The method used is the Life Cycle Assessment (LCA) method. LCA helps decision-makers to measure the environmental impact based on the production process so that they can identify environmental impacts based on their activities [8,9–11]. It is essential as a basis for efforts to reduce environmental impacts. LCA has four stages, namely goal and scope definition, life cycle inventory (LCI) collection, life cycle impact assessment (LCIA), and life cycle interpretation phase [12,13].

2. Methodology

2.1. The production process of tofu in Sugihmanik

The primary raw materials for the tofu production process are soybeans and water. The tofu production process consists of soaking, grinding, cooking, clumping, printing, and frying. The clumping process requires materials to precipitate the protein. Tofu products generally use vinegar, but in tofu production in the SME, Sugihmanik Village uses whey. Whey ferments the remaining water from the clotting process, which is left for 1-2 nights. The bacterial fermentation of wastewater from the clotting process produces lactic acid, which can coagulate protein. The energy used consists of gasoline to fuel soybean grinders, rice husks, and corn cobs for cooking and frying. The final product of SMEs is fried tofu, so it requires cooking oil for the finishing process. The tofu production process also produced non-product output (NPO) in the form of liquid waste, solid waste in tofu dregs, roasted husk ash, roasted corn cobs, and used cooking oil.

2.2. Life Cycle Assessment

Life cycle assessment (LCA) is an internationally recognized standard for estimating the environmental impact of a product, process, or activity [8]. LCA helps decision-makers measure the environmental impact based on the production process to identify environmental impacts based on their activities [9]–[11]. Measurements are interpreted in terms of potential impact and evaluated in categories [14]. It is essential as a basis for efforts to reduce environmental impacts. LCA has four stages, namely goal and scope definition, life cycle inventory (LCI) collection, life cycle impact assessment (LCIA), and life cycle interpretation phase [12-13].

3. Result and discussion

3.1. Input-output diagram of the tofu production process

The tofu production process in Sugihmanik generally has the same process and technology, namely the traditional method. The production process is carried out in small sizes, which is known as cooking. One-time cooking usually uses 5 kg of soybeans which are then processed into two trays of molds. Each SME produces soybeans between 100 - 150 kg or 20 - 30 cooking). The input-output diagram is made based on 30 samples of observations in Sugihmanik Village SMEs. Figure 1 is an input-output diagram of the tofu production process.

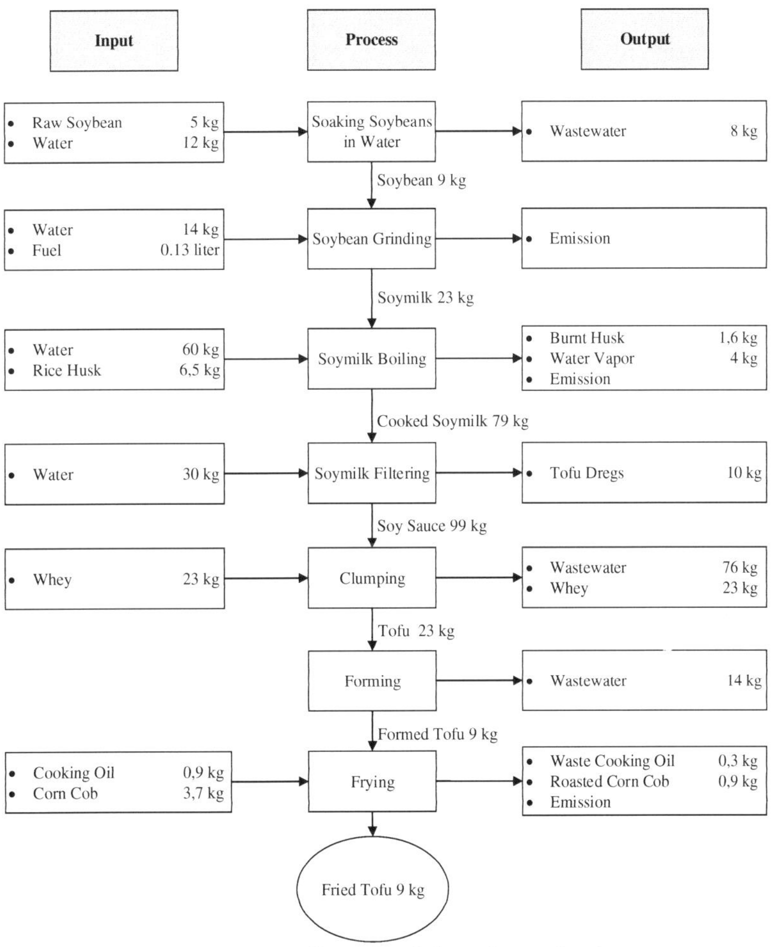

Figure 1. Input-output diagram of the tofu production process.

3.2. The result of the Life Cycle Assessment

The research goal is to identify environmental impacts and costs arising from the production process of tofu SMEs in Sugihmanik Village. The scope studied is Gate to Gate, which produces tofu from soaking to frying. The system does not review the soybean supply process from the supplier to the supplier and does not evaluate the distribution and marketing process. LCA evaluation is based on material and energy consumption and the resulting output, both in product and non-product output. Material consumption is calculated in units of mass (kg) and volume (litres).

The second stage is the Life Cycle Inventory (LCI) phase which explains the inputs and outputs from the beginning to the end of the tofu production process. Table 1 describes the life cycle inventory for one cooking.

IOP Conf. Series: Earth and Environmental Science **894** (2021) 012004 doi:10.1088/1755-1315/894/1/012004

Table 1. Life cycle inventory of tofu production process.

Process	Input			Output		
	Material	**Unit**	**Massa**	**Material**	**Unit**	**Massa**
Soaking	Raw soybean	kg	5	Soybean	kg	9
	Water	kg	12	Wastewater	kg	8
Grinding of Soybean	Water	kg	14	Soymilk	kg	23
	Fuel	Litres	0.13			
	Soybean	kg	9	Emission		
Heating of Soymilk	Water	kg	60	Cooked soymilk	kg	79
	Rice Husk	kg	6.5	Burnt husk	kg	1,6
	Soymilk	kg	23	Water vapor	kg	4
				Emission		
Filtering	Water	kg	30	Soy sauce	kg	99
	Cooked soymilk	kg	79	Tofu dregs	kg	10
Clumping	Whey	kg	23	Tofu	kg	23
	Soy sauce	kg	99	Wastewater	kg	99
Forming	Tofu	kg	23	Formed tofu	kg	9
				Wastewater	kg	14
Frying	Cooking oil	kg	0.9	Waste cooking oil	kg	0.3
	Corn Cob	kg	3.7	Roasted corn cob	kg	0.9
	Formed tofu	kg	9	Fried tofu	kg	9
				Emission		

The next stage is the Life Cycle Impact Assessment (LCIA) phase. This phase aims to evaluate the environmental impact resulting from the tofu production system. The LCIA phase in this study uses the Eco-Cost 2017 v1.5 method on the SimaPro software with 12 impact categories. Eco-costs is a measure to state the amount of environmental impact of a product based on the costs incurred to reduce the pollution it causes. Environmental costs should be considered as a hidden obligation or 'external costs' in environmental economics [15]. This phase has several stages consisting of characterization, normalization, weighting, and a single score. The characterization stage identifies production factors and groups them into several categories of environmental impacts based on the method used. Table 2 shows the impact characterization values at the stages of the production process using SimaPro software.

The normalization stage aims to assess an activity that contributes to environmental impacts. The normalization value is the multiplication between the characterization and normalization values so that the impact categories have the same units—normalization factor for each impact category using Eco-Cost 2017 v1.5 method. The weighting stage is a stage of assigning weights to each category of environmental impact. The weighting factor has different values depending on the method used and the level of importance of the impact category. In this study, the weighting factor using the Eco-Cost 2017 v1.5 method, where all impact categories have a weight value of 1. Thus, the normalization and weighting stages have the same impact category value. The single score stage aims to classify the value of the impact category of each process or activity. Based on the value of this stage, it can be seen which process contributes the most to the environmental impact and damage. Table 3 is the result of a single score after the normalization and weighting stages in euro, and Table 4 is the result of a single score in IDR. Figure 2 is the result of a single score in a bar chart. The biggest impact category is on climate change and acidification. Climate change is a change in the earth's climatic conditions that are becoming hotter. Global warming is caused by increasing greenhouse gases, namely CO_2, CH_4, N_2O, in the troposphere. These gases are like the greenhouse effect, which reflects radiation from the earth back to the earth. Sources of greenhouse gases are produced from burning fuel, burning biomass, piles of garbage, draining peatlands, cement, ethanol, and even from the plants themselves. At tofu SMEs in Sugihmanik, CO_2 is produced from the cooking and frying process wherein the process uses rice husks, corn cobs, and electrical energy for water pumping machines and cooking oil which produces CO_2 gas. From the results of the characteristics, it is known that the value of the impact of climate change on the tofu production process is 47.93 kg CO_2 equivalent per batch.

Tabel 2. The result of the characterization stage.

Impact Category	Unit	Soaking	Grinding	Heating	Filtering	Clumping	Forming	Frying
Climate Change	kg $C0_2$ eq	5.04	3.63	$1.57x10^1$	7.60	2.78	0.00	$1.32x10^1$
Acidification	kg SO_2 eq	$2.05x10^{-2}$	$2.00x10^{-2}$	$1.00x10^{-1}$	$3.28x10^{-2}$	$1.26x10^{-2}$	0.00	$6.00x10^{-2}$
Eutrophication	kg PO_4 eq	$3.00x10^{-3}$	$8.40x10^{-4}$	$5.42x10^{-3}$	$1.73x10^{-3}$	$2.72x10^{-2}$	$3.04x10^{-3}$	$3.00x10^{-2}$
Photochemical Oxidant	kg C_2H_4 eq	$2.00x10^{-4}$	$1.40x10^{-4}$	$6.40x10^{-4}$	$2.92x10^{-4}$	$1.32x10^{-4}$	0.00	$3.10x10^{-4}$
Fine Dust	kg $PM_{2.5}$ eq	$9.00x10^{-4}$	$8.50x10^{-4}$	$3.57x10^{-3}$	$1.77x10^{-3}$	$4.36x10^{-4}$	0.00	$4.40x10^{-4}$
Human Toxity	Cases	$5.00x10^{-8}$	$5.00x10^{-8}$	$2.00x10^{-7}$	$9.83x10^{-8}$	$1.52x10^{-8}$	0.00	$9.00x10^{-9}$
Ecotoxicity (Freshwater)	$PAF.m^3.day$	$1.16x10^3$	$1.13x10^3$	$4.84x10^3$	$2.38x10^3$	$3.28x10^2$	0.00	$4.75x10^2$
Metals Scarcity	Euro	$1.25x10^{-1}$	$5.00x10^{-2}$	$2.00x10^{-1}$	$1.05x10^{-1}$	$1.68x10^{-2}$	0.00	$1.90x10^{-3}$
Oil & Gas Depletion excl	kg Oil eq	$3.90x10^{-2}$	$4.00x10^{-2}$	$2.00x10^{-1}$	$8.46x10^{-2}$	$1.27x10^{-2}$	0.00	$5.00x10^{-4}$
Waste	MJ	0.00	0.00	$1.72x10^1$	0.00	0.00	0.00	$1.01x10^1$
Land-Use	Bio factor	0.00	0.00	0.00	0.00	0.00	0.00	0.00
Water Stress Indicator	WSI factor	$2.58x10^{-2}$	$1.00x10^{-2}$	$1.00x10^{-1}$	$1.60x10^{-2}$	$8.70x10^{-2}$	0.00	$2.00x10^{-4}$

Acidification occurs due to air pollution, which causes a decrease in the pH of the water and makes the water acidic. This acidification can cause acidification of water in both the sea and rivers, damage forests, and health problems for humans. Acidification also causes an increase in SO_2 (sulfur dioxide) in the earth's atmosphere, so that it can cause acid rain. A small portion of SO_2 is caused by nature, such as volcanoes, and most of it is caused by burning fossil fuels to generate electricity, vehicles, manufacturing processes, oil refineries, and other industries. In these tofu SMEs, the most significant impact of acidification is generated from the cooking process due to electricity to pump water generated from burning fossil fuels. Burning corn cobs in the tofu frying process also has an enormous impact. The total value of the environmental impact of the acidification category from the tofu production in Eko Budi enterprise is 0.21 Kg SO_2 equivalent.

Table 3. The result of a single score in euro.

Impact Category	Unit	Soaking	Grinding	Heating	Filtering	Clumping	Forming	Frying	Total
Climate Change	Euro	0.584	0.421	1.800	0.882	0.322	0	1.528	5,537
Acidification	Euro	0.179	0.141	0.600	0.287	0.110	0	0.500	1.817
Eutrophication	Euro	0.013	0.004	0.020	0.007	0.113	0.013	0.112	0,282

Impact Category	Unit	Soaking	Grinding	Heating	Filtering	Clumping	Forming	Frying	Total
Photochemical Oxidant Formation	Euro	0.002	0.001	0.010	0.003	0.001	0.000	0.003	0.021
Fine Dust	Euro	0.031	0.029	0.100	0.060	0.015	0.000	0.015	0.251
Human Toxity	Euro	0.046	0.043	0.200	0.091	0.014	0.000	0.008	0.401
Ecotoxicity (Freshwater)	Euro	0.006	0.006	0.027	0.013	0.002	0.000	0.003	0.057
Metals Scarcity	Euro	0.125	0.049	0.200	0.105	0.017	0.000	0.002	0.498
Oil & Gas Depletion excl Energy	Euro	0.031	0.032	0.100	0.070	0.010	0.000	0.000	0.244
Waste	Euro	0.000	0.000	0.200	0.000	0.000	0.000	0.109	0.309
Land-Use	Euro	0.000	0.000	0.000	0.000	0.000	0.000	0.000	0.000
Water Stress Indicator	Euro	0.026	0.008	0.100	0.016	0.087	0.000	0.000	0.237
Total		1.040	0.733	3.414	1.530	0.690	0.013	2.280	9.653

Table 4. The result of a single score in IDR.

Impact	Unit	Soaking	Grinding	Heating	Filtering	Clumping	Forming	Frying	Total
Climate Change	IDR	9,970	7,181	31,090	15,046	5,503	-	26,086	94,875
Acidification	IDR	3,058	2,399	10,546	4,903	1,886	-	8,525	31,316
Eutrophication	IDR	216	60	386	123	1,932	217	1,919	4,852
Photochemical	IDR	41	25	113	52	23	-	55	309
Fine Dust	IDR	536	491	2,070	1,029	253	-	258	4,637
Human Toxicity	IDR	780	726	3,098	1,544	239	-	140	6,527
Ecotoxicity	IDR	110	107	458	225	31	-	45	976
Metals Scarcity	IDR	2,133	844	3,582	1,790	286	-	32	8,668
Oil & Gas	IDR	532	541	2,314	1,156	173	-	7	4,723
Waste	IDR	-	-	3,178	-	-	-	1,855	5,033
Land-Use	IDR	-	-	-	-	-	-	-	-
Water Stress	IDR	440	130	1,433	273	1,481	-	4	3,761
Total		17,814	12,503	58,269	26,141	11,808	217	38,927	165,6799

This study obtained an eco-cost value of €9.7076 to produce one batch of tofu with a yield of 9 kg. Based on the Euro to Rupiah exchange rate on August 2, 2021, it is known that it is IDR 17,066.87, which is equivalent to IDR 165,679. SMEs produce an average of about 20 batches of tofu, so it has an eco-cost value of around IDR. 3,313,580/day/SME.

From Figure 3, it is known that the most significant single score is in the cooking and frying process. The cooking process is caused by the extensive use of electricity for water pumps that use fossil fuels (conventional) as a producer of electrical energy and rice husk fuel for boiler engines. In the frying process, the environmental impact becomes large due to the use of corncob as fuel. This study recommends energy conservation efforts in the production process of tofu using biogas as a substitute for rice husks and corn husks.

IOP Conf. Series: Earth and Environmental Science **894** (2021) 012004 doi:10.1088/1755-1315/894/1/012004

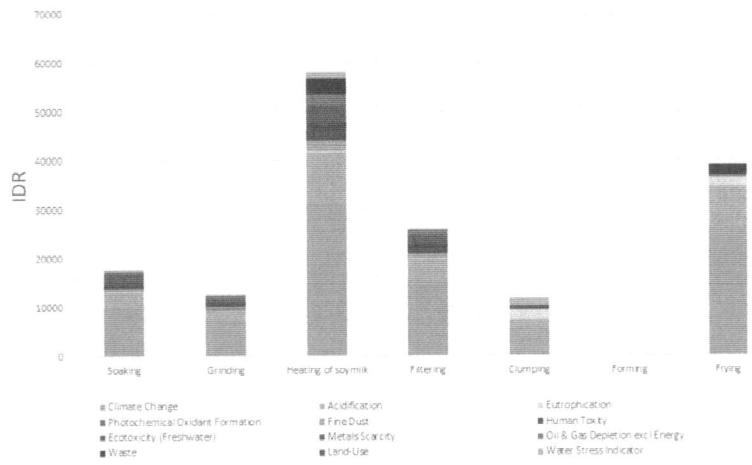

Figure 2. Single score.

Table 5. The comparison of environmental impact between existing conditions and proposed.

Impact Category	Characterization			Single score (euro)		Single score (IDR)	
	Unit	**Existing**	**Proposed**	**Existing**	**Proposed**	**Existing**	**Proposed**
Climate Change	kg CO2 eq	47.93	41.90	5.56	4.86	94875.38	82939.24
Acidification	kg SO2 eq	0.21	0.16	1.84	1.43	31316.44	24369.62
Eutrophication	kg PO4 eq	0.07	0.05	0.28	0.12	4852.44	2026.17
Photochemical Oxidant Formation	kg C2H4 eq	0.00	0.00	0.02	0.02	309.15	318.49
Fine Dust	kg PM2.5 eq	0.01	0.01	0.27	0.26	4637.30	4483.69
Human Toxicity	Cases	0.00	0.00	0.38	0.38	6526.90	6525.96
Ecotoxicity (Freshwater)	PAF.m^3.day	10,326.43	10,219.39	0.06	0.06	976.38	966.26
Metals Scarcity	Euro	0.51	0.55	0.51	0.55	86.67.57	9391.97
Oil & Gas Depletion excl Energy	kg Oil eq	0.35	0.35	0.28	0.28	4722.83	4781.14
Waste	MJ	27.32	0.10	0.30	0.00	5032.87	18.43
Land-Use	Bio factor	0.00	0.00	0.00	0.00	0.00	0.00
Water Stress Indicator	WSI factor	0.22	0.25	0.22	0.25	3761.38	4328.98
Total				9.71	8.22	165678.64	140149.96

Suparni [16] stated that biogas could be alternative energy to replace firewood in the cooking process. The use of biogas can result in a cleaner cooking process because it does not produce smoke and reduces greenhouse gas emissions due to wood fuel. Biogas itself can be produced from the processing of tofu liquid waste which is produced in large quantities. To see the effect of biogas, this study has recalculated if energy conservation has been carried out. For the amount of biogas used, the equation for the conversion of biogas with rice husks and corn cobs is carried out. 1 kg of corn cobs is equivalent to 0.21675 m^3 biogas [17]. Meanwhile, 1 kg of rice husk is equivalent to 0.25 m^3 of biogas [18].

Furthermore, the processing is carried out with SimaPro replacing the energy of rice husks and corn cobs with biogas.

From Table 1, it can be seen that if the energy source of corn husks and corncobs is changed to biogas, the eco-cost value from IDR 165,679 to IDR 140,150 per batch so will decrease by around IDR. 25,529 per batch. If SMEs cook 20 batches per day, the eco-cost will be reduced by around IDR 510,574 per day. Sugihmanik Village has around 30 tofu SMEs, so switching energy sources to biogas will reduce the eco-cost by around IDR 15,317,208 per day or around IDR 459,516,240 per month.

Tofu SMEs already know the importance of tofu wastewater treatment so as not to damage the environment. Tofu SMEs who are members of an association have purchased land planned to build a wastewater treatment plant. However, in reality, they have not treated the waste and immediately dumped it into the river. The main problem lies in the limited cost of constructing WWTPs, which are high and do not understand the technology. It requires the participation of universities and local governments. The pattern of collaboration needs to be applied in overcoming environmental problems that are considered disturbing to the surrounding community.

4. Conclusion

This study has found that the highest impact categories occur in the climate change and acidification categories. The most impactful activities are the cooking and frying processes. Due to the very high level of water consumption where the extraction uses an electric pump, the process requires energy from rice husks and corn cobs. To reduce the environmental impact, it is necessary to conserve water and energy by building WWTPs and biodigesters. This study also evaluates the environmental impact of energy conversion using biogas. This study has not yet designed an appropriate WWTP design for water conservation. Further studies need to be carried out appropriate WWTP design and evaluation of BOD/COD level after the WWTP and environmental impact measurement with life cycle assessment if the WWTP has been built.

Acknowledgment

Many thanks, the authors say to Diponegoro University, which has funded community service with the program "IPTEK untuk Desa Binaan Undip 2021."

References

[1] Zheng L, Regenstein J M, Teng F and Li Y 2020 *Comprehensive Reviews in Food Science and Food Safety* **19(6)** 3683–3714
[2] Yang Y *et al* 2021 *Food Chemistry* **346** no October 2020
[3] BPS 2018 Perkembangan Data Usaha Mikro, Kecil, Menengah dan Usaha Besar. In Www.Depkop.Go.Id (Vol. 2000, Issue 1)
[4] Faisal M, Gani A, Mulana F and Daimon H 2016 *Asian J Chem* **28(3)** 501–507
[5] Suryoko S, Hadi S P and Purwanto 2019 *E3S Web of Conferences* **125(201)** 9
[6] Carucci G, Carrasco F, Trifoni K, Majone M and Beccari M 2005 *Journal of Environmental Engineering* **131(7)** 1037–1045
[7] Kurniawati S D, Supartono W and Suyantohadi A 2019 *IOP Conf Ser Earth Environ Sci* **365(1)**
[8] Bessou C, Cynthia L, Alice V, Hadrien H, Henri V, Jean-Pierre C 2016 *International Journal of Life Cycle Assessment* **21(3)** 297–310
[9] Hartini S, Sari D, Aisy N and Widharto Y 2020 *E3S Web of Conferences 202* **10004** p 1–9
[10] Hartini S, Wicaksono P A, Prastawa H, Hadyan A F and Sriyanto 2019 *IOP Conference Series: Materials Science and Engineering* **598(1)**
[11] Prastawa H and Hartini S 2019 The influence of product design on environmental impacts using life cycle assessment **030015** no June p 030015
[12] Internationaol Organization for Standardization 2006 *Environmental management -- Life cycle assessment - Principles and framework* p 20 [Online]
[13] Ramos A and Rouboa A 2020 *Environmental Impact Assessment Review* **85** no April p 106469

IOP Conf. Series: Earth and Environmental Science **894** (2021) 012004 doi:10.1088/1755-1315/894/1/012004

[14] N Duarte da Silva Lima, I de Alencar Nääs, Garcia R G and D Jorge de Moura 2019 *Journal of Cleaner Production* **237**
[15] Vogtlander J G, Scheepens A E, Bocken N M P and Peck D 2017 *Journal of Remanufacturing* **7(1)** p 1–17
[16] Rahayu S S, Budiyono B and Purwanto P 2018 *E3S Web of Conferences* **31** p 2017–2019
[17] Li W *et al* 2015 *Bioresource Technology* **194** p 276–282
[18] Djoefrie B and Nastiti D M 2019 *Pencegahan, Pengendalian, dan Pemanfaatan Limbah Organik.* (PT Penerbit IPB Press)

IOP Conf. Series: Earth and Environmental Science **894** (2021) 012005 doi:10.1088/1755-1315/894/1/012005

Study to reduce greenhouse gas emissions at waste landfill in Medan City

I Suryati[1]*, A Farindah[1], I Indrawan[2]

[1]Environmental Engineering, Universitas Sumatera Utara, Indonesia
[2]Civil Engineering, Universitas Sumatera Utara, Indonesia

isratl@usu.ac.id

Abstract. Landfill is a place where waste reaches the final stage. The piles of waste can generate greenhouse gas emissions that cause global warming the potential of climate change. The greenhouse gas emission generates from the piles of waste is CH_4 emission. The research purpose is to count CH_4 emission in the waste landfill in Medan city located in Terjun, projection CH_4 emission for ten years later is 2020-2029 and decisive the effort reduction of CH_4 emission. The scenarios of reducing CH_4 emission in Terjun waste landfill reduce the potential CH_4 emission for ten years later. The calculation of CH_4 emission from the piles of waste in Terjun waste landfill using FOD method (First Orde Decay) by IPCC (Intergovernmental Panel on Climate Change) in 2006. In 2019, CH_4 emission in Terjun waste landfill was 12,350.750-ton CH_4 and had an uplift in 2029 can reach 17,143.087-ton CH_4. There are two scenarios for reducing CH_4 emission in the Terjun waste landfill; the first is the processing of waste in the source (composting), and the second is reducing the waste by using incineration technology Terjun landfill. The first scenario (composting) can reduce CH_4 emission by 14.80%. The second scenario can reduce by 63.37% the CH_4 emission in Terjun waste landfill. The chosen alternative scenario for reducing CH_4 in the Terjun waste landfill is the first scenario, the processing of waste in the source (composting).

1. Introduction

The increasing number of residents in Medan also increases the amount of waste generated in Medan. Piles of both organic and inorganic waste will cause the waste to undergo weathering, which triggers greenhouse gas formation, namely CH_4 gas. Every 1 ton of solid waste will produce about 50 kg of methane gas [1]. Based on the 2006 Intergovernmental Panel on Climate Change (IPCC) report, the waste sector also contributes greenhouse gas to the atmosphere, specifically from landfills, contributing between 3 – 4% of global greenhouse gas emissions [2].

Greenhouse gas is a gas in the atmosphere that functions to absorb infrared radiation and determines the atmosphere's temperature. The existence of various human activities can increase the concentration of greenhouse gas emissions in the atmosphere. It can trigger the problem of global warming and climate change. The generation of CH_4 gas in the landfill usually starts 6-12 months after being placed, increases after the landfill's closure and gradually decrease over 30-50 years. One ton of domestic waste can produce 300 m^3 of biogas. CH_4 gas is one of the main greenhouse gasses against global warming. CH_4 gas concentration in the atmosphere has increased by about 0.6% per year and more than doubled over the last two centuries [3].

The purpose of this study is to calculate the CH_4 gas emissions produced at Terjun Waste Landfill, projecting the CH_4 gas emissions produced by Terjun Waste Landfill for the next ten years and determining the efforts to reduce CH_4 gas emissions at Terjun Waste Landfill. Indonesia is committed to reducing greenhouse gas emissions by 26% and 41% if it gets help from abroad. The increasing greenhouse gas emissions in the atmosphere will cause global warming that triggers climate change, so it is necessary to carry out an inventory of greenhouse gas emissions which will produce data on the estimated burden of greenhouse gas emissions which is used as the primary basis in determining efforts to mitigate greenhouse gas emissions.

2. Methodology

2.1. Data collection

The primary data for this research is the characteristics of the waste and the calorific value of the waste. Waste taken in the form of waste aged 27 years, 20 years and ten years at Terjun Waste Landfill to test the characteristics of the waste in the form of physical characteristics of the waste such as water content, volatile content, ash content and calorific value was carried out in the laboratory. The purpose of taking waste aged 10-27 years is so that the results of the characteristic test can represent the characteristics of the waste in Terjun Waste Landfill.

The secondary data of this study is the population of Medan in 1993-2012 obtained from the Central Statistics Agency of Medan. The amount of waste generation for Medan based on the Ministry of Public Works is 0.8 kg/p/day. Waste composition data were obtained from previous studies [4].

2.2. Waste generation projection calculation

The calculation of the projected waste generation at Terjun Waste Landfill is carried out using the formula. The formula for calculating the projected waste generation at Terjun Waste Landfill is [5]:
Amount of projection year waste =

$$Previous\ year\ waste\ (1 + avarege\ of\ \ waste\ growth)^{projection\ year - previous\ year} \quad (2.1)$$

$$\text{The average waste growth } = ((final\ value / initial\ value)^{1/t} - 1)\ x\ 100\% \quad (2.2)$$

2.3. Calculation of CH_4 emissions at Terjun Waste landfill

Calculation of CH_4 emissions is carried out based on the provisions of the IPCC [6]. This study uses the FOD (First Order Decay) method. This method is more accurate because it can calculate methane recovery, where methane gas produced from solid waste in the landfill can be used as environmentally friendly energy. The equation used to calculate methane gas emissions in the waste sector at the landfill in this study refers to the 2006 IPCC guidelines.

$$DDOCm\ rem_T = DDOCmd_T . e^{-k(13-M)/12} \quad (2.3)$$

$$DDOCm\ dec_T = DDOCmd_T .(1 - e^{k(13-M)/12}) \quad (2.4)$$

$$DDOCma_T \ = \ DDOCm\ rem_T + (DDOCma_{T-1} . e^{-k}) \quad (2.5)$$

$$DDOCm\ decomp_T = DDOCm\ dec_T + DDOCma_{T-1} .(1 - e^{-k}) \quad (2.6)$$

$$CH_4\ Generated_T = DDOCmdecomp_T . F. 16/12 \quad (2.7)$$

$$DDOCm = W.\ DOC\ .\ DOCf\ .\ MCF \quad (2.8)$$

$$Emisi\ CH_4 = [\sum x\ Timbulan\ CH_{4\ \ x,T} - R_T]\ .\ (1 - OX_T) \quad (2.9)$$

Information:

T	= Inventory year
DDOCm remT	= Mass of DDOCm deposited, which does not decompose in year T that remains until the end of the year (Gg)
DDOCmdT	= DDOCm at the end of the year (T-1) (Gg)
DDOCm decT	= DDOCm in year T which has decomposed at the end of year T (Gg)
DDOCmaT	= DDOCm accumulated in landfill at the end of year T (Gg)

IOP Conf. Series: Earth and Environmental Science **894** (2021) 012005 doi:10.1088/1755-1315/894/1/012005

DDOCmaT-1 = DDOCm accumulated in the landfill at the end of the year (T-1) (Gg)
DDOCm decompT = DDOCm decomposed in year T (Gg)
M = Month when reaction started, equal to +7 average waiting time (months)
k = Reaction rate constant (y-1)
CH_4 *Generated*$_T$ = CH_4 formed in year T as a result of the decomposition of organic components that are stored in the trash (DDOC) (Gg)
F = Fraction of CH_4 generated in landfill (0,5)
16/12 = CH_4 / C molecular weight ratio
DDOCm = Mass of DOC which stored in waste in a biodegradable landfill (Gg)
W = Mass dumped garbage (Gg)
DOC = Carbon degraded organic (Gg C/Gg Garbage)
DOCf = Fraction of biodegradable DOC (for Indonesia 0,5)
MCf = CH_4 correction factor for aerobic decomposition (0,8)
R_T = Amount of CH_4 recovered in year T (Gg)
OX = Oxidation factor in year T (Fraction).

Parameters that use IPCC default numbers are oxidation factor (OX) and DOC (Degradable Organic Carbon) data.

3. Results and discussion

3.1. Waste generation at Terjun Waste Landfill

There is no data on waste generation at Terjun waste landfill from the beginning of the operating year until 2012. It was because there were no scales at that time and the administrative system at Terjun Waste Landfill was also not adequate. In 2013, we started weighing the waste transported by trucks that enter the waste landfill. Every truck that enters will be recorded first before entering the waste landfill. The amount of waste generated at Terjun Waste Landfill in 1993-2019 can be seen in Figure 1.

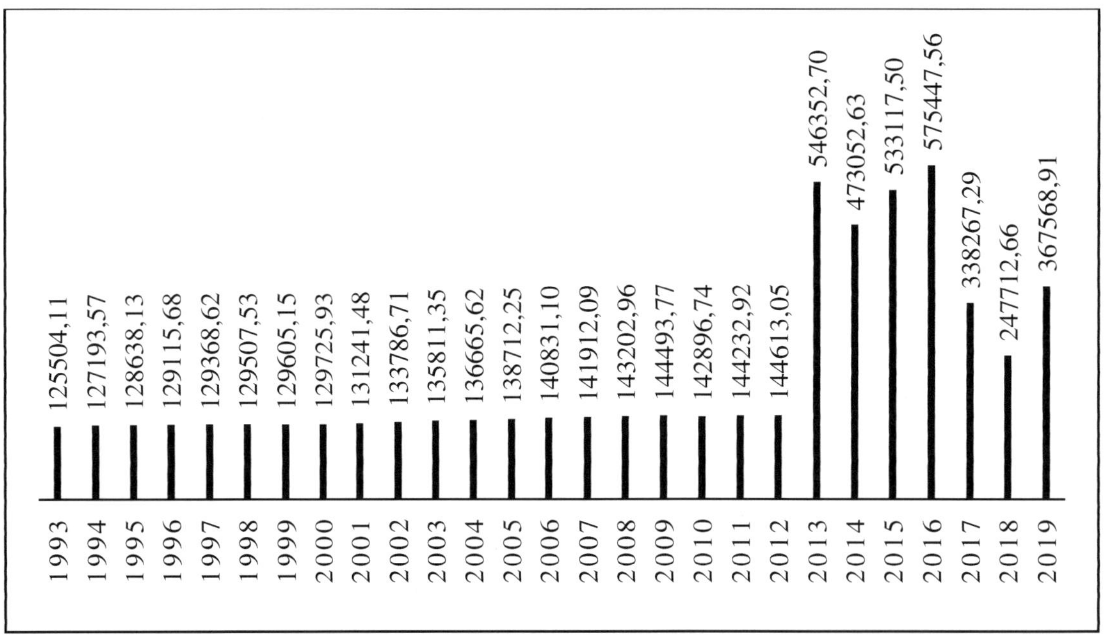

Figure 1. Waste generation (tons) at the Terjun Waste Landfill in 1993-2019.

3.2. Projection of waste generation at Terjun Waste Landfill
Projection of waste generation can be seen in Figure 2.

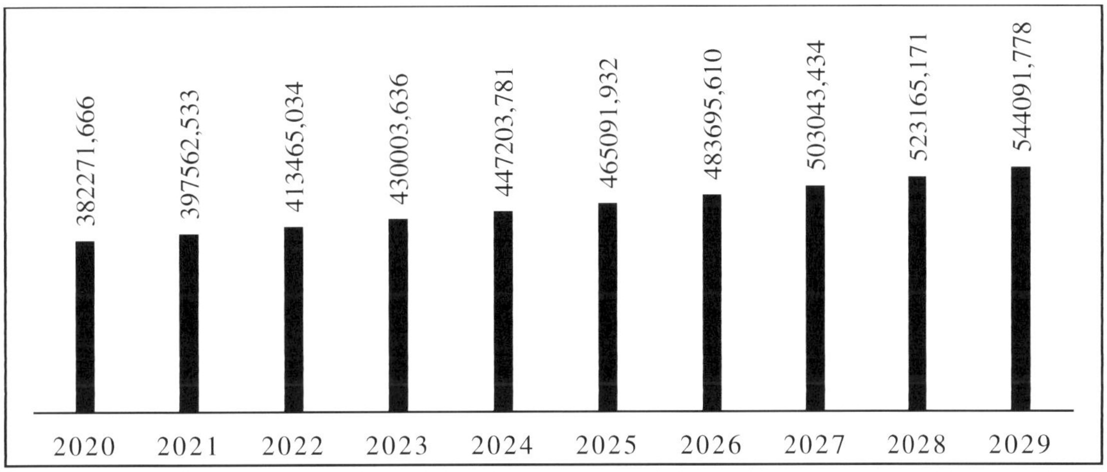

Figure 2. Projection of waste generation (tons) at Terjun Waste Landfill.

Based on Figure 2, waste generation always increases every year. With the increasing number of residents in Medan, more waste in Medan will be disposed of to the Terjun waste landfill. The percentage increase in waste generation from 2020-2029 is 7.79%. The amount of waste generated by a city is primarily determined by all community activities in that city. These activities include trading activities, housing, offices, industrial activities, agriculture and others. One of the things that affect the waste generation in a city is the number and density of the population. The amount of waste generated is directly proportional to the population. The more the population, the more waste generated. Likewise, for population density [7].

Other factors influence waste generation apart from a large number of residents. The level of income of the population affects the generation of waste in Bukittinggi. Residents with high incomes (HI) generate higher waste than people with middle and low incomes (LI). Waste generation for high-income residents produces 0.28 kg/person/day, middle-income residents generate 0.22 kg/person/day and waste generation generated by low-income residents produces 0.10 kg/person/day. It is because people with high incomes (HI) do more activities that produce waste. The method of handling waste by high-income residents is different from that of low-income residents. Residents with high incomes (HI) usually throw away what is no longer needed. In comparison, people with low incomes (LI) can still use the waste they produce, such as food waste used for animal feed or yard waste used as fertilizer [8]. The amount of generation is directly proportional to community life. The higher the level of life, the greater the generation of waste [9].

3.3. CH_4 emissions at Terjun Waste Landfill
The calculation of CH_4 emissions begins with calculating the year the Terjun waste landfill started operating, 1993. In 1993, the CH_4 emission in the Terjun waste landfill was 0 tons. It took 13 months to produce the CH_4 anaerobic process from the waste heap in the waste landfill. Methane gas is formed from the anaerobic decomposition of organic matter by methane bacteria (methanogens). The aerobic decomposition process lasts for approximately 2-3 weeks until the oxygen supply in the waste is reduced. After the O_2 gas is exhausted, it is followed by anaerobic acid decomposition. Anaerobic bacteria produce CO_2 and organic acids. This process occurs for 1-2 years, and the temperature of the waste has decreased. Furthermore, the decomposition process enters the final stage, namely anaerobic methane. In this process, the bacteria produce CH_4 and CO_2 [10].

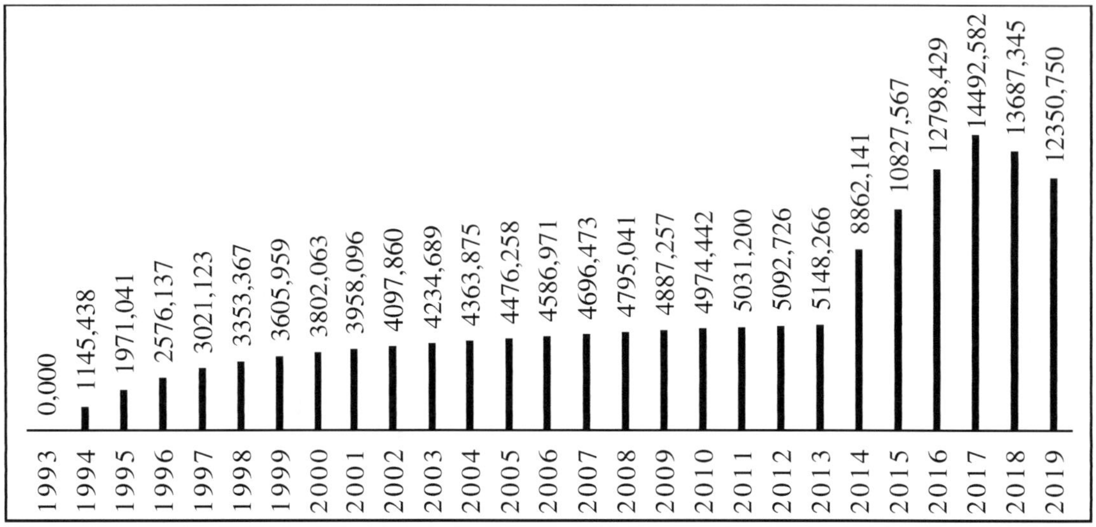

Figure 3. Total CH_4 emissions 1993-2019 (tons CH_4).

After calculating CH_4 emissions at Terjun Waste Landfill from the beginning of operations until 2019, a projected CH_4 emission calculation is carried out for the next ten years. The calculation is also the same as the previous calculation of CH_4 emissions. The following is an example of calculating CH_4 emissions at the Terjun Waste Landfill in 2020. Total Terjun waste landfill emissions from 1993 to 2029 can be seen in Figure 4.

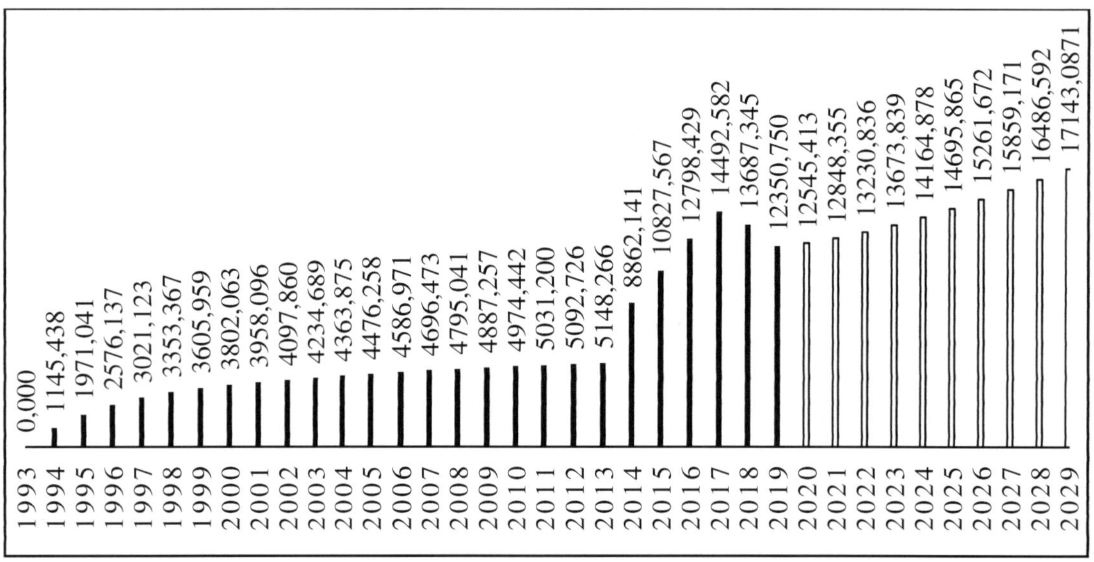

Figure 4. Total CH_4 emissions at Terjun Waste Landfill in 1993-2029 (tons of CH_4).

Based on Figure 4, the amount of CH_4 emissions produced by the waste at Terjun Waste Landfill in 1993 was 0 tons of CH_4. It took 13 months for the formation of CH_4 from the waste heap. Therefore, CH_4 emissions in 1994 amounted to 1,145.438 tons of CH_4 or 28,635.947 tons of CO_2e. In 2014, CH_4 emissions at Terjun Waste Landfill experienced a significant increase, amounting to 8,862.141 tons of CH_4 or 221,553.13 tons of CO_2e. The amount of waste that went to Terjun Waste Landfill in 2013 increased after Namo Bintang Waste landfill was officially closed. The increase in the amount of CH_4 emissions is due to the increasing amount of waste generation.

Terjun Waste Landfill still uses an open dumping system and has more than 10 meters of waste thickness. The magnitude of the potential for CH_4 gas produced at the waste landfill is not only influenced by the increase in waste generation. The thickness of the waste also affects the formation of CH_4 through air access and the density of the waste. Open dumping waste landfill system with a thickness of > 5 meters of waste, more CH_4 gas production due to the upper layer of waste preventing air access from entering the lower layer and creating anaerobic conditions so that more CH_4 gas is produced [11].

The composition of the waste also affects the production of CH_4 gas in waste landfills. The more organic waste there is, bacteria decompose, producing more gas [12]. The composition of the waste will determine the nutrient content that has a vital role in gas formation. Organic waste disposed of in waste landfills is the most significant factor affecting the quality of gas emissions because it contains organic carbon that is easily degraded [13].

3.4. CH_4 emission reduction scenarios

3.4.1. Waste treatment at source (composting). In this scenario, the data needed is the number of residents of Medan who are willing to do composting and the amount of organic waste that is reduced to compost. Based on the questionnaire results, only 58% of the residents of Medan are willing to sort their waste at the source (composting).

The reduction in organic waste that is processed by aerobic composting is 40%-60% [14]. The reduced waste in this scenario is organic in food waste and leaf/garden waste. The percentage of waste reduction in this scenario is 60%. The amount of waste that was reduced at Terjun Waste Landfill due to composting was 26.61%. Next, calculate CH_4 emissions from waste that is not composted and disposed of at Terjun Waste Landfill. The purpose of calculating CH_4 emissions from the waste is to find out how much CH_4 emissions are after composting activities at the source. The calculation of CH_4 emissions is similar to baseline CH_4 emissions, but the difference is the amount of waste reduced because it has been reduced from the source. The comparison of CH_4 emissions in Terjun waste landfill before and after composting can be seen in Figure 5.

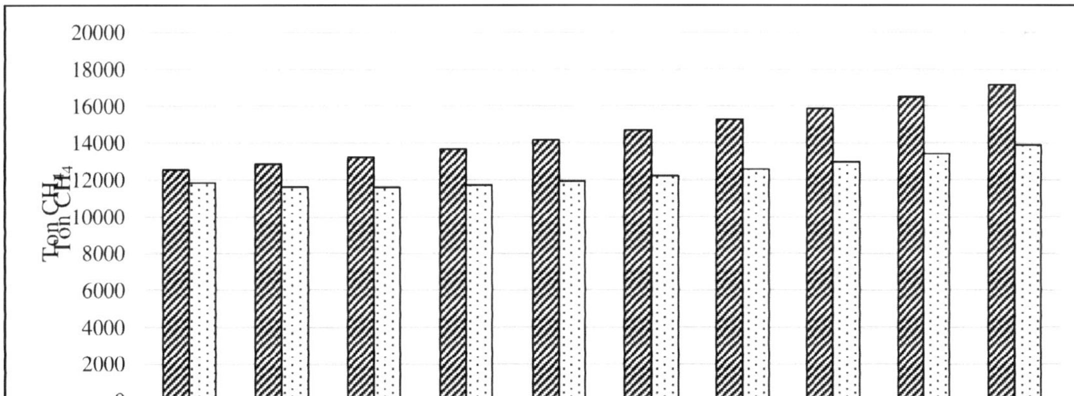

	2020	2021	2022	2023	2024	2025	2026	2027	2028	2029
Before	12545,41	12848,35	13230,83	13673,83	14164,87	14695,86	15261,67	15859,17	16486,59	17143,08
After	11840,28	11616,27	11591,80	11708,55	11927,39	12222,14	12575,29	12975,23	13414,28	13887,49

Figure 5. Comparison of the amount of CH_4 emissions (tons of CH_4) before and after composting at the source in 2020-2029.

Based on Figure 5, the number of emissions at Terjun Waste Landfill will decrease if composting is carried out at the source. It is due to the reduction of organic waste every year. In Figure 5, the CH_4 emissions in 2020 before composting was 12,545.413 tons of CH_4 or equivalent to 313,635.322 tons of

CO_2e. After composting, it became 11,840.284 tons of CH_4 or equivalent to 296,007.097 tons of CO_2e. The per cent reduction in CH_4 emissions from 2020-2029 with composting at the source is 14.80%. If there is an increase in the willingness of the residents of Medan to do composting at the source every year, then the amount of reduced waste will increase, and the percentage of CH_4 reduction can also increase every year.

Processing of waste by composting will reduce CH_4 emissions. Without processing, organic material in a waste landfill will produce CH_4 emissions of 2,326,007 tons/year. Meanwhile, with composting activities, CH_4 emissions are reduced to 9,255-18,344 kg/year. Indonesia's experience in reducing CH_4 emissions from composting activities carried out by the World Bank through the Compost Subsidy Program in 2004-2005 in 16 cities in the western part of the island of Java which succeeded in reducing CH_4 emissions by 11,049-14,916 m^3 or equivalent to 254,129-343,074 tons of CO_2 [15].

Composting 1,9 tons of waste can produce 1 ton of compost, while one ton of waste is stored in a landfill can produce 0.20-0.27 m^3 of CH_4 emissions. Thus, producing 1 ton of compost can reduce CH_4 emissions by 0.21-0.29 tons of CH_4 or equivalent to 5-7 tons of CO_2e [16].

3.4.2. Waste reduction with incineration technology at Terjun Waste Landfill. The results of testing the calorific value of Terjun waste landfill are very high, namely 12,009,834.67 kJ/kg. The usefulness of knowing the calorific value of waste is to find out whether the waste is good or not if it is managed by incineration. Waste can be processed by incinerator without the need for additional fuel with a minimum calorific value of 2,200 kcal/kg or the equivalent of 9,196 kJ/kg [17]. Various studies state that the calorific value required for WtE from the incineration process whose energy will later be utilized is 7,000 kJ/kg or more and should not be below 6,000 kJ/kg [18]. Therefore, the waste in Terjun Waste Landfill is very suitable if it is managed by incineration because it has a very high calorific value. The process of processing waste by incineration can reduce waste by 80%-90% [19]. In this scenario, the reduction of waste by incinerator is 90%. The comparison of CH_4 emissions before and after the waste is processed by the incinerator can be seen in Figure 6.

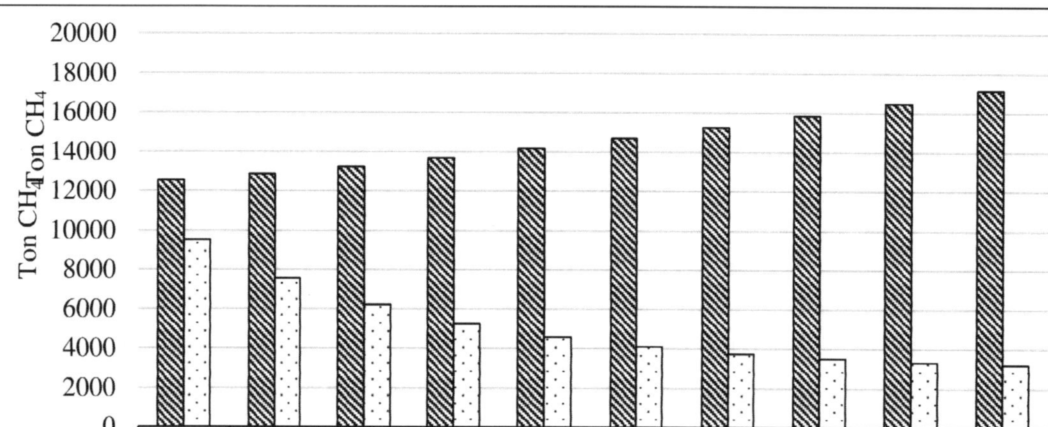

	2020	2021	2022	2023	2024	2025	2026	2027	2028	2029
Before	12545,41	12848,35	13230,83	13673,83	14164,87	14695,86	15261,67	15859,17	16486,59	17143,08
After	9526,192	7572,828	6212,836	5258,878	4584,433	4103,882	3759,150	3510,719	3331,601	3203,280

Figure 6. Comparison of the amount of CH_4 emissions before and after the waste is reduced by the incinerator in 2020-2029.

Figure 6 shows the amount of CH_4 emissions in 2020 before the waste reduction with the incinerator was 12,545.413 tons of CH_4 or equivalent to 313,635.322 tons of CO_2e, but after the incinerator reduced the waste, it became 9,526.192 tons of CH_4 or equivalent to 238,154.800 tons of CO_2e. The percentage of CH_4 emission reduction from 2019-2028 by reducing waste with an incinerator is also 63.37%.

Incineration technology is very feasible at Terjun Waste Landfill because, in terms of the characteristics and calorific value of the waste, it is very feasible to be managed by the incineration process.

3.4.3. Alternative scenario. After obtaining the total CH_4 emissions from each scenario, the magnitude of the CH4 emissions reduction can be known. The graph of the reduction of CH_4 emissions at Terjun Waste Landfill in 2020-2029 can be seen in Figure 7.

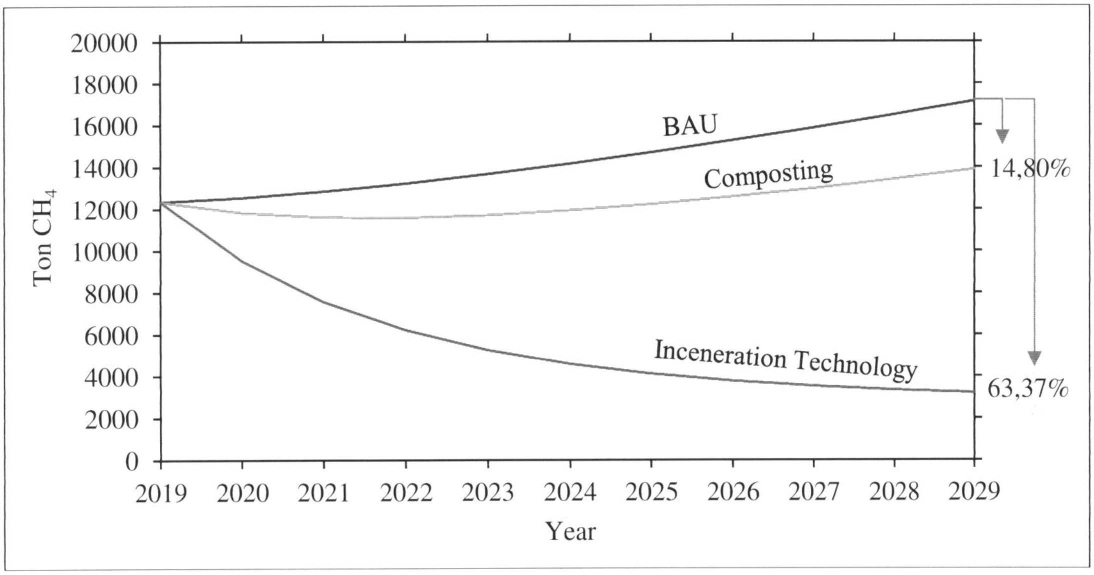

Figure 7. Graph of CH_4 emission reduction in Terjun Waste Landfill in 2019-2029.

Based on Figure 7, the most significant reduction in CH_4 emissions is in scenario 2, namely incineration technology. Incineration technology can reduce waste, a source of CH_4 emissions in Terjun Waste Landfill, using an incinerator by 63.37% so that the CH_4 emissions in Terjun waste landfill are reduced a lot. Besides reducing waste, the steam produced from burning waste using an incinerator can be used as renewable energy. However, from an environmental point of view, it is also dangerous because incinerators are feared to cause air pollution gases that endanger human health and the environment if the temperature is not optimal when operating. The use of this incinerator also requires a huge investment cost, requires a large area of land, and its operation is not easy; it must be maintained.

In scenario 1, composting can only reduce CH_4 emissions at Terjun Waste Landfill from 2020- 2029 by 14.80%. At least the reduction of CH_4 emissions in Terjun Waste Landfill from composting activities is caused by several factors. First, every year the amount of waste is increasing, but the willingness of the residents of Medan to do composting at home is not too much, only 58%. If the population's willingness to do composting increases every year, it is possible that the reduced amount of CH_4 emissions in the Terjun Waste Landfill could continue to increase. Second, not all the waste used as compost can be composted, only food waste and garden/garden waste. However, not all food waste and garden/garden waste can be composted. Only 40%-60% of the waste can be processed into compost.

Based on the description above, the alternative scenario chosen to reduce CH_4 emissions in Terjun Waste Landfill in the next ten years is scenario 1, namely processing waste at the source (composting). The selection of this scenario is assessed based on the impact on environmental quality, land requirements, investment costs, and operation.

4. Conclusion

The amount of CH_4 emissions produced at Terjun Waste Landfill in 1994 was 1,145.438 tons of CH_4, or equivalent to 28,635.947 tons of CO_2e from the total waste of 34.978 Gg. In 2019 the CH_4 emissions increased to 12,350.750 tons of CH_4 or equivalent to 308,768.750 tons of CO_2e from a total waste of 367.569 Gg. Projected CH_4 emissions produced by Terjun Waste Landfill in the next ten years continue to increase. By 2029 emissions of CH_4 reach 17,143.087 tons of CH_4 or equivalent to 428,577.177 tons of CO_2e from the total waste 544.092 Gg. Efforts to reduce CH_4 emissions in the waste sector at Terjun Waste Landfill by processing waste at the source (composting) and reducing waste with incineration technology at Terjun Waste Landfill. In 2029, scenario 1 (composting) can reduce CH_4 emissions in Terjun Waste Landfill to 13,887.492 tons of CH_4 or equivalent to 347,187.288 tons of CO_2e. The percentage of CH_4 emission reduction in 2019-2028 from composting activities is 14.80%. Meanwhile, scenario 2 (waste reduction with an incinerator) can reduce CH_4 gas emissions at Terjun Waste Landfill in 2029 to 3,203.280 tons of CH_4 or equivalent to 80,081.991 tons of CO_2e. The percentage of CH_4 emission reduction in 2020-2028 from waste reduction with incinerator is 63.37%.

References

[1] R A A Rahma et al 2015 *J Ekosains* **7** 106
[2] Khatulistiwa M R et al 2015 *J Tek Ling Lahan Basah.* 2
[3] Purwanta W 2009 *J Tek Ling* **10** 5
[4] Khair H et al 2019 *J Urb Env Techno* **3** 42
[5] Jaisyullah U A 2017 *Thesis of Environmental Engineering* (Surabaya: Intitut Teknologi Sepuluh November)
[6] IPCC 2006 *IPCC Guidelines for National Greenhouse Gas Inventories* (Japan: IGES)
[7] Putri C N 2018 Thesis of Environmental Engineering (Medan: Universitas Sumatera Utara)
[8] Damanhuri E and Padmi T 2018 *Pengelolaan Sampah Terpadu Edisi Kedua* (Bandung: Institut Teknologi Bandung)
[9] Ruslinda Y et al 2012 *J Tek Ling* **9** 5
[10] Indarto A M 2007 *Thesis of Environmental Science* (Semarang: Universitas Sebelas Maret)
[11] Prabowo S et al 2019 *J Bioeksperimen* **5** 22-31
[12] Sari A M 2018 Thesis of Environmental Engineering (Makasar: Universitas Hasanuddin)
[13] Santosa S dan Soemarno 2014 *J Indonesian Green Techno* **3** 36
[14] Peraturan Menteri Pekerjaan Umum Nomor 03 2013 *Penyelenggaraan Prasarana dan Sarana Persampahan Dalam Penanganan Sampah Rumah Tangga dan Sampah Sejenis Sampah Rumah Tangga* (Jakarta: Menteri Pekerjaan Umum Republik Indonesia)
[15] Wahyono S 2016 *J Tek Ling* **17** 39
[16] Suprihatin et al 2002 *J Tek Ind Pertanian.* **18** 59
[17] Cointreau S 2006 *Occupational and Environmental Health Issues of Solid Waste Management* (Washington DC: The World Bank Group)
[18] GIZ 2017 *Waste to Energy Options in Municipal Solid Waste Management-A Guide for Decision Makers in Developing and Emerging Countries* (Deutsch: Gesellschaft für Int)
[19] Rini T S et al 2020 *J Rese Technol* **6** 104

IOP Conf. Series: Earth and Environmental Science **894** (2021) 012006 doi:10.1088/1755-1315/894/1/012006

Water quality and management of Citepus River, Bandung, Indonesia

Y M Yustiani[1], F K Tarigan[1]*, M Nurkanti[2], S Wahyuni[1]

[1]The Department of Environmental Engineering, Universitas Pasundan, Jl. Dr. Setiabudhi 193, Bandung 40153, Indonesia

[2]Department of Biology Education, Universitas Pasundan, Jl. Tamansari 6-8, Bandung, 40116, Indonesia

fadhlan.173050019@mail.unpas.ac.id

Abstract. Citepus River is one of the urban rivers that pass through Bandung City, Indonesia. Visually, the Citepus River water shows relatively poor condition. This study aims to examine the water quality of the Citepus River, the results of which are used to recommend river management efforts. This research was conducted using secondary data obtained from the Bandung City Environmental Agency for the 2018-2020 time duration. Assessment of river water quality is carried out by comparing water quality monitoring results with the applicable quality standards. In addition, indexation was also carried out using the Storet method. The study results show that almost all parameters exceed the Class 2 of maximum water quality standard based on Government Regulation No. 82 of 2001 concerning Water Quality Management and Pollution Control. The results of calculations using the Storet method stated that the water quality in the Citepus River from the upstream segment to the downstream segment was classified as class D, which means heavily polluted, with the highest score in the middle segment at -54 and the lowest score in the upstream segment at -42, while the downstream segment, the score obtained is -52. Based on these results, management efforts can be recommended by determining the capacity, identification and inventory of wastewater sources, determining requirements and procedures for waste disposal, prohibiting waste disposal into rivers, monitoring river water quality, and monitoring wastewater quality.

1. Introduction

Water is a resource that must be preserved and maintained in terms of quality and quantity. Water resource potential differences will affect the difference in river water quality and quantity in each region so that the problems that arise related to water resources are also different in management and planning [1].

Rivers have an essential role in society. Human activities such as industrial and household waste cause deterioration in water quality, especially in urban areas [2]. The activities of residents around the river cause the decline in river water quality due to being polluted by various wastes. Pollution that often occurs in rivers can affect the balance of the ecosystem and decrease aquatic biota [3]. Several sources of water pollution in household waste, agriculture, livestock, industrial or other types of waste come from direct use water bodies, such as toilets and water transportation [4]. Pollution in an ecosystem that is high enough will poison the organisms in it. The decline in species diversity can be considered as one of the signs of pollution [5]. Based on Government Regulation Number 22 of 2021 concerning the

IOP Conf. Series: Earth and Environmental Science **894** (2021) 012006 doi:10.1088/1755-1315/894/1/012006

Implementation of Environmental Protection and Management Article 1 paragraph 33, it is explained that a river basin, from now on referred to as a watershed, is a land area unit that is an integral part of the river and its tributaries which function to accommodate, store and drain water.

Citepus Watershed is a sub-watershed of the Upper Citarum Watershed with a drainage area of 23.15 km2 (2315 Ha). The sub-districts included in this watershed are Sukasari, Sukajadi, Cicendo, Andir, Astana Anyar, Bojongloa Kaler and Bojongloa Kidul sub-districts. Citepus rivers have length 11.25 kilometers ends in Kampung Bojong Citepus, Desa Cakuang Wetan. The Citepus River does not end directly into Citarum River but first enter other rivers such as the Cirangrang River, Cikamanuman River, Cibojong River, which ends directly into Citarum Rivers.

Therefore, some efforts to control pollution have been carried out. The purpose of this study is to examine the parameters that can affect the quality of Citepus River water; the result can be used as recommendations for efforts to manage the water quality of the Citepus River. This study aims to identify the condition of physical, chemical, and biological parameters that can affect water quality in Citepus River and determine the proper effort in managing water quality in Citepus River.

2. Methodology

According to government regulation, water quality analysis was carried out by comparing Citepus River water quality data obtained from the Bandung City Environment Service with river water quality standards. Locations of the monitoring point represent upstream, middle stream and downstream segments.

Regulation No. 82 of 2001 concerning Determination of water quality status using the Storet Method according to the Decree of the Minister of the Environment 115 of 2003 concerning Guidelines for Determining the Status of Water Quality in order to see trends in water quality as a whole, water quality management and water pollution control. The way to determine the status of water quality is to use the value from the US-EPA (Environmental Protection Agency) by classifying water quality into four classes, i.e. Grade A: Very good, score = 0 means it meets the quality standard; Grade B: Good, score = -1 to -10 means lightly polluted; Grade C: Medium, score = -11 to -30 means moderately polluted and Grade D: Poor, score -31 means heavily polluted.

Identification of the river quality management was also conducted. The river water improvement efforts were evaluated based on a 3-years calculated Storet Index. The evaluation will show the effectiveness of the activities on river water quality rehabilitation.

3. Result and Discussion

3.1. *Citepus River water quality*

Secondary data for the results of research on the water quality of the Citepus River in 2018 to 2020 was obtained from the Bandung City Environmental Service and carried out in 3 segments, i.e. upstream (Jl. Pajajaran), middle (Jl. Kebon Jati), downstream (Jl. Pagarsih).

Table 1. Citepus River water quality (2018-2020).

No	Parameter	Units	Water Quality Standard**	Year	Upstream	Middlestream	Downstream
1	pH	-	6-9	2018	7.22	7.68	7.49
				2019	7.55	7.34	7.20
				2020	6.5	6.62	6.49
2	BOD	mg/l	3	2018	26.98*	29.23*	21.63*
				2019	28.97*	37.34*	28.29*
				2020	12.76*	7.68*	14.81*
3	COD	mg/l	25	2018	72.53*	77.52*	60.09*
				2019	122.13*	138.56*	119.44*
				2020	41.17*	24.01*	44.89*

No	Parameter	Units	Water Quality Standard**	Year	Upstream	Middlestream	Downstream
4	Cr(+6)	mg/l	0.05	2018	0.0032	0.0041	0.0032
				2019	0.0038	0.0052	0.0040
				2020	0.0436	0.043	0.0173
5	Nitrite	mg/l	0.06	2018	0.2395*	2.0416*	0.0034
				2019	0.0321	0.0592	0.0168
				2020	16.77*	15.820*	4.5826*
6	Grease	mg/l	1	2018	0.94	0.94	1.00
				2019	0.94	3*	2*
				2020	0.94	0.94	0.94
7	MBAS (Detergent)	mg/l	0.2	2018	0.91*	0.4211*	0.04254*
				2019	0.1795	0.2627*	0.2093*
				2020	0.8982*	0.4446*	0.8066*
8	total p	mg/l	0.2	2018	1.1763*	1.2115*	0.6621*
				2019	0.4219*	0.7258*	0.6122*
				2020	0.2358*	0.3307*	0.1857

*) above maximum standard
**) Class II; river water to be used as irrigation, recreational function, farming

The table above shows the water quality of the Citepus River, most of the water quality parameters of the Citepus River show numbers that exceed the quality standards that have been set for river water quality. The quality standard applied in this study is Government Regulation No. 82 of 2001. From the quality standard that has been set, the classification of river water quality is taken for Class II; namely, water whose designation can be used for water recreation infrastructure/facilities, freshwater fish farming. Livestock, water to irrigate crops, and other uses require the same water quality as those uses.

3.2. Citepus River water quality index

The analysis using the Storet Index method was carried out to determine the status of the water quality of the Citepus River in the upstream, middle and downstream areas. The data processing results with the Storet method can be seen in Tables 2, 3 and 4.

Table 2. Status of water quality by storet method in the upstream segment of the Citepus River.

No	Parameter	Units	Water Quality Standard	River Water Concentration			Storet Index Total score
			Class II	Maximum	Minimum	Average	
1	pH	-	6-9	7.55	6.5	7.09	0
2	BOD	mg/l	3	28.97*	12.76*	22.9	-10
3	COD	mg/l	25	122.14*	41.17*	78.61	-10
4	Cr(+6)	mg/l	0.05	0.043	0.0032	0.016	0
5	Nitrite	mg/l	0.06	16.77*	0.032	5.682	-4
6	Oil and Grease	mg/l	1	0.94	0.94	0.94	0
7	MBAS detergent	mg/l	0.2	0.91*	0.179	0.663	-8
8	Total Phosphate	mg/l	0.2	1.176*	0.235*	0.611	-10
			Score				-42

Table 3. Status of water quality by storet method in the middle stream of the Citepus River.

No	Parameter	Units	Water Quality Standard	River Water Concentration			Storet Index Total score
			Class II	Maximum	Minimum	Average	
1	pH	-	6-9	7.68	6.62	7.21	0
2	BOD	mg/l	3	37.34*	7.68*	24.75*	-10
3	COD	mg/l	25	138.56*	24.02*	80.04*	-10
4	Cr(+6)	mg/l	0.05	0.043	0.0041	0.0174	0
5	Nitrite	mg/l	0.06	15.820*	0.059	5.973*	-10
6	Oil and Grease	mg/l	1	3*	0.94	1.63*	-8
7	MBAS detergent	mg/l	0.2	0.445*	0.263*	0.376*	-10
8	Total Phosphate	mg/l	0.2	1.211*	0.331*	0.756*	-6
			Score				-54

Table 4. Status of water quality by storet method in the downstream segment of the Citepus River.

No	Parameter	Units	Water Quality Standard	River Water Concentration			Storet Index Total score
			Class II	Maximum	Minimum	Average	
1	pH	-	6-9	7.49	6.49	7.06	0
2	BOD	mg/l	3	28.29*	14.81*	21.58*	-10
3	COD	mg/l	25	119.44*	44.89*	74.81*	-10
4	Cr(+6)	mg/l	0.05	0.0173	0.0032	0.0082	0
5	Nitrite	mg/l	0.06	4.5826*	0.0034	1.5243*	-8
6	Oil and Grease	mg/l	1	2*	0.94	1.47*	-8
7	MBAS detergent	mg/l	0.2	0.8066*	0.04254	0.3528*	-8
8	Total Phosphate	mg/l	0.2	0.6621*	0.1857	0.4867*	-8
			Score				-52

Table 2, 3 and 4 show that almost all parameters are above the maximum standards. There are only pH and Cr(+6) that meet the standard. Table 5 display the status of the Citepus River water quality.

Table 5. Pollution quality of the Citepus River.

Segment	Score	Grade	Information
Upstream	-42	D	Heavily polluted
Middlestream	-54	D	Heavily polluted
Downstream	-52	D	Heavily polluted

The analysis results using the Storet Index method found that overall, from upstream to downstream, the Citepus River had experienced heavy pollution. It can be seen in Table 5 that the lowest score is in the middle segment with a score of -54, and the highest is in the upstream segment with a score of -42; the lower the score indicates the condition of river water quality violates the quality standard. The presence of a low score at the centre of the sampling point indicates that the highest level of pollution occurs in the middle segment [6]. Of all the segments studied, they fall into class D. According to the

US-EPA, which classifies water quality classes, a score of -31 is classified as class D, which means heavily polluted.

Poor river water quality, with a low Storet index score and polluted quality status, is either lightly, moderately or heavily polluted and often found in Indonesia's urban rivers [7]. In addition, other surface waters affected by domestic waste also have poor quality status [8], [9].

3.3. Citepus River management
River water quality management is one part of river hydrology restoration [10]. Water quality management is carried out by controlling water pollution by maintaining water functions to meet quality standards [11]. The process of preventing the occurrence of pollution is better than dealing with the pollution that has occurred. Such prevention efforts can be carried out following Indonesian Government Regulation No. 38 of 2011 concerning Rivers, in article 27 it is stated that the prevention of river water pollution can be done through (1) Determination of pollution load capacity; (2) Identification and inventory of sources of wastewater entering rivers; (3) Determination of requirements and procedures for wastewater disposal; (3) Prohibition of dumping waste into rivers; (4) Monitoring river water quality; (5) Supervision of wastewater entering the river.

River water quality management must also be carried out by various parties, namely the government, the private sector, academics and the general public. The participation of all parties will facilitate efforts to rehabilitate the water quality of the Citepus River for the better and restore its functions as determined, i.e. Class II. The river water quality information system is also an essential means of disseminating data on the Citepus River, especially in the information digitalization era [12].

4. Conclusion
The results of secondary data research obtained from the Bandung City Environmental Agency from 2018-2020 show that almost all parameters above maximum standard of Class 2 water quality standards based on Government Regulation No. 82 of 2001 and No. 22 of 2021 concerning Water Quality Management and Pollution Control. From the results of the analysis using the Storet Index method, it was found that the Citepus River entered from the upstream segment to the downstream segment into class D, which means heavily polluted, with the highest score in the middle segment of -54 and a score of -54 the lowest was in the upstream segment of -42, while the downstream segment the score obtained was -52. Based on PP No.38/2001, prevention efforts can be carried out by determining the capacity, identification and inventory of wastewater sources, stipulating requirements and procedures for waste disposal, prohibiting waste disposal into rivers, monitoring river water quality, and controlling wastewater.

References
[1] Sudarmadji H P and Widyastuti M 2013 *Pengelolaan Sumberdaya Air Terpadu* (Yogyakarta: Gadjah Mada University Press)
[2] Yustiani Y M, Wahyuni S and Dewi SNF 2019 *International Journal of Geomate* **16(56)** p 118-122
[3] Yustiani Y M 2016 *Pollution Research* **35(3)** p 475-481
[4] Baherem B, Suprihatin S and Indrasti N S 2014 *Jurnal Pengelolaan Sumber Daya Alam dan Lingkungan* **4(1)**
[5] Singh V, Shukla S and Singh A 2021 *Open J Plant Sci* **6(1)** p 011-014
[6] Arnop O, Budiyanto B and Saefuddin R 2019 *Naturalis* **8(1)** p 16-24
[7] Gannes Cs, and Wisnu W 2021 *Jurnal Teknologi* **14(1)** p 28–38
[8] Aristawidya M, Hasan Z, Iskandar, Yustiawati and Herawati H 2020 *Limnotek Perairan Darat Tropis di Indonesia* **27(10)** p 27-38
[9] Asuhadi S and Manan A 2018 *Jurnal Kelautan Nasional* **12(2)** p 109-119
[10] Yustiani Y M, Hasbiah AW, Matsumoto T and Rachman I 2019 *IOP Conference Series: Earth and Environmental Science* **245** 1 012033

[11] Pohan D A S, Budiyono, Syafrudin 2016 *Jurnal Ilmu Lingkungan* **14(2)** p 63-71
[12] Yustiani Y M, Lidya L, Matsumoto T, Rachman I and Komariah I 2017 *International Journal of Engineering and Technology* **9(1)** p 137-142

IOP Conf. Series: Earth and Environmental Science **894** (2021) 012007 doi:10.1088/1755-1315/894/1/012007

Environmental ethics analysis on household waste management behaviour

V Zulfa[1*], U Hasanah[1], P A Utami[1]

[1]Department of Family Welfare Education, Faculty of Engineering, Universitas Negeri Jakarta, Indonesia, 13220.

vaniazulfa@unj.ac.id

Abstract. This study aims to analyze the effect of environmental ethics knowledge on household waste management behaviour in Jatibening Village, Pondok Gede District, Bekasi City. The research data collection was conducted from October to November 2020. The data was collected using a questionnaire from 127 homemakers. Respondents were selected using a simple random sampling method. The data that has been collected is processed and analyzed using descriptive statistics and inferential statistics, namely simple linear regression. The simple linear regression analysis results indicate that Environmental Ethics Knowledge has a significant positive effect on Household Waste Management Behavior ($\beta = 0.215$, $p < 0.05$). These results indicate that increasing knowledge about environmental ethics can improve the behaviour of homemakers in managing household waste. Thus, it is hoped that homemakers can increase their knowledge of environmental ethics to improve their behaviour in managing household waste.

1. Introduction

For humans, the environment is everything in the environment, whether in the form of living things, inanimate objects, real or abstract objects, including other humans, and the atmosphere formed due to interactions between elements in nature. Besides, the environment is also one of the essential elements on earth. It causes the environment to have a unique role in the survival of all living things in the world [1]. Humans cannot live without the resources provided by nature, which is why humans must protect nature as best they can. One of the actions that can be taken to protect nature is to overcome environmental problems, such as waste problems. Garbage is leftover material that is no longer used, disliked/, or something that must be thrown away, which generally comes from activities carried out by humans. In activities, humans produce or produce waste. Humans produce many wastes, and it is necessary to carry out waste management to convert waste into materials that have economic value or process waste into materials that are not harmful to the environment [2]. In this case, if there is no community ability to manage waste, then the waste can cause environmental problems [3].

Waste management can be defined as a field that deals with the control of waste generation, storage, collection, transfer and transportation, processing and disposal of waste in a manner that is by the best principles relating to public health, economy, engineering, protection of nature, beauty and other environmental considerations as well as considering the wider community. Improper waste management can lead to piles of garbage. It can create a potential effect on environmental changes. Ignorance of the problem of waste management results in degradation of environmental quality, which does not provide comfort for life, thus reducing the quality of public health. Environmental quality degradation is triggered by behaviour patterns of people who are not environmentally friendly, such as throwing garbage in water bodies, causing garbage to accumulate in waterways, and causing various other inherited problems. This condition often occurs in densely populated areas in urban areas [4]. For example, environmental problems that occur in Bekasi City are caused by improper waste management. Namely, there are 88 illegal waste disposal points closed by the Environment Agency of Bekasi City, West Java, and the remaining 32 points for illegal dumping are spread evenly in 12 local districts. The condition of the 32 illegal waste disposal points is very worrying, so it is urgent to implement land management. One of them is located in the Jatibening sub-district, which is included in the Pondok Gede District area (Head of the Bekasi City Environmental Service, Jumhana Luthfi in suara.*com* 18 September 2018). Besides, the waste in the area will be disposed of at the Sumur Batu Final Disposal Site with an area of 3.5 hectares with a maximum capacity of 650 tons per day [5].

Some residents refuse to dismantle illegal TPS because some RTs have not yet provided a garbage storage area, including the RT 04 RW 03, Jatibening Village, Pondok Gede District, Bekasi City. In dealing with this problem, the Head of the Bekasi City Environment Agency is activating some programs, one of which is a waste bank that applies the principles of Reuse, Reduce, and Recycle (3R).

Based on the results of a preliminary study, the behaviour of homemakers in RT 04/RW 03 Jatibening Village in managing household waste is not good. It is because there are 6 out of 12 homemakers who have thrown their trash in an illegal garbage dump. Furthermore, 10 out of 12 homemakers do not manage household waste by type. Besides, 7 out of 12 homemakers manage their household waste by burning it. In this case, the knowledge about the environment is owned to manage the environment properly. One of the research results on waste management behaviour shows that environmental knowledge has a significant positive effect on waste management behaviour. It means that environmental knowledge greatly determines behaviour in terms of good waste management [6]. Thus, environmental knowledge is essential for every individual to have.

Understanding the environment (eco-literacy) owned by a person is not sufficient to shape one's behaviour in preserving the environment, so other supporting factors are needed [1]. Besides, humans are expected to instil morals or ethics in themselves because having good ethics towards the environment will foster participation in environmental care wherever they are [7]. It can be concluded that the knowledge that is expected to be possessed by each individual is knowledge of environmental ethics. These results indicate a problem in environmental ethics knowledge. Therefore, this study aims to analyze the effect of environmental ethics knowledge on household waste management behaviour.

2. Methodology

2.1. Research design

This research is an associative quantitative study that aims to analyze the effect of environmental ethics knowledge on household waste management behaviour. Meanwhile, the research method used is a survey method with a correlational approach. This research was conducted at Jatibening Village, Pondok Gede District, Bekasi City.

2.2. Instrument

Environmental ethics knowledge instruments are measured from three dimensions, namely nominal environmental literacy, functional environmental literacy, and operational environmental literacy. This instrument consists of 23 statements with two response options, namely true and false, provided that the correct answer is given a score of one (1) and the wrong answer is given a score of zero (0). The scores obtained are added and converted in the form of an index. After that, the data were categorized into three categories, namely low (index <60), medium (index 60-80), and high (index> 80).

Instruments household waste management behaviour is measured in three dimensions, namely thinking back (rethink), replace environmentally friendly materials (replace), and repair damaged goods (repair). This instrument consists of 9 statements with four response options, namely: Always (A), Often (O), Rarely (R), and Never (N). With ratings of 4, 3, 2, 1 for positive statements and 1, 2, 3, 4 for negative statements. The scores obtained are added and converted in the form of an index. After that, the data were categorized into three categories, namely low (index <60), medium (index 60-80), and high (index> 80). The p This research is an associative quantitative study that aims to analyze the effect of environmental ethics knowledge on household waste management behaviour. Meanwhile, the research method used is a survey method with a correlational approach. This research was conducted at Jatibening Village, Pondok Gede District, Bekasi City.

2.3. *Sample and sampling technique*
This study was housewives in RT 04/RW 03 Jatibening Village, Pondok Gede District, Bekasi City. Data obtained from the local RT shows the number of homemakers there are 200 people. The sampling technique in this study used *probability sampling techniques,* with a *simple random sampling technique*. In determining the number of samples, the researcher used Isaac and Michael's formula table. Based on the table, a sample of 200 housewives with an error rate of 5%, a sample of 127 housewives was obtained.

2.4. *Data analysis*
That had been collected were analyzed using descriptive statistics and inferential statistics. Descriptive statistics are used to analyze the description of the research variables, such as the minimum value, maximum value, average value, standard deviation, categories, and percentages. The inferential statistic used is a simple linear regression test used to analyze the effect of environmental ethics knowledge on household waste management behaviour. Before carrying out a simple linear regression test, several prerequisite tests must be done, namely the normality and linearity tests. The results of the Lilliefors normality test show that the research data is usually distributed. The linearity test results also show that the data is linear.

3. Results and discussion

3.1. *Characteristics of respondents*
The housewives involved in this study were between 20 years and 79 years old. The highest percentage of age was 30-39 years (35%). The housewives involved in this study had taken primary education to tertiary education. The highest percentage of education was completed in high school (39%). The sample of homemakers involved in this research is not working (homemakers) and working as Civil Servants (PNS), Permanent Employees, and Entrepreneurs/Traders. The highest percentage is not working (housewives) (89%). The housewives involved in this study had no income and ranged from <IDR 1,000,000 to> IDR 5,000,000. The highest percentage of respondent's income is not yet income (89.8%).

3.2. *Environmental ethics knowledge*
Environmental ethics knowledge is included in the elements of environmental literacy knowledge. Environmental literacy has the same concept as knowledge of environmental ethics, which consists of knowledge, aspects of attitude, and concern for the environment (Kamil *et al.*, 2019). Environmental literacy owned by each individual is different, depending on the ability to understand and the intensity of individual interactions with the environment. Related to integrating the constituent elements of environmental literacy, individual abilities can be divided into several stages/levels as follows [9]: nominal environmental literacy, functional environmental literacy, and operational environmental literacy.

The data shows that the respondents answered correctly (99.2%) for the statement that garbage dumped in rivers or waterways can cause flooding. It refers to the results of previous research that flood disasters can be said to be non-natural disasters caused by human behaviour, including throwing garbage inappropriately or throwing garbage into the riverbank [10]. Respondents answered correctly (41.7%) for the statement that coughs, flu, and migrants are diseases caused by bad waste management. It refers to the results of previous research that solid waste can cause several problems if not properly managed, such as trash can become a breeding ground for insects or nuisance animals known as disease vectors that can cause infectious diseases (such as diarrhoea, cholera, typhus), pollution. Air (such as foul odours, dust, poisonous gases, and combustion fumes) and water pollution (such as changes in the colour and smell of river water) [11]. Thus, the statement that coughs, flu, and migrants are diseases caused by improper waste management is not quite right.

Respondents answered correctly (89.0%) for the statement that accumulated garbage did not affect public health. Accumulation or disposal of waste in the wrong place can cause various sources of disease that affect human health [12]. Thus, the statement that accumulated waste does not affect public health is not quite right. Furthermore, respondents answered correctly (76.4%) for the statement that using recycled products is not considered as an act of maintaining environmental balance. It refers to the results of previous studies that efforts to use natural resources efficiently, recycling, and *eco-green are* carried out to maintain the balance of nature [13]. Thus, the statement that using recyclable products is not included in maintaining environmental balance is not quite right. Furthermore, respondents answered correctly (96.9%) that waste management is an activity that can have a positive impact on the environment. Good waste management will positively impact the surrounding community [14].

Respondents answered correctly (96.1%) for the statement that proper waste management can provide benefits in balance and proportion to current and future generations. They refer to the results of previous research that the waste management system provides benefits in terms of environmental health and socio-economic community (Sarfiah and Juliprijanto, 2017). Thus, it can be said that waste management activities can positively impact the environment and society. Furthermore, respondents answered correctly (76.4%) for the statement that current waste management does not affect waste management in the future. Awareness and a sense of responsibility to protect and care for the environment are driving factors promoted in every individual. On a planning scale, waste management that is not controlled from now on will impact the next generation in the future [16]. Thus, the statement that waste management carried out at this time does not affect waste management in the future is incorrect.

Respondents answered correctly (96.9%) that technological progress can affect the waste management process to make it more effective and efficient. It refers to previous research that in achieving levels of effectiveness and efficiency in dealing with waste problems, it is necessary to choose the suitable method and technology and active participation from the community and related government agencies (Ray, 2017). Thus, choosing the right technology in dealing with problems can affect the process of waste management activities to run effectively and efficiently. Furthermore, respondents answered correctly (51.2%) for the statement that sophisticated technology is believed to be able to solve the waste problem. Waste can come from technological advances/ so that people tend to be more consumptive and follow technological developments [18]. Thus, the statement that sophisticated technology is believed to solve the waste problem is not quite right.

Respondents answered correctly (92.9%) for the statement that separating wet and dry waste is the most uncomplicated waste management process that can be done at home. The simple thing in waste management starts with separating wet and dry waste so that waste that can still be used can be reused [19]. Furthermore, respondents answered correctly (66.1%) that waste management can be carried out in the community first, then it can be applied in the home environment. It refers to previous research that states that waste management is carried out within households and outside households. Homemakers carry out domestic waste management. Meanwhile, outside the household, the management is carried out by the RT/RW management. Thus, proper waste management carried out by homemakers can be done first in the house, then it can be done outside the home with other communities [20].

Respondents answered correctly (93.7%) for the sanctions statement that was less firm and did not provide a deterrent effect for people who did not carry out waste management. Relevant agencies are expected to be more active in supervising and imposing strict sanctions on people who violate applicable regulations to cause a deterrent effect [21]. Furthermore, respondents answered correctly (93.7%) that the implementation of waste management requires legal regulations governing these activities, which are formed in law. In this case, the law governing waste management is Law No. 18 of 2008. The results are presented in Table 1.

Table 1. Distribution of respondents by category nominal dimension of environmental literacy.

Environmental Literacy Nominal	Total	
	Total	Percent
Low (Index <60)	5	3.9
Medium (Index 60-80)	53	41.7
High (Index> 80)	69	54.3
Total	127	
Minimum-Maximum Value	54.0-100	
Average ± Standard deviation	82.3 ± 13.3	

The second dimension of environmental, ethical knowledge is functional environmental literacy. This dimension measures the housewife's ability to understand the nature and interactions between human social systems and natural systems and to be able to analyze the problems that occur in these interactions. The data shows that the respondent answered correctly (90.6%) that humans have an interdependent relationship with nature which requires humans to protect nature, such as by carrying out waste management. It is by the objectives of *ecological citizenship*, according to Crane (2008), that *ecological citizenship* can raise citizen awareness that humans have an interrelated relationship with nature/ so that humans can build good relationships with the natural environment. Furthermore, respondents answered correctly (88.2%) that the community was littered because they did not know how to manage waste. It refers to previous research that behaviour in action based on science is more durable and becomes a habit. It is due to knowing the risk of the action being taken. The better the knowledge a person has, the better the behaviour in managing the environment [22].

Furthermore, respondents answered correctly (91.3%) for the statement of sorting the waste according to nature's ability to break down waste, such as leaf waste separated from plastic waste. Waste selection is carried out by separating waste based on its type, organic or inorganic waste. Leaf litter and plastic waste are different types of waste [23]. Furthermore, respondents answered correctly (95.3%) for the statement that frequent flooding was caused by people who like to litter. Furthermore, respondents answered correctly (90.6%) that the lack of human resources is a problem in managing waste. It refers to previous research that the problems in waste management that often occur, including behaviour and lifestyle, still tend to increase the rate of waste piles and limited resources [24].

Respondents answered correctly (87.4%) for the statement that limited budget funds were a problem in carrying out waste management. They are referring to previous research that the government must have a budget for funding waste management. In this case, it can be said that a budget is needed as needed in carrying out waste management activities [25]. Results of the study are presented in Table 2.

Table 2. Distribution of respondents by category dimension of environmental literacy functional.

Functional Literacy Environment	Total	
	Total	Percent
Low (Index <60)	5	3.9
Medium (Index 60-80)	15	11.8
High (Index> 80)	107	84.2
Total	127	
Minimum-Maximum Value	33.0-100	
Average ± Standard deviation	90.5 ± 14.6	

The third dimension of environmental ethical knowledge is functional environmental literacy. This dimension measures how much the ability of housewives to identify organizations or groups that contribute to environmental change. The data shows that the respondent answered correctly (89.8%) for the statement that the organization/group managing the waste provided training in waste management. Furthermore, the respondents answered correctly (89.0%) for the statement of the organization/group of waste management to help increase public awareness in carrying out waste management activities. Community participation in waste management and the existence of a waste management organization will have a positive social impact. One of the positive impacts is expected to increase the knowledge and capacity of the community in managing the environment [26]. Furthermore, the respondents answered correctly (71.7%) that the organization/group of waste managers that provides training in waste management is only responsible for implementing the training. It refers to previous research that the implementing element or *(implementor),* both organizations or individuals, is responsible for obtaining implementation and supervision of the waste management implementation process [27]. Thus, the statement of the waste management organization/group that provides training in waste management, only being responsible for implementing the training, is inaccurate. Furthermore, the respondents answered correctly (99.2%) that the organization/group of waste management assists the government in implementing programs regarding waste management. The research results are presented in Table 3.

Table 3. Distribution of respondents based on dimensions of operational environmental literacy category.

Operational Environment Literacy	Total	
	Total	Percent
Low (Index <60)	7	5.5
Medium (Index 60-80)	50	39.3
High (Index> 80)	70	55.1
Total	127	
Minimum-Maximum Value	50.0-100	
Average ± Standard deviation	87.4 ± 15.0	

The study results indicate that most homemakers know environmental ethics in the dimensions of functional environmental literacy in the high category (84.2%). In Table 6, the results show that most homemakers have knowledge of environmental ethics in the operational environment literacy dimension in the high category (55.1%). The highest knowledge of environmental ethics based on average scores is in the dimension of functional environmental literacy, where homemakers can understand the nature and interactions between human social systems and natural systems and can adequately analyze problems that occur in these interactions. It refers to previous research that humans and the environment are integral natural systems forming ecosystems that influence each other. As a creature endowed with a mind, humans are the most perfect and noble creatures. The human obligation in interacting with nature is to use nature responsibly by environmental ethics by protecting and preserving it [28].

Based on the average value, the lowest average score for knowledge of environmental ethics is in the dimension of nominal environmental literacy, in which the ability of homemakers to understand and appreciate natural systems is sufficient. Referring to previous research, humans were created to appreciate nature because humans need nature as a place to live. If humans do not respect nature, it will cause disasters detrimental to humans themselves [29]. Thus, homemakers can increase their nominal environmental literacy by understanding every element and principle of nature to perfect their knowledge of environmental ethics.

3.3 Household waste management behaviour

Behaviour that can support environmental improvement efforts is behaviour in managing household waste. A method 3R in waste management, namely thinking back (rethink), replace environmentally friendly materials (replace), and repair damaged goods (repair).

The first dimension of household waste management behaviour is to think again (rethink). This dimension measures how precise the housewife is in determining the products to buy and use. The data shows that respondents always answered (85.0%) and often (14.2%) for buying products based on needs, such as products that must be fulfilled for daily life. These results indicate that the respondents agree with the activity of buying products to fulfil their daily needs. It refers to previous research that someone making a purchase meets their needs (Supangkat and Supriyatin, 2017). Furthermore, respondents always answered (40.9%) and often (26.0%) for buying products based on previous consumer assessments, such as the number of positive responses to the product. These results indicate that the respondent agrees before making a purchase activity to ensure that there is a positive response from previous consumers to the product. It refers to previous research that one of the factors in the tendency of consumers to buy products is because the product has a good image in society [31].

Furthermore, respondents always answered (80.3%) and often (14.2%) for statements that buy products prioritizing health elements, such as products that have been officially registered at BPOM. These results indicate that respondents who agree in making purchasing decisions can first pay attention to the health elements. This statement is in line with the statement that intelligent consumers will always prioritize health when deciding to buy a product [32]. The research results are presented in Table 4.

Table 4. Distribution of respondents by category dimensional thinking back *(Rethink).*

Rethink	**Total**	
	Total	**Per cent**
Low (Index <60)	4	3.1
Medium (Index 60-80)	19	14.9
High (Index> 80)	104	81.8
Total	127	
Minimum-Maximum Value	50.0-100	
Average ± Standard deviation	88.4 ± 11.5	

The second dimension of household waste management behaviour is to replace environmentally friendly materials *(replace).* This dimension measures the level of compliance of homemakers to replace items that are not environmentally friendly with environmentally friendly goods. The data presented in Table 9 shows that the respondent always answered (51,2 %) and often (27.6%) stated that they do not bring their shopping bags when going to the market. These results indicate that respondents agree not to carry shopping bags when travelling. Referring to previous research, plastic shopping bags usually obtained from stores can make it easier for buyers to move groceries (Furqaan and Santoso, 2017). Furthermore, respondents always answered (46.5%) and often (28.3%) for statements using plastic straws when drinking. These results indicate that the respondents agreed to use a plastic straw. It refers to previous research that another form of plastic waste is plastic straws that are easy to use and dispose of. Most restaurants serve plastic straws with drinks. It makes people reluctant to consume directly from glass and choose to use a straw [34].

Furthermore, respondents always answered (50.4%) and often (30.7%) for the statement that they brought their cutlery when travelling. These results indicate that the respondents agreed to bring cutlery when travelling. Furthermore, respondents always answered (48.8%) and often (33.1%) for the statement that they brought their food containers when travelling. These results indicate that the respondents agreed to carry food containers with them when travelling. The results of the study are presented in Table 5.

Table 5. Distribution of respondents based on the dimension category of replacing environmentally friendly materials *(replace).*

Replace	**Total**	
	Total	**Per cent**
Low (Index <60)	14	11.0
Medium (Index 60-80)	39	30.7
High (Index> 80)	74	58.2
Total	127	

Replace	**Total**	
	Total	**Per cent**
Minimum-Maximum Value	38.0-100	
Average ± Standard deviation	80.8 ± 16.4	

The third dimension of household waste management behaviour is repairing damaged items *(repair)*. This dimension measures the level of the housewife's ability to repair damaged goods so that the product's life can be extended. The data shows that the respondent always answered (41.7%) and often (27.6%) for the statement of collecting used paper that is still fit for use. These results indicate that the respondents agreed to collect used paper. Furthermore, respondents always answered (54.3%) and often (23.6%) for the statement they collected used plastic, which was still suitable for use. These results indicate that the respondents agreed to collect used plastic that can still be used. The results of the study are presented in Table 6.

Table 6. Distribution of respondents by category dimensional fix damaged goods *(repair)*.

Repair	**Total**	
	Total	**Per cent**
Low (Index <60)	23	18.1
Medium (Index 60-80)	34	26.7
High (Index> 80)	70	55.1
Total	127	
Minimum-Maximum Value	25.0-100	
Average ± Standard deviation	78.8 ± 21.1	

The study results indicate that most homemakers apply household waste management behaviour by *(rethink)* in the high category (81.8%). The highest household waste management behaviour based on the average value is on the dimension *(rethink)*, where the accuracy of the housewife in determining which products to buy and use is good. Referring to previous research that human consumption is carried out to meet needs both in the form of primary and secondary needs so that it is expected to be wiser in choosing which needs are the most important of an item and which have benefits and uses so as not to waste actions (Astuti, 2013). Thus, it is essential to think again (rethink) in determining which products to buy and use. Based on the average value, the lowest average value for household waste management behaviour is in the dimension of repairing damaged goods *(repair)*, in which the accuracy of homemakers in repairing damaged goods is sufficient. It refers to previous research that if the damage to goods is categorized as severely damaged and cannot be repaired, it is better to buy new items [36]. Thus, some housewives think it is better to buy new things than spend much money repairing damaged goods.

3.4. Relationship between knowledge of environmental ethics and behaviour of household waste management

The correlation test results showed that knowledge of environmental ethics had a significant positive relationship with household waste management behaviour ($r = 0.215$, $p < 0.05$). These results indicate that the better the knowledge of environmental ethics owned by housewives, the better the behaviour of homemakers in managing household waste. Referring to the results of previous research shows that environmental knowledge has a significant positive effect on waste processing behaviour. It means that environmental knowledge determines behaviour in terms of sound waste processing.

3.5. The effect of environmental ethics knowledge and household waste management behaviour

The influence of environmental ethics knowledge on household waste management behaviour was analyzed using a simple linear regression test. Before carrying out a simple linear regression test, a prerequisite test for data analysis was carried out, namely the Lilliefors normality test. The prerequisite test results show that data on environmental ethics knowledge and household waste management behaviour is usually distributed. Furthermore, the data were analyzed using a simple linear regression test. The analysis results show that knowledge of environmental ethics has a significant positive effect on household waste management behaviour ($\beta = 0.215$, $p < 0.05$). These results indicate that the higher the knowledge of environmental ethics owned by housewives, the higher the behaviour of homemakers in managing household waste.

The magnitude of the influence between the variable knowledge of environmental ethics (X) on the behaviour of household waste management (Y) can be seen from the value of the R square, which is equal to 0.046 x 100%, or it can be said that the variable knowledge of environmental ethics has an effect of 4.6% on the behaviour variable of household waste management. Stairs. The effect obtained from the two variables is low. It is because knowledge of environmental ethics can significantly influence the behaviour of household waste management if a caring attitude towards the environment supports it. They refer to previous research that the existence of indirect knowledge significantly increases behaviour by increasing attitudes, which can then improve behaviour. No matter how much knowledge a person has, if a positive attitude does not support it, the effect on behaviour will be small [37].

A person's attitude in waste processing has something to do with education level and income. The level of education informal education can affect a person's attitude towards waste processing. The higher the level of education of a person, the higher the attitude of concern for waste processing and vice versa. Meanwhile, the greater a person's income, the higher the attitude of concern for household waste processing and vice versa [6].

Based on the respondent's description of the respondent's education data, most of the respondents were high school graduates. Thus, it can be said that the respondent's education level is sufficient. Meanwhile, for the respondent's income data, it is not yet income. It is because most respondents are not working (homemakers). Thus, low income will slightly affect a person's caring attitude so that the effect on behaviour in managing waste is low. However, the index value of each dimension of the two variables is in the high category.

4 Conclusion

The results showed that environmental ethics knowledge was included in the high category (66.9%). Meanwhile, the behaviour of household waste management is in the high category (64.5%). The simple linear regression analysis results showed that environmental ethics knowledge had a significant positive effect on the behaviour of household waste management at ($\beta=0.215$, $p < 0.05$). These results indicate that the higher the knowledge of environmental ethics owned by housewives, the higher the behaviour of homemakers in managing household waste. Thus, knowing environmental ethics can improve one's behaviour in managing household waste.

References

[1] Nadiroh N, Hasanah U and Zulfa V 2019 *Indones. J. Geogr.* **51(2)** 42–49
[2] Fadhilah A, Sugianto H, Hadi K, Firmandhani S W, Woro T and Pandelaki E E 2011 *Modul* **11(2)** 62–71
[3] Yogiesti V, Hariyani S and Sutikno F R 2010 *Tata Kota dan Drh.* **2(2)** 95–102
[4] Setianto H 2018 *Jurnal Dosen Universitas PGRI Palembang* **8**
[5] Ariefana P 2018 *Bau, Bekasi Tutup Ratusan Tempat Pembuangan Sampah Liar (suara.com)*
[6] Putra R P, Saam Z and Tantoro S 2019 *Ilmu Lingkung.* **13(2)** p 196–205
[7] Ariwidodo E 2014 *Nuansa* **11(1)** p 1–20
[8] Kamil P A, Putri E and Ridha S 2019 *Georafflesia* **4** no April p 127–138
[9] Loubser Swanepoel and Chacko 2001 *South African J. Educ* **21(4)**
[10] Silalahi B 2017 *J. Ilm. Keperawatan IMELDA* **3(1)** p 208–217
[11] Hutabarat B T F and Siagian R I O I 2015 *J. Kedokt. Komunitas dan Trop.* **3(1)** 41–47
[12] Syafri I, Rendra P P R and Sulastri M 2020 *J. Pengabdi. Kpd. Masy.* **4(2)** 46–49

[13] Pane K A and Suryono 2012 *J. Arsit.* **1(1)** 52–59
[14] Khaira M, Hasanah U and Hayati I 2020 *J. Pengabdi. Kpd. Masy.* **2(2)** 187–195
[15] Sarfiah S N and Juliprijanto W 2017 *J. Ris. Ekon. Pembang.* p 165–184
[16] Manurung R A 2013 *J. Wil. dan Lingkung.* **1(3)** 227–244
[17] Ray R I W 2017 *Indones. J. Public Adm.* **3(1)** 18–34
[18] Hakim M Z 2019 *Amanna Gappa* **27(2)** 111–121
[19] Fitriyah L, Akbar A and Wibowo B 2010 *Diversifikasi Briket Berbahan Dasar Sampah Organik Sebagai Alternatif Baru Bahan Bakar Bagi Masyarakat*
[20] Mulyadi A, Husein S and Saam Z 2010 *J. Ilmu Lingkung.* **3(2)** 147–162
[21] Juniardi A, Asrinawaty and Ilmi M B 2020 *J. Publ. Kesehat. Masy. Indones.* **7(1)** 10–15
[22] Sukerti N L G, Sudarma I M and Pujaastawa I B 2017 *ECOTROPHIC J. Ilmu Lingkung. (Journal Environ. Sci.* **11(2)** 148–155
[23] Saragih R I *et al* 2020 *METHOMIKA J. Manaj. Inform. dan Komputerisasi Akunt.* **4(1)** 47–55
[24] Amanda A 2019 *J. S-1 Ilmu Adm. Publik* p 1–18
[25] Wahyudi A, Abdillah R A and Nuthqi A I I 2020 *Pros. Simp. Nas.* p 1018–1045
[26] Subekti S 2009 p 24–30
[27] Rahman Y 2016 p 4–22
[28] Tulalessy Q D 2018 *J. Trit. Pendidik.* **1(1)** 51–56
[29] Efrida 2016 *J. Ekspresi Seni* **18(1)** 62–77
[30] Supangkat A H and Supriyatin 2017 *J. Ilmu dan Ris. Manaj.* **6(9)** p 1–19
[31] Edbert S, Tumbel A and Tumbuan W A 2014 *J. Ris. Ekon. Manajemen, Bisnis dan Akunt.* 2(3) 1035–1044
[32] Rosha Z and Khaidir A 2019 *J. Manaj. Univ. Bung Hatta* **14(1)** 13–19
[33] Furqaan N M and Santoso K H 2017 **6(1)**
[34] Syafrikurniasari N and Widiani P 2020 *J. Lugas* **4(1)** 17–26
[35] Astuti E D 2013 *Psikoborneo* **1(2)** 79–83
[36] Pahlevi R, Imron A and Kusumaningrum D 2016 *J. Manaj. Pendidik.* **25(1)** 88–94
[37] Saputra S N A and Mulasari S A 2017 *Kes Mas J. Kesehat. Masy.* 11(1) 22–27

Integrating ecology and zero runoff in a vertical village residential design in West Jakarta

M Prihandrijanti[1], H M Kristanto[1] and Y D Apritasari[1]

[1]Architecture department, Universitas Agung Podomoro, Jakarta, 11470, Indonesia

maria.prihandrijanti@podomorouniversity.ac.id

Abstract. Diminishing availability of urban land in metropolitan cities such as Jakarta resulting in high-density settlements, which impacted the quality of life of the urban village communities. One of the government's efforts to solve this problem is by building subsidized apartments. However, this effort mainly focused on the land scarcity and the required housing aspects, while ecological and environmental roles toward the life quality of the dwellers are often set aside though very influential. On the other hand, Jakarta is very vulnerable to natural disasters related to water, such as flooding caused by stormwater runoff which cannot be accommodated by the insufficient capacity of the surrounding environment. Therefore, the development of urban settlements in Jakarta has to pay more attention to integrate ecology and zero runoff management from the design conception phase. A vertical village residential design in West Jakarta that integrates these two aspects through ecological architecture elements (nature, water, human and environment) was proposed as a role model design for vertical village housing. Various water design features have been creatively designed and integrated into the landscape and building elements resulting in zero runoff and an ecologically significant built environment.

1. Introduction

As a metropolitan city, Jakarta attracts many people from all around Indonesia to come and try to seize the opportunities offered by various businesses and infrastructure developments. Since 1980 the population of Greater Jakarta has doubled from 11.9 million to 28 million. Due to rural-urban migration, an estimated 250,000 people relocate to the city [1]. Many of these people belong to the low-income group who do not have a decent place to live in Jakarta. Their financial limitations and diminishing availability of urban land in Jakarta resulted in the development of high-density settlements, which impacted the quality of life of the urban village communities. One of the government's efforts to solve this problem is by building subsidized apartments. These efforts have contributed to reduce the housing backlog and provide an option of affordable housing. However, there are still some issues about the program qualities, especially in ecological and environmental roles toward the life quality of the dwellers, which are often set aside though very influential.

The pressures from urbanization and population growth also impacted land coverage and adequate stormwater drainage, worsening the city's vulnerability to flooding. The high urbanization rate has decreased the ability of water bodies to accommodate stormwater. In addition, the topography of Jakarta increases the flood risk. Jakarta is located on a low plain on the northwest coast of Java, and the greater

Jakarta area is enclosed by several mountains, which form the upstream catchment areas of the 13 rivers flowing through Jakarta. The majority of forests that existed upstream have been transformed into residential and agricultural areas over the past several decades, which resulted in the upstream catchment areas sending their stormwater to Jakarta.

Meanwhile, the city's river catchment has difficulty preventing flooding because of sedimentation in the existing rivers, illegal settlements along the riverbanks and poor waste management, which caused the waterways to be blocked during heavy rainfall, thus even worsening the flood risk. Most of the city is prone to muddy and flooded conditions, especially during the rainy season (typically from October to April). Figure 1 shows a map of Jakarta and the floodwater levels during the February 2007 flood event per subdistrict level. As seen, locations with the highest food level (dark red) are adjacent to the Ciliwung, Pesanggrahan and Sunter Rivers [2]. Figure 2 contains a map of 3 significant flood events in Jakarta in 1996, 2002 and 2007. It can be seen that more recent floods took place in larger areas and consequently impacted more people.

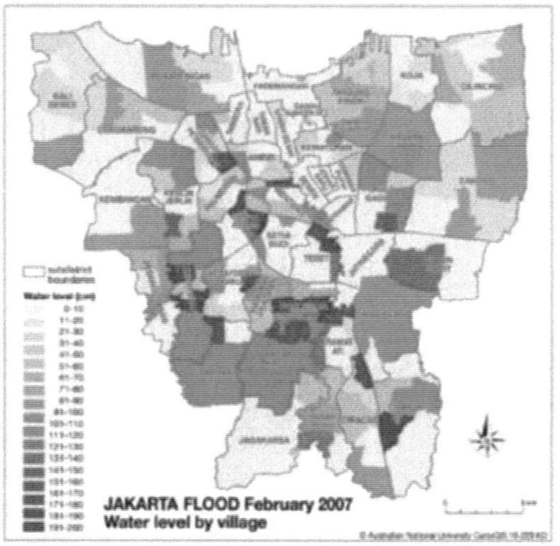

Figure 1. Map of Jakarta's flood disaster in 2007 [2].

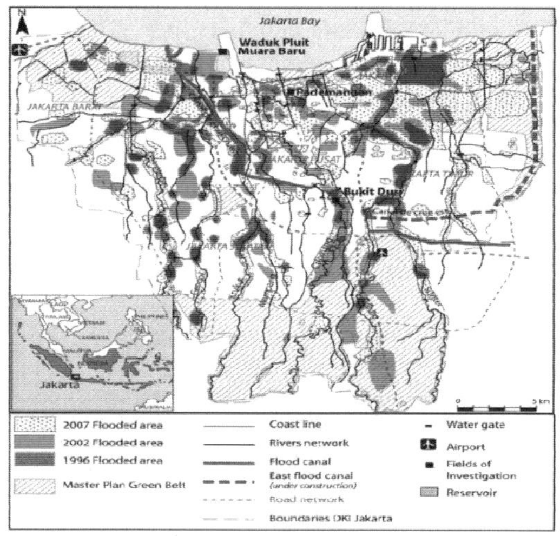

Figure 2. Map of 3 flood events in Jakarta [3].

A recent flood event from 31 December 2019 until 1 January 2020 caused flooding across the Greater Jakarta area, including some areas in West Java and Banten provinces. Hundreds of districts and sub-districts were flooded. The event is worse in terms of water volume and scale compared to the 2007 flood. In 2007 around three-quarters of Jakarta's particular region was immersed, and almost half of a million people were affected.

Impacts of flooding events happening in Jakarta give the city government urgency to prioritize flood risk management in urban settlements. By understanding the causes and effects of flood impacts and designing, investing in and implementing measures to minimize the impacts must become part of development thinking and be embedded into broader development goals [1]. However, the conventional strategy commonly implemented by governments to mitigate the impacts of flooding was to send stormwater away from built areas by the construction of storm sewers/channels or detention basins which have become increasingly difficult because of minimal land availability, especially in high-density urban settlements in Jakarta's village, and the high proportion of paved surfaces and roofs as well as the reduction of green open spaces in these areas. Studies were done in several vertical housing (rumah susun or rusun) in Jakarta, such as in Rusunawa Muara Baru, Rusun Cinta Kasih Cengkareng, and Rusunawa Jatinegara Barat [4] mentioned that stormwater management mainly was still done by channeling rainwater to the nearest water body or city's stormwater channel.

The low ability of the land to accommodate, hold and store stormwater caused high runoff and increased flood risk, making the conventional strategy not following the natural water cycle because the stormwater would end up in the sea and mix with seawater, thus requiring more extended time to supply the needs of freshwater for agriculture and the people living inland, especially in the cities. It represents a linear system instead of a circular system as the water cycle should be. In the water cycle, the precipitation is supposed to infiltrate the soil, refill aquifers or be stored in surface inland water bodies for various uses. On the other hand, less than half of Jakarta's population has access to piped water which means that most residents and activities must rely on groundwater for their freshwater demand. This draining of groundwater aquifers contributes to land subsidence. Jakarta was stated to subside up to 15 cm/year, making it one of the fastest sinking cities globally. Therefore, the development of urban settlements in Jakarta needs to pay more attention to integrating ecology and zero runoff management from the design conception phase. This study aims to design vertical village residential in West Jakarta that integrates these two aspects through ecological architecture elements (nature, water, human and environment) to be a role model design for vertical village housing.

1.1. Vertical village
The problems with the urban village are mainly about the lack of infrastructures and facilities to support the high-density population living there, mostly living in semi-permanent or non-permanent housing with minimum sanitation services and minimum natural lighting and natural lighting ventilation, thus lowering their quality of life. On the other hand, these urban villages, better known as "kampung" in Indonesia, have unique, intense social and cultural interaction characteristics. That is why these kampungs or urban villages need particular attention, mainly through design and re-organization or maybe revitalization in certain parts, including architectural solutions, especially in areas that have become slum areas because this will significantly affect the life quality of the people who live there. The role of architecture becomes essential to create a solution to answer those problems and tackling the problem of limited land availability by designing a vertical village with ecological architecture approach.

The vertical village is a more humane idea of vertical housing for low-income communities, which maintains the community's character, activities, and culture without eliminating their local wisdom but can still contribute to the environment's development. It can also accommodate facilities to improve the community's wellbeing by supporting the informal economy and entrepreneurship. In a vertical village, the housing function could represent the village (kampung) image in a "vertical form" and solve the decreasing availability of urban space, such as Jakarta, which has the highest population density in Indonesia. However, there are differences between the common kampung typology and the vertical village typology, so the design of vertical village housing should accommodate and represent the existence of the verticalized kampung without losing its essence, nature or character.

1.2. Ecological architecture
There are two shaping forms in ecological architecture, namely the biotic and abiotic environments, interacting with each other in the ecosystem. The ecological architecture integrates four ecosystem elements, i.e., nature, water, humans and the environment, to create a harmonious and friendly ecosystem [5]. Van der Ryn and Cowan [6] have a similar concept in their design principles of ecological architecture by emphasizing solutions grown from the place and people, design with nature, and make nature visible in the design. These elements will be used as the vertical village design reference. In general, ecological architecture has a balancing role in harmonizing human and nature and human and its built environment through the design parameters that relate to harmonizing and balancing the created space's functions.

Ecological architecture means the combination between architecture and environment, which regard or pay special attention to the harmony between man and his environment. Several strategies to integrate ecological architecture in the design of vertical village housing emphasize the integration of surrounding (the design environment context) ecological conditions, climate's influence on building, and building function-oriented toward the behaviors of urban village communities (such as solid social interactions).

Table 1. Elements and principles of ecological architecture.

Elements	Principles
Building materials use	• Minimize non-environmentally friendly materials • Prioritizing use of sustainable and reusable materials • Improving efficiency of materials use
Building renewable energy use	• Use solar panel • Minimize energy use from ventilation and lighting through passive design • Optimize building energy use
Environmentally friendly waste management	• Eliminate air, water and soil pollution • Use organic, reusable or biodegradable materials • Apply Zero Runoff concept in site planning
Improving adaptive functionality	• Take into account materials and water flow in the ecosystem that will be shaped/formed in the design area • Preserve biodiversity created in the environment

The above principles listed in Table 1 are further applied as ecological architecture design parameters among others in these aspects:

1. Sustainability of green areas in design location for infiltration
2. Choice of design site that does not disturb nature's sustainability
3. Encourage use of natural materials
4. Improve ventilation as passive design response
5. Implement Zero Runoff in design area
6. Use wall materials that minimize Urban Heat Island Effect

2. Methodology

This research was done using qualitative and quantitative approaches such as direct observation in case study areas, literature review, interviews with relevant stakeholders, collecting secondary data to support analysis, and simulation software. Results from this analysis were evaluated and then applied to the design process. Before the study area was chosen, a suitability analysis was done to select the location from 3 site alternatives, then micro and macro site analysis was performed on the selected site. A volumetric study was also conducted for three alternatives of masses for the chosen site to obtain the design mass, which will be developed and designed according to the intended design concept and strategies. The design concept can be seen in Figure 3.

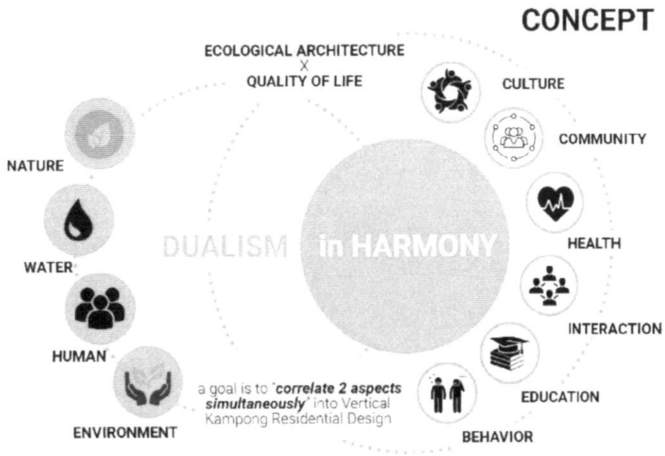

Figure 3. Design concept used in this research.

2.1. Study area

The object of this research was RW 016 Kapuk (West Jakarta), which has the worst slum status in DKI Jakarta and compared the conditions/characteristics with one of the existing and similar urban vertical housing complexes in Jakarta, namely Rumah Susun (Rusun) Conver Kemayoran.

RW 016 Kapuk is the most densely populated area in Jakarta having in average around 300 family cards (KK or Kartu Keluarga) in one Rukun Tetangga (RT) and there are 25 RTs in RW 016 (total 9433 KK). Most of its residents (about 80%) are migrants working as factory workers or in informal sectors.

(a) (b)

Figure 4. Existing conditions at study areas: RW016 Kapuk (a) and Rusun Conver Kemayoran (b). Source: personal documentation.

Rusun Conver Kemayoran is located in Jalan Conver 3 no 101 RT 12 RW 16 Kebon Kosong, Kemayoran district, Central Jakarta. It has a total area of 11,845 m2. In the beginning, it was built for the low-income urban community, disaster victims and relocated families, but then the residency has changed to middle-up economy classes until now. In this study, Rusun Conver Kemayoran is used to be a precedence case study as the vertical village that is already existing and be a reference for the vertical village that will be made in RW 016 Kapuk.

2.2. Ecological conditions

Table 2. Comparison of existing conditions at study areas.

Aspect	Rusun Conver Kemayoran	RW 016 Kapuk
Area	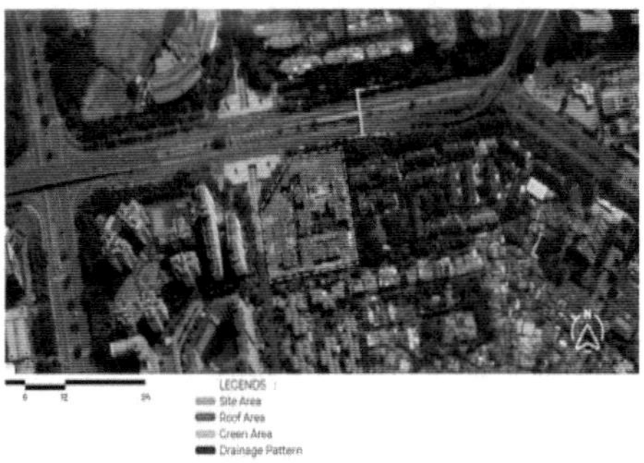	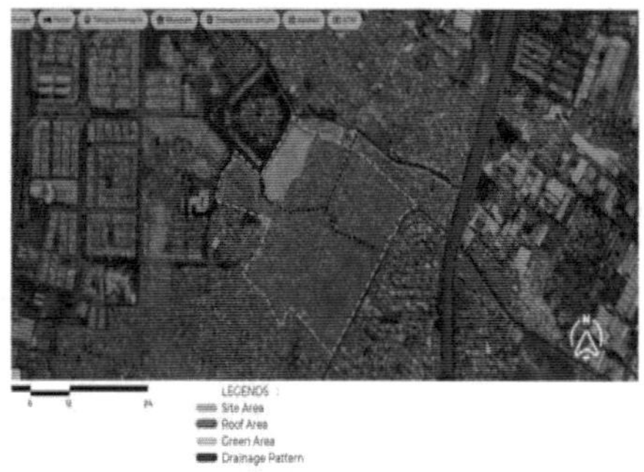
	Total area: 11,845 m^2 Street area: 1728 m^2 (15 %) Green area: 4621 m^2 (39 %) Roof area: 5496 m^2 (46 %) Water body area: -	Total area: 220,000 m^2 Street area: 12,535 m^2 (5.7 %) Green area: 38,935 m^2 (17.7 %) Roof area: 169,070 m^2 (74.4 %) Water body area: 4,920 m^2 (2.2 %)
Building shape and structure	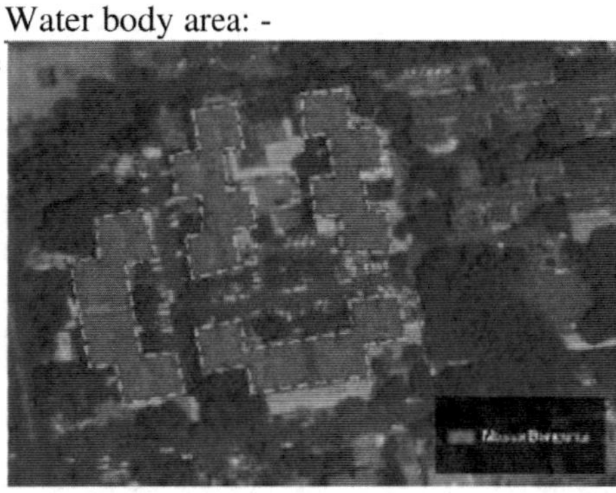	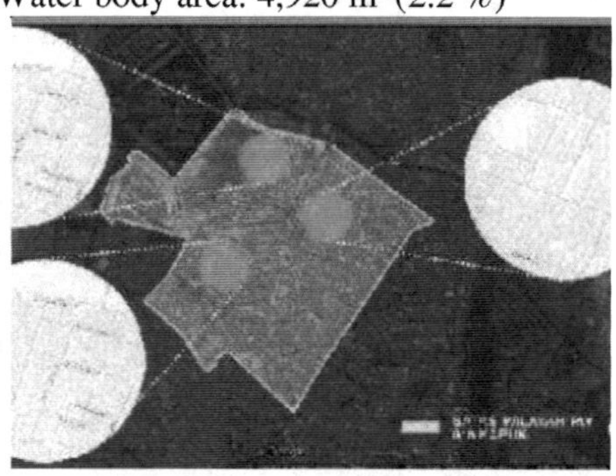
	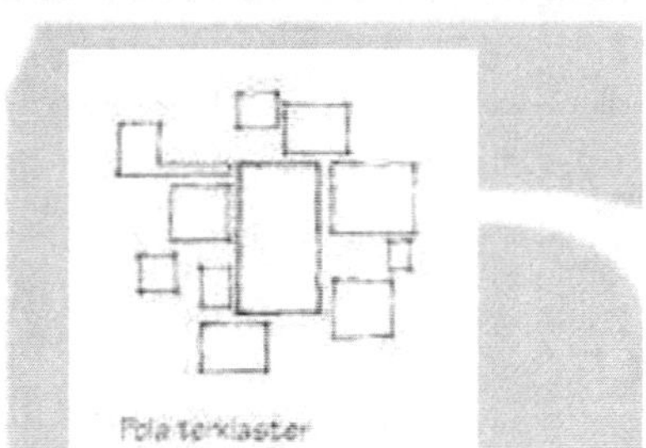	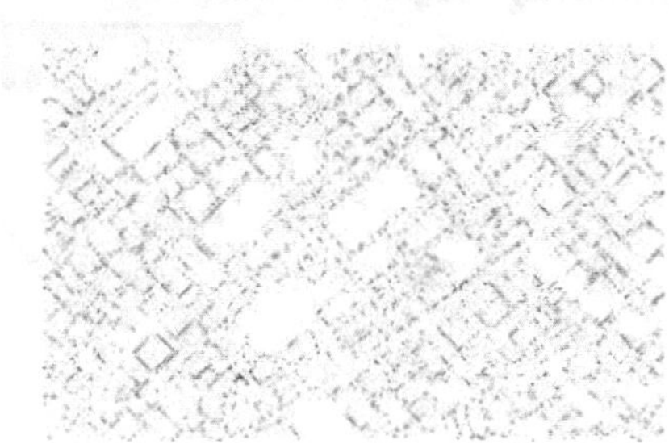
	Clustered layout forming a void used as o pen spaces in ground floor	Irregular grid layout of each building for ming small corridors for circulation

Aspect	Rusun Conver Kemayoran	RW 016 Kapuk
Lighting and ventilation system	 Each unit has enough openings for light and ventilation	 Non-permanent houses and very crowded condition, mostly have minimal openings
Response to climate	Buildings have responded to climate	Most houses have not responded to climate
Drainage system	 Closed drainage channels and good layout (available in every block). Drainage channels discharge to bigger channel (1 m diameter) under the pedestrian walkway and then go to the city's sewer in front of the rusun.	 Open drainage channels, smell bad, very dirty and muddy most of the time, even when it did not rain. In the rainy season very often flooded (>50 cm deep)
Design considering nature	 Already considered green spaces	 Very crowded and minimal green spaces

3. Result and discussion

The design concept for this vertical village housing came up from the problems happening in high-density settlements, which impacted the quality of life of the urban village communities. The government's effort to solve this problem by building "rusun" (vertical housing) has not given enough attention to ecological and environmental sustainability, particularly the impacts from disturbed water cycle on the environment and life quality of the communities.

Implementation of ecological architecture concept in this research was done through simulations, calculations and analysis of the four elements (nature, water, human and environment) resulted in Table 3.

Table 3. Ecological architecture concept application into building.

No	Elements	Design application
1	Kampung structure interpretation	Housing design tried to interpret the horizontal structure of kampung into vertical structure.
2	Building form follows context	Building form responded to the context pragmatically.
3	Use of renewable energy	Solar panels are used as alternative energy source for lighting and irrigation pumps (automatic drip system).
4	Ecological corridors in building	Place various kinds of vegetations in every floor to create a living ecosystem on each floor.
5	Natural ventilation	Natural ventilation for each massing/block from continuous void stretching from the lowest to highest floor.
6	Climate response	Orientation, form of massing and openings were simulated to respond to the climate on site.
7	Modular housing	Easier to construct, more economical and aesthetically better-looking building.
8	Natural lighting	Natural lighting from voids and balconies creates shading during the day.
9	Universal design	Public ramps were incorporated as a response to human aspect and as a form of universal design for everyone.
10	Communal area	Communal area on the ground floor has cross-programming characteristics to suit various occasions needed by the residents, e.g. weddings or group gatherings.
11	Steel structure for columns and beams	Steel structure has good characteristics and aesthetics, as well as recyclable.
12	Urban green area	Green areas have lots of voids between densely populated settlements around the design site.

The modular form was chosen to represent interpretation and response to the context of "kampung." The modular form also created a better view for each unit and shading for other units in the lower level. It was developed from the simulation of horizontal urban villages in vertical form and aimed to obtain more effective forms of natural lighting and ventilation. The simulation was also done to determine practical mass orientation to maximize natural daylighting during the day. Design calculation can also maximize the roof space to set up solar panels, contributing to 20-25% of electricity consumption in the design site.

Roof space can also be used for urban farming and communal area for socialization. The green open space in the middle of the site is intended to create bonding between masses and provides a better view from each unit. The terraces formed on each floor also add to the central green open space in the middle. The accessibility and connectivity in the building were also considered ecologically friendly and healthier for everyone by encouraging people to walk and interact with each other between blocks. Provision of the communal area is an interpretation of urban village sociocultural and implementing the ecological concept. The landscape area was designed to become a communal area or improve the aesthetics and function as an infiltration and rainwater management area. The integration of ecological architecture and zero runoff in this vertical village housing design can be seen in Figure 5.

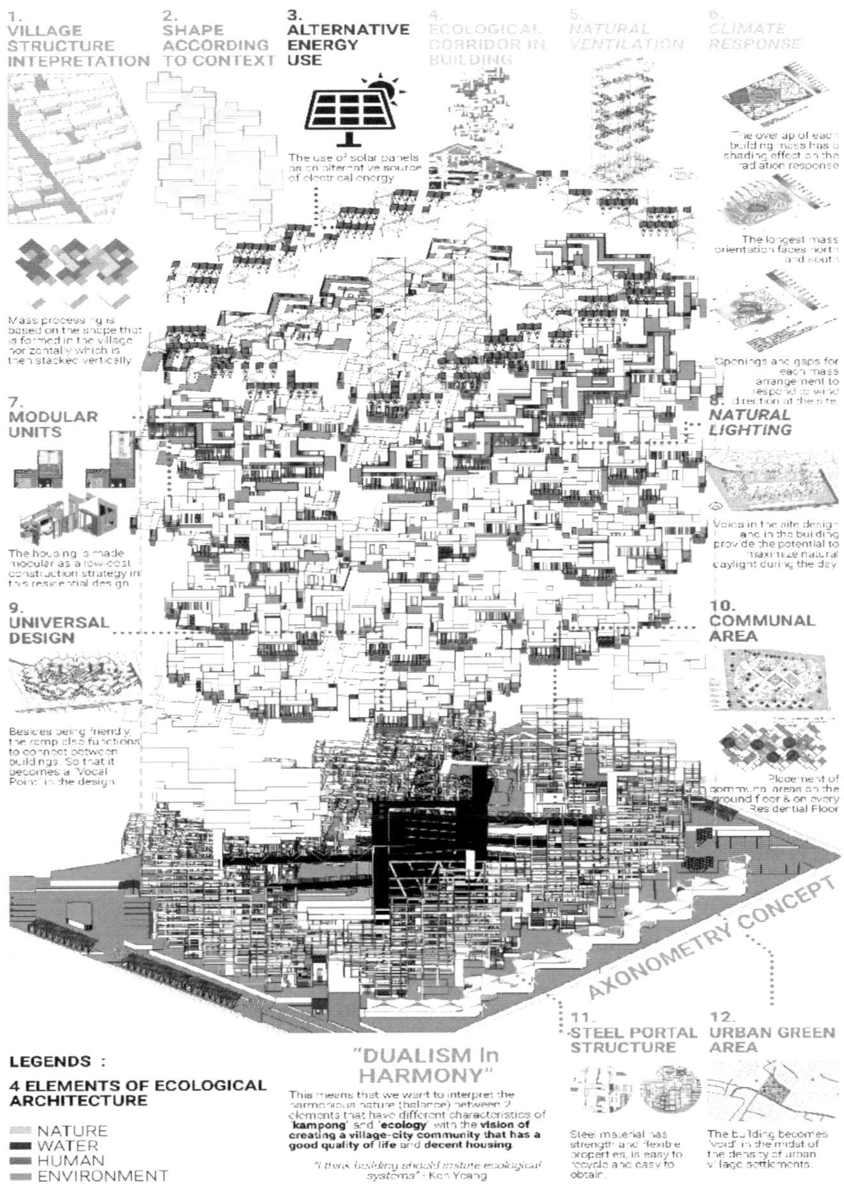

Figure 5. Design of vertical village housing using ecological architecture concept.

The ecological corridors were embedded in the buildings through terraces (green facades), rooftop gardens, parks, and grass blocks in parking lots. Total green area (excluding roof garden and grass blocks) amounts to 4725 m2 (KDH = 50% of total area), more extensive than the regulation for building's KDH (Koefisien Dasar Hijau = 20%). These green spaces naturally improve the potential of Zero Runoff on the site. By linking ecological landscapes in the vertical village, rainwater does not have to flow on the streets or through the drainage channels but the green and blue landscapes/corridors as it should be. The application of Zero Runoff in this design is also supported with 11 infiltration wells receiving a calculated runoff volume of approximately 250 m3 from non-roof areas and a retention pond

IOP Conf. Series: Earth and Environmental Science **894** (2021) 012008 doi:10.1088/1755-1315/894/1/012008

having a volume of around 370 m3 coming from the roof area (Figure 6); thus, the Zero Runoff system was designed to handle a total of 620 m3 runoff and divert it from flooding the surrounding communities.

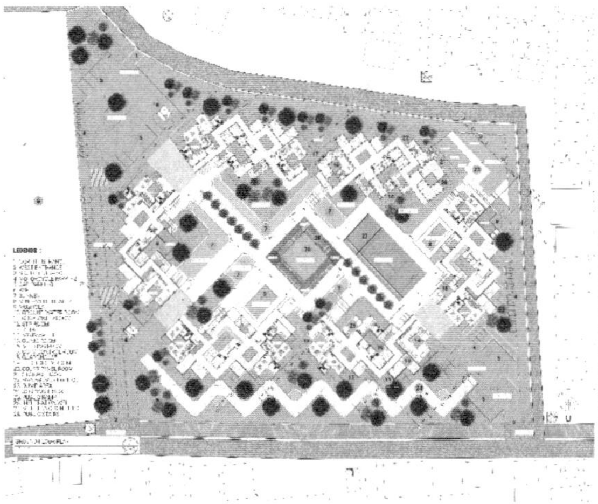

Figure 6. Zero runoff implementation in vertical village design.

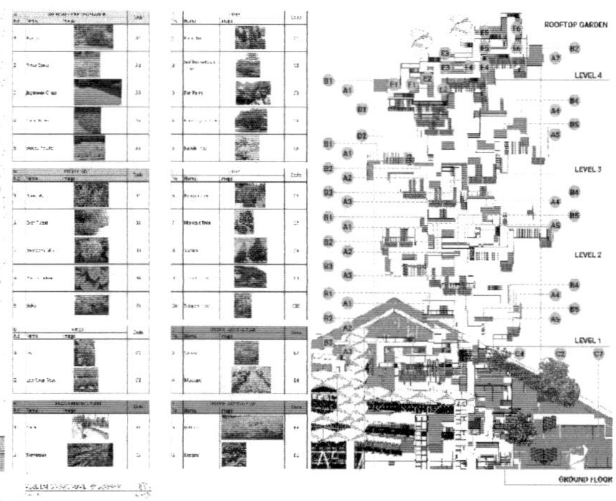

Figure 7. Vegetation axonometry in vertical village design.

Figure 7 shows the various types of vegetation planted on-site and the potential of increasing the biodiversity in the area, creating a rich ecosystem that improves health, well-being, and environmental quality and can be utilized as a green building envelope to reduce the Urban Heat Island effect.

Through the integration of ecology (including ecological architecture) and sustainable water management, which include Zero Runoff rainwater management, we can help nature to work more effectively, promote natural hydrologic and ecosystem processes, conserve and reuse freshwater, improve public health and well-being, and prevent disaster and losses from flooding at the same time.

4. Conclusion

Limited land availability and high urbanization rate have impacted the life quality of urban village communities in Jakarta, especially from ecological aspects. On the other hand, amplifying the development of built and impervious areas coupled with increasingly unpredictable and extreme weather causes urban villages in Jakarta to become more prone to disaster and losses from flooding.

The multidimensional analysis done in this study has shown that integrating ecology with Zero Runoff in the design of vertical village housing has great potential to be a solution to rectify these problems and simultaneously improve the quality of life and environment.

References

[1] Jha A K, Bloch R and Lamond J 2012 *Cities and Flooding: A Guide to Integrated Urban Flood Risk Management for the 21st Century* (Washington DC: The World Bank)

[2] Cobián Á J A and Resosudarmo B P 2019 *Environ. Econ. Policy Stud.* **21** 555–77

[3] Texier P 2008 *Disaster Management and Prevention* **17(3)** 358-72

[4] Sunarharum T M, Sloan M and Susilawati C 2013 *The 9th Annual Conf. of Int. Institute of Infrastructure Renewal and Reconstruction* (Brisbane)

[5] Notario D, Suprata F, Canti M and Karmawan LU 2021 *Mitra: Jurnal Pemberdayaan Masyarakat* **5(1)** 1-10

[6] Damayanti S, Moersidik SS and Sarwono SW 2014 *Lingkungan Tropis* **8(1)** 1-12

[7] Leepel RJ, Utomo SW and Suganda E 2017 *J. Manusia & Lingkungan* **24(3)** 116-124

IOP Conf. Series: Earth and Environmental Science **894** (2021) 012008 doi:10.1088/1755-1315/894/1/012008

[8] Widiyani NMW Rusunawa sustainability assessment and the impact to residents' capabilities [dissertation on the internet]. Grenoble (FR): Université Grenoble Alpes; 2017. [cited 2021 Aug 20]. Available from: https://dumas.ccsd.cnrs.fr/dumas-01590375/document
[9] Wijayanti SH, Pandia WSS and Sutarno HY 2019 *ASEAN J. of Community Engagement* **3(2)** 266-86
[10] Yeang K 2008 *Ecodesign: A Manual for Ecological Design* (Chichester: Wiley)
[11] Van der Ryn S and Cowan S 2007 *Ecological Design: 10th Anniversary Edition* (Washington DC: Island Press)

Locality values in thermal comfort embodied in traditional Indonesian houses: a literature review

P U Pramesti[1]*, M I Hasan[1], M Ramandhika[1]

Vocational School, Universitas Diponegoro, Indonesia

previ.pramesti@gmail.com

Abstract. This study aims to explore how traditional buildings of Indonesia adapted and responded to the current climate. Scientific publications regarding traditional houses from various regions in Indonesia were gathered to be examined the locality values in thermal comfort. The examination of the scientific publications highlighted some housing elements and arrangements that influence the thermal comfort of the houses. The building elements that affect the thermal comfort encompass lifted house on stilts; appropriate size and amount of windows; ventilated roof; selected natural material; divider arrangement; and the existence of veranda. Furthermore, as exemplified in Bali, the proper spatial arrangement between house buildings and open space influences traditional Indonesian houses' thermal comfort. In conclusion, dealing with climate in a tropical area with high humidity can be demonstrated in those locality values without consuming excess energy for a cooling system. The significance of this study is beneficial for the development of tropical houses, which is expectedly to consider those locality values.

1. Introduction

Artificial lighting, heating, cooling, and ventilation are used heavily in modern Indonesian structures. It implies that mechanical equipment consumes more energy, which is not sustainable. Thermal comfort is a state in which a person's psychological, physiological, and behavioural patterns are at ease while they are in an environment with a specific temperature. According to theory, humans can adjust to thermal changes in three ways: behavioural pattern adaptation, physiological adaptation, and psychological adaptation. Several heat sources contribute to the rise in temperature in a room; natural heat sources, such as the sun and geothermal, are the first to be considered. Biological heat sources, such as humans and animals, are the second source of heat. An electric mechanical heat source, such as machinery, lights, and other equipment, is the last heat source. The metabolic rate, ventilation rate of the garments worn, room temperature, room air humidity, and the speed of airflow on the skin surface are all factors that affect a person's thermal comfort [1].

Traditional structures are known as a source of information for sustainable, energy-efficient, and climate responsive design, despite modern buildings consuming much energy. One of the most noticeable differences between modern and traditional buildings in Indonesia is providing thermal comfort to occupants. The diversity of Indonesian cultures and customs are renowned. Its traditional dwellings reflect this diversity as well. The majority of traditional Indonesian houses conform to rules based on local customs or culture [2]. However, other houses are guided entirely by demands and technical considerations. Traditional houses, on the other hand, are noted for their excellent thermal performance. It is the reason for responding to the environment, materials and techniques available in the area, and the local way of life [3]. The houses are notable for using passive design principles to enhance thermal comfort. As previously said, traditional buildings are influenced by the local climate, and traditional Indonesian houses must thus be influenced by the country's hot and humid climate.

Indonesia is positioned between -11° and 6° latitude. Thus, it has a tropical climate. Due to Indonesia is an archipelago lying between the Pacific Ocean and the Indian Ocean, its climate is hot and humid. High average temperatures, little day-night and seasonal differences, high humidity, and copious rainfall are characteristics of a hot and humid climate, as are high and relatively diffuse solar radiation [4]. Overheating may not be the most serious issue in this climate, but it is exacerbated by high humidity, limiting evaporation potential [5].

Traditional buildings dependability and existence have always been a source of architectural learning over the years. Modern architectural approaches can be filtered through local knowledge and wisdom. It utilizes local materials and harmonization with the surroundings results in an adaptive design consistent with previous user behaviour.

2. Methodology

Scientific publication in literature review is commonly conducted to generate a holistic understanding of a variable from various authors, research locations, and authors' points of view [6,7]. The scientific literature on thermal comfort in traditional Indonesian houses was gathered to investigate how people treated internal and external circumstances in their houses to accomplish thermal comfort in the past. A total of twelve articles were chosen for synthesis, including one book, four journals, and seven proceedings. This research paper examines case studies of traditional Indonesian dwellings from various parts of the country. Table 1 shows a summary of the 12 publications.

3. Result and discussion

Table 1. Outline of literature review.

Author	Type of publication	Location	Findings in thermal comfort	Keyword
Izziah et al. 2021	Proceeding	Aceh	Cool air is blown from beneath of then enter the house through timber floor [8]	Stilt house
Samra and Imbardi. 2020	Proceeding	Riau	Sufficient ventilation to maximize air circulation [9]	Windows size
Siahaan. 2020	Proceeding	Nias	Sloping and ventilated roof design maximizes air circulation; timber material responds to the local climate; stilt house to decrease humidity from soil[10]	Sloping and Ventilated roof; material; Stilt house
Idham. 2018	Journal	Central Java and Yogyakarta	Wall height is lower than ceiling to allows hot air to circulate upwards; house orientation towards the sea to allocates natural cooling system [11]	Wall and ceiling arrangement; house orientation
Wiryomartono. 2014	Book	West Java	Naga community built their houses 35-40cm high from the ground and remain space under the house empty for ventilation.[12]	Stilt house

Author	Type of publication	Location	Findings in thermal comfort	Keyword
Naing and Hadi 2020	Journal	South Sulawesi	Selected timber materials capable of maintaining air humidity[13]	Material
Asriningpuri. 2020	Proceeding	Greater Jakarta	The existence of veranda as the response to the local climate.[14]	Veranda
Yudantini. 2021	Proceeding	Bali	Proper percentage between building mass and open space allows smooth air circulation to buildings [15]	Open space
Kane,Mishra, and Dutta. 2016		Lampung Java Toraja	Cross ventilation that occurs in the house is caused by the air movement through the windows in the building[16]	Window
Anwar. 2019	Proceeding	Southern Sumatera	Change their behaviour and activity to fulfil their comfort[17]	Behaviour and activity
Latif et al. 2019	Journal	Bugis	A roof that has a steeper slope or a larger attic volume reduces heat; Material variables exert a considerable influence on heat induction from the external environment. Houses that do not have ceilings, utilizing non-insulating materials on the roof, walls and ceilings are the sources of excessive internal temperatures.[18]	Slooping and ventilated roof; Material; ceiling;
Lahji et al. 2011	Journal	Flores	Using local materials such as fibres for roof, woods and bamboos for walls and floors, woods[19]	Material

The thermal comfort of the static model and the thermal comfort of the adaptive model can be mentioned in the formation of thermal comfort in traditional buildings in Indonesia, according to the literature review results. Static thermal comfort is defined as thermal comfort that is established based on the idea that the thermal conditions in a room remain constant regardless of the temperature and differences in people's ability and behaviour to adapt to the thermal environment. Several static models of thermal comfort are developed based on the above review and applied to traditional buildings, including stilt houses; window size to maximize air circulation; ventilated roof design to maximize air circulation; wall and ceiling arrangements; house orientation; material; the existence of veranda; open spaces; ceiling. If the static model fails to achieve thermal comfort, an adaptive model can be used, based on the idea that the temperature outside the building affects the temperature inside the building, and humans can interact with and adapt to different temperatures. The adaptive thermal comfort model is based on one's expectations of thermal environmental conditions based on historical memories, behaviour patterns, and thermal environmental conditions. According to the literature assessment, one of the adaptive models in traditional structures is adjusting their behaviour and activity to meet their comfort needs.

3.1. Stilt house

Stilt houses can be found in a variety of traditional Indonesian structures around the country. Several factors influenced people's decision to create stilt houses as their residential structures in the past. Izziah et al. noted in their article on the connection between traditional Acehnese houses and the current state of the modern Acehnese community that the stage's architecture tries to shield the stage's inhabitants from wild creatures. The development of such houses addresses the needs of the climate, particularly in terms of climate. Raising the house's floor to 2.5 meters allows cold, damp air from the basement to be drawn into the house's rooms via wooden floors spaced at regular intervals. In his book about traditional settlements in Indonesia, Wiryomartono explains that the Naga community in West Java built their houses 35-40cm high from the ground and left space under the house empty for ventilation. In his paper, Siahaan mentions another criterion for the installation of dwellings on stilts in Nias. The mild form of the house, which is located in hilly and forest locations, can protect the people from being attacked by wild animals. In terms of thermal comfort, these conditions avoid flooding and humidity, which can damage building materials [20].

3.2. Window

According to the SNI (Indonesia National Standards) of the General Department's regulations, a room in the Residential houses must have ventilation of at least 5% of the room's floor area and windows of at least 10% of the floor area the room [21]. The building significantly impacts the amount of work required to use wind for room conditioning. The size of the opening can be modified according to the wind flow requirements. The entrance and outlet area comparison affects the speed of the wind entering the room. The air velocity in the room will be lower than outside when the intake is greater than the outlet. The air velocity inside the room will be higher than outside when the inlet is smaller than the outlet. The effort to employ the wind in the air room conditioning is heavily influenced by the direction of the opening. The direction of motion and pattern of air in the air space will be determined by the guide at the intake, and the varying shapes of the guides will result in distinct airflow patterns. The leaf windows and louvres are the part utilized to guiding in the window opening [22]. According to Samra and Imbardi's article, Riau traditional home architecture is a vernacular building that can be altered or created over time in response to changing needs and tropical circumstances on Sumatra's east coast [9]. The structure is constructed specifically for the tropics, so fresh air circulation is controlled through vents built into the walls. The concept is best suited for hot and humid tropical climates. Penetrations are also discovered in kitchen walls and floors to maximize fresh air delivery and reduce concentrations of cooking pollutants. Air circulation is a type of shaft ventilation in which the apertures are on opposing sides of the room. A similar case happened in Lampung traditional houses. The air movement in Lampung's traditional houses allows for cross ventilation. Air movement is primarily at or below window level, with the average amount of wind inside the room being lower than outside and the average amount of wind inside the room being lower than outside. If the wind speed outside is slightly higher, turbulence can cause discomfort to occupants due to its pushing influence on incoming airflow [16].

3.3. Material

The heat of the sun enters the structure by conduction and solar radiation. The amount of sunlight that passes through the envelope. The type of material employed determines how the building envelope reacts to environmental conditions as the building's exterior layer. The building material acts as a barrier between the outside temperature and solar radiation and the internal temperature. The materials used in construction are influenced by the climate. The qualities and thickness of the material and the colour of the material's outer surface are all aspects to consider [23]. A material's thermal conductivity is a measurement of its ability to transfer heat. Heat transfer occurs at a low pace in materials with poor thermal conductivity; however, heat transfer occurs at a high rate in materials with high thermal conductivity [24].

The availability of resources in the surrounding environment influences material selection in traditional buildings. Due to technological and transportation limitations, local ancestral wisdom is used in a traditional Indonesian dwelling. As a response to the tropical climate, the Nias traditional house researched by Siahaan featured a sloping roof. Furthermore, as the most common building material, wood can respond to changes in temperature and humidity. Woven leaves of thatch are used as a roof covering to shield the building from the blazing sun. Humidity is anticipated by positioning the column on a stone foundation and not directly touching the earth to avoid decay/weathering [20].

As detailed in the essay by Naidah Naing and Karim Hadi, similar considerations also exist in Bugis traditional dwellings. The materials utilized, such as ironwood, Merbau, and spinach wood, are flexible, resilient, natural electrical insulators, capable of maintaining humidity at seasonal temperatures, and non-flammable [25].

The wave size on the zinc roofing material was sufficient to have a considerable effect on the intensity of heat entering the building, according to research conducted by Latif et al. on traditional Bugis dwellings. The average internal temperature of dwellings with little corrugated tin roofs is higher than those with sizeable zinc roofs. When it comes to wall materials, the highest temperature is achieved when zinc, wooden boards, and plywood are used, and the lowest temperature is achieved when gamacca is used.

The Gurusina indigenous people's local knowledge is supplemented by local culture using local wisdom materials to feel comfortable and safe in a traditional dwelling that reflects three characteristics of thermal comfort: temperature, humidity, and wind flow. Traditional Flores houses are built with environmentally friendly sheathing materials that may be recycled quickly (bamboo, palm fibre, wood). The state of the house accomplished the optimal thermal comfort in the PMV analysis conducted by Lahji et al. on the typical Flores house employing these materials. The thermal comfort level of inhabitants in a comfortable place thermal space is predicted using PMV analysis, which is affected by outdoor and indoor temperatures, occupant activities, and outerwear. Measurements of temperature, humidity, and wind speed and interviews with occupants on the perceived thermal comfort provide simulation data on field measurements.

3.4. Orientation

In order to profit naturally from heating and cooling systems, orientation must be in harmony with other aspects. According to Soetiadji in Dwi Kustianingrum's research [26], the orientation of a shape is its location concerning the base plane, cardinal directions, or the observer's point of view. Soetiadji also said that the type of orientation results from the influence of orientation on anything, resulting in the building. Because the equatorial region receives much radiation, it is the hottest place on the planet. According to Hamdani, Bekkouche, Benouaz, and Cherier [27], in the case of housing, the building's orientation to the sun has a considerable impact on the air temperature in the room.

The majority of the vernacular dwellings in Java areas still adhere to the traditional housing layout and orientation. The Queen of the South Seas Nyai Roro Kidul and Dewi Sri are revered as the Javanese's goddesses of land and fertility. As a result, the dwelling always faces south, and all functions other than offerings are prohibited in the central hall. Air circulation is also greatly aided by proper home orientation. Acclimatization is performed through the direction of the building in the smaller internal space of a beach home. The orientation of a house in the southern region of Java is beneficial for acclimating coastal dwellings and inland dwellings that tend to take into account the South Sea breeze. The orientation of northern dwellings is related to the roadway rather than the breeze. North Java's coastal breezes are not as strong as those in the south. Hence they have less of an impact on indoor air quality. Meanwhile, orientation is not a high priority in this area. More excellent rooms, larger openings, and more porous envelopes assure air movement.

In North Nias, according to the Omo Hada settlement pattern, all houses face north-south, with the length of the house facing the road. This position provides natural daylight and ventilation, which is ideal for any home.

3.5. Sloping, ventilated roof and ceiling

In Nias traditional house, the interior of the structure, protected from moisture, with the roof window

(Lawa-Lawa) and lattices on the building's walls, which provide lighting and natural airing and allow sunlight into the building. The building's high roof space helps to ensure that air circulation is optimized.

Bugis traditional house has a similar condition because the heat buildup process occurs faster on a triangular roof with a steeper slope or a bigger attic volume. It will be hotter than a roof with a steeper slope or a larger attic capacity. During the hot afternoon, houses without ceilings have a high average interior temperature. According to Ong [28], solar radiation absorbed by the roof and immediately discharged to the occupants causes the high temperature in the home's interior that is not shielded by the ceiling. As a result, ceilings constructed of wood are suitable for homes in humid tropical regions. In this situation, the ceiling height observed in research ranges from 200 to 265 cm. The room temperature drops as the ceiling rises. This finding is consistent with several prior investigations, which found that the air is stratified from the bottom to the top due to the high temperature. If the room is low, hot air will gather towards the top, bringing the hot air closer to the adequate human space.

3.6. Veranda and outdoor spaces
The house's one-of-a-kindness is Because Betawian descendants are faithful to their customs or traditions, notably in caring for ancestor ashes. The home has "meja abu" on the front area following the veranda ("serambi"). Based on Asriningpuri's research, veranda use to restrain the sun glair or radiation of the sunlight that enters the main house in frontage to preserve the indoor temperature from the back sideuse the service area. Meanwhile, in Balinese traditional houses, outdoor spaces with traditional landscapes have the goal of being environmentally friendly while maintaining comfort for its residents, regulating the flow of air into the room, and adjusting lighting according to its function, as well as acoustic function that can be found in the outdoor of the housing. The building's composition pattern generates a plaza in the centre called natah, which serves as a point of direction, circulation, an emergency point in an earthquake, air circulation, and natural illumination for the structures.

3.7. Behaviour and activity
Traditional structures feature a passive design that uses less energy. Traditional structures are built to collect much solar energy and use it sparingly. As a result, the zero-energy building design approach has been applied to conventional structures. The climate today is not the same as it was when the limas home was constructed. As a result, the user's behaviour and activity may vary to accommodate their comfort. Because limas was built before electricity, the rooms relied on natural light from the windows. When electricity is available, artificial lights are used in each room during the day and at night. Only when natural lighting is insufficient are the lights used for daily activities. The windows were opened for natural lighting during the day for visual comfort. Artificial lighting is installed in all rooms and is only turned on at night. The natural lighting is adequate for the activities of the residents. The limas have electric fans in the bedroom and living area for further comfort, with the afternoon and mid-afternoon were the most popular times to use a fan.

4. Conclusion
Indonesian traditional houses were designed in response to the region's climatic conditions. Specific house arrangements generate a thermal comfort system in a pearl of local wisdom; the genius loci of local wisdom values are less applied to current housings in Indonesia. The increase of timber materials and demand for modern-look houses fade these values away. However, another value lay in the traditional Indonesian houses, such as the arrangements of openings and house layout, which are still applicable to current architectural needs. The generous amount of windows applied in traditional Indonesian houses applies to current housing needs, which is in line with the green architecture issue to maximize natural ventilation and reduce the use of air conditioners. The verandah in the local wisdom of traditional Indonesian houses is also applicable to the current architectural realm due to tons of its thermal and acoustic comfort function. Finally, this study reveals the local wisdom values of traditional Indonesian houses and contributes to building design knowledge with responses to the environment.

References

[1] Alta 2019 *Alta Integra* 1
[2] Toe DHC, Kubota T 2015 *Sol Energy* **114** 229
[3] Jayasudha P, Dhanasekaran M, Devadas MD, and Ramachandran N 2014 *Indian J Tradit Knowl* **13(4)** 762
[4] Coch H 1998 *Renew Sustain Energy Rev* **2(1–2)** 67
[5] Szokolay SV 2008 *Journal of the American College of Radiology* **8** 259
[6] Hasan MI, Noor Prabowo B, and Bava Mohidin H H 2021 *J Des Built Environ* **21(1)** 21
[7] Othman Z, Aird R, and Buys L. 2015 *Front Archit Res* **4(1)** 12
[8] Izziah, Meutia E, Sahputra Z, and Irwansyah M 2021 *IOP Conf Ser Mater Sci Eng* **1087(1)**
[9] Samra B, and Imbardi I 2020 *IOP Conf Ser Earth Environ Sci* **469**
[10] Siahaan F 2020 *IOP Conf Ser Earth Environ Sci* **452**
[11] Idham NC 2018 *Front Archit Res* **7(3)** 317
[12] Wiryomartono B 2014 *Perspectives on Traditional Settlements and Communities. Perspectives on Traditional Settlements and Communities*
[13] Naing N, and Hadi K 2020 *Int Rev Spat Plan Sustain Dev* **8(3)** 1
[14] Asriningpuri H 2020 *IOP Conf Ser Earth Environ Sci* **402**
[15] Yudantini NM 2021 *IOP Conf Ser Earth Environ Sci* **738(1)**
[16] Kane SN, Mishra A, and Dutta AK 2016 *J Phys Conf Ser* **755(1)**
[17] Anwar WFF 2019 *Journal of Physics: Conference Series* **1198**
[18] Latif S, Hamzah B, Rahim R, and Mulyadi R 2019 *Tesa Arsit* **17(1)** 61
[19] Lahji K, Susetyarto MB, and Walaretina R 2021 *IS LIVAS* 63
[20] Siahaan F 2020 *IOP Conf Ser Earth Environ Sci* **452(1)**
[21] SNI 2010 *SNI 03-6390-2010*
[22] Arifah AB, Adhitama MS, and Nugroho AM 2017 *J Mhs Arsit* **5(4)** 1
[23] Pramesti PU, Ramandhika M, Hasan MI, and Werdiningsih H 2021 *IOP Conf Ser Earth Environ Sci* **623(1)**
[24] Lubis RN 2020 *Repositori Institusi Universitas Sumatera Utara*
[25] Naing N, and Hadi K 2020 *Int Rev Spat Plan Sustain Dev* **8(3)** 1
[26] Kustianingrum D, Muhamad YA, Rizqika M, Wijaya AN, and Pramana AD 2016 *J Reka Karsa* **4(1)** 1
[27] Hamdani M, Bekkouche SMA, Benouaz T, and Cherier MK 2012 *Energy Procedia* **8** 636
[28] Ong KS 2011 *Energy Convers Manag* **52(6)** 2405

IOP Conf. Series: Earth and Environmental Science **894** (2021) 012010 doi:10.1088/1755-1315/894/1/012010

Visualizations of Berea sandstone pores using neutron tomography

R Setiati[1*], I Sumirat[2], S G Sukaryo[2], P S Adisoemarta[1], F Akbar[2], TS Marpaung[1], and Bharoto[2]

[1]Master of Petroleum Engineering, Faculty of Earth and Energy Technology, Universitas Trisakti, Jakarta, Indonesia
[2]Center for Advanced Materials Science and Technology, Serpong, Tangerang Selatan, Indonesia

rinisetiati@trisakti.ac.id

Abstract. Sandstone is a reservoir rock commonly found by oil and gas companies. Oil and gas usually trap inside the sandstone pores. A Destructive-Testing (DT) method or chemical solution method is usually used to measure the porosities of the sandstone sample. On the other hand, neutron computed tomography (NCT) can visualise and quantify all the porosities of the sandstone non-destructively. Neutron tomography is an imaging technique that employs neutron generated by a nuclear research reactor. The NCT produces cross-sectional images of the object that was used to visualise the Barea sandstone porosities distribution. The results show that each rock sample possesses connected and concentrated pores in the middle part of the rock with total porosity of ± 20%.

1. Introduction

Indonesia possesses an abundant level of minerals where oil and natural gas are the primary mineral resources. Sandstone is one of the most commonly found minerals in oil and gas companies. As oil and gas reservoir rock, sandstone possesses pores that can be filled by oil and gas. Generally, to produce a visualization of a rock, Destructive Testing (DT) is used as an option, for example, the Wood's Metal method. Wood's Metal is a mixed metal that commonly consists of Cadmium (Cd) and Lead (Pb) used to stick metals together. This method is utilized by injecting Wood's Metal into the rock pores. After that, the rock is dissolved into a chemical solution which will destroy the rock and leave the hardened Wood's Metal that shows the rock pores distribution [1]. On the other hand, by utilizing neutron tomography, the distribution of the pores can be visually produced by utilizing specific software. This method is categorized as the NDT method, where it is not necessary to destroy or dissolve the rock sample. By utilizing neutron tomography, we can also determine the value of rock porosity by using specific software [2].

Neutron radiography facility – BATAN is already enhanced for tomographic data acquisition and has been utilized to characterize and facilitate NDT on industrial materials in agriculture, medicine, archaeological geology, and building construction [3]. Neutron radiography is also used to characterize material porosity such as concretes on cultural heritage buildings. The latest utilization of neutron tomography is in the oil industry [4]. A neutron is a sub particle of an atom that possesses no charge or is neutral, not reacting with other electrical charges, and can move freely towards any other materials until it reaches the nucleus. Because of that, neutrons are non-destructive and possess a high penetration value [5]. By utilizing neutron penetration as a radiation test, imaging is possible as a non-destructive analytical tool to determine porosity and water content in porous media, including sandstone, concrete, and so on [6]. The successful neutron penetration through the sandstone sample and the performance of special software, it is possible to predict the percentage of tar/oil/water content in the sandstone sample. Therefore, the use of Neutron Tomography as Neutron Imaging in visualizing the pores of sandstone is appropriate [7].

Unlike x-rays, neutrons interact with the nucleus of the atom, which is why they are sensitive to the different isotopes of matter. One interesting example is hydrogen (and it compounds, such as water and hydrocarbons), which weakly attenuates x-rays while being highly opaque to neutrons. Similar considerations can be made for some salts (the chlorine in NaCl is highly visible) or lithium (for battery applications). Conversely, metal such as Lead or Gold are far more transparent to neutrons than to X-rays [8].

Neutron tomography is an imaging technique that utilises neutron as its source to produce an internal structure of an object by showing the cross-sectional image of the object. Neutron tomography technique utilises neutron of a research reactor with 15MWatt thermal energy. Neutron will react on sandstone sample and form shadows on scintillator Lithium (Li). The scintillator will change neutron into blue light and then transfer it into a CCD camera. The sandstone itself is placed on a Goniometer turntable, and the image is recorded at a very turning degree. From the image, sinogram and image reconstruction are produced. Sinogram is an all-angle composition of an object projection on a single slice. Sinogram act as an input for the following process, which is the image reconstruction process [9].

The novelty of this research is that there are still few studies on the visualisation of sandstone pores using the Neutron Tomography tool for the oil and gas industry, especially in Indonesia. BATAN's nuclear reactor is the only nuclear reactor in Indonesia. Therefore, this study uses a Neutron Tomography tool that utilises neutrons from the G. A. Siwabessy Multipurpose Reactor - Serpong Nuclear Area [2]. In a previous study by Bharoto et al. [1], Limestone cores obtained directly from oil and gas reservoirs were used, while in this study, Sandstone cores in the form of Berea cores were used. Limestone has more varied materials when compared to Berea Sandstone, so Limestone has pores that are also more varied (large and small) than Berea sandstone which has almost uniform pores. Therefore, this study was conducted to visualise the pores of Berea sandstone using a Neutron Tomography tool. The reason for using this tool is because this Neutron Tomography tool uses the Non-Destructive Testing (NDT) method rather than the Destructive Testing (DT) method, which can damage the sample. This study also uses three cores instead of one to ensure the data obtained are correct.

2. Methodology

This research was conducted by utilising neutron tomography belonging to the National Nuclear Energy Agency of Indonesia (BATAN), located at G.A. Siwabessy Multi-Purpose Reactor – Serpong Nuclear Area [2]. On this neutron tomographic sandstone pore visualisation test, we utilised three sandstone samples representing three repetitions to improve data accuracy. The utilised samples are Berea Core X. Berea Cores are ideal as samples because of their average pore structure uniformity and relatively low clay content [10]. The Core X is cut into three parts and named Core X4, Core X3, and Core X7. These three cores have diameters of respectively 2.458 cm; 2.462 cm; and 2.460 cm and heights of respectively 3.747 cm; 3.793 cm; and 3.770 cm.

Figure 1. Images of Core X4, Core X3, and Core X7.

These core samples were then saturated with brine until their pores were filled with brine. Brine is a solution that contains sodium chloride and aquadest. After that, these three samples were tested on a neutron tomography owned by BATAN. Figure 1 shows the images of Core X4, Core X3 and Core X7.

Neutron tomography data was collected by turning the object from 0° to 360° during neutron irradiation, and the imaging data were taken at every 1° interval. Each sample projection was captured on a Charge-Coupled Device (CCD) camera and stored on a computer with LabView software's 16-bit Tagged Image File Format (TIFF). Before the tomography reconstruction, every noise on these images was removed by using image processing software called ImageJ. After that, the image reconstruction process was generated by using Octopus 8.8 software. The output of this data processing is a cross-sectional image of the object along its vertical axis. By utilising VG Studio Max 3.0 image visualisation software, the produced cross-sectional image (voxel data) was shown on a 3-dimensional image [1]. These images were analysed to obtain porosity value by using available software. The neutron tomography (RN1) image can be seen in Figure 2, and the neutron tomography scheme can be seen in Figure 3.

Figure 2. Neutron tomography (RN1).

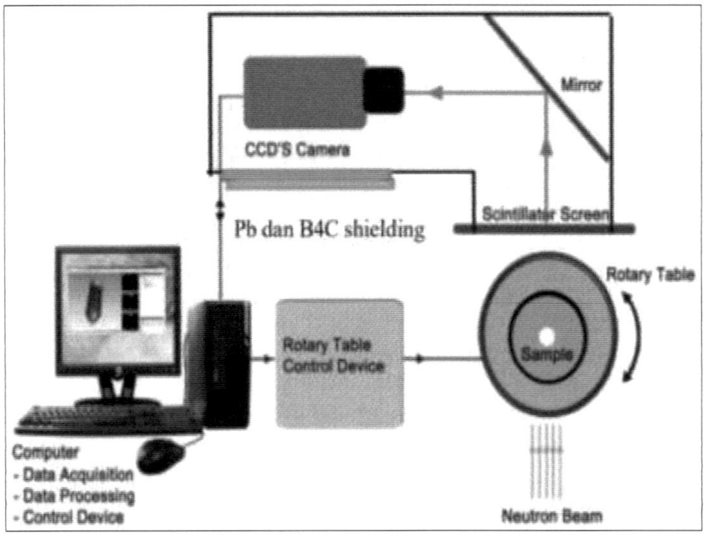

Figure 3. Neutron tomography system scheme.

3. Results and discussion

The produced images through Octopus 8.8 software are sinogram images cleaned from noises and outliers using ImageJ software. These noises and outliers are unwanted because they can disrupt the primary data required in this research. Because of that, to obtain a clearer sinogram image, these noises and outliers must be eliminated. The horizontal cross-section image of Core X4, X3, and X7 was respectively taken on slices 135, 1,243, and 1,235. These three horizontal cross-sectional images can be seen in Figure 4.

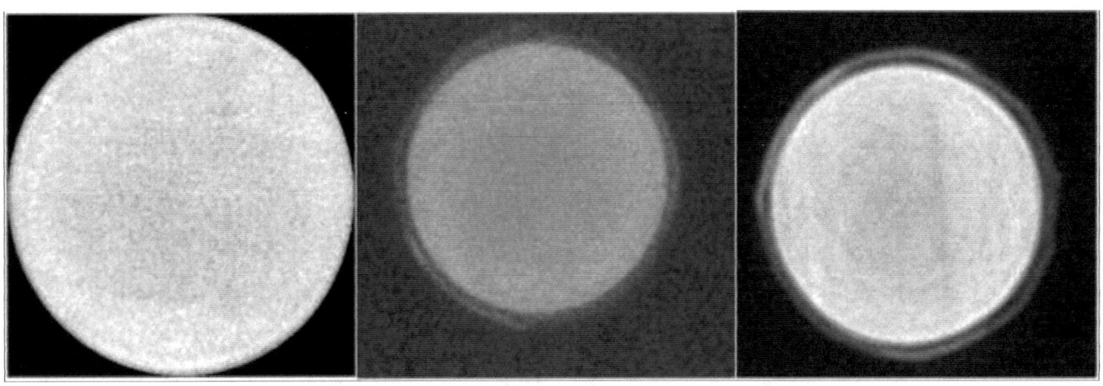

Figure 4. Sinogram of core X4, X3, and X7 (from left to right).

As we can see in Figure 4, the black dots on the core are the core pores. Meanwhile, the white areas are core materials (matrix). Brine is a sodium chloride solution which is partially composed of water containing hydrogen atoms. The hydrogen atom greatly attenuates the thermal neutrons so that the neutrons will be absorbed in the brine and therefore appear a contrast or darker colour than the others in the image, so that it can be distinguished between the pores of the rock and the material (matrix). The same thing was also obtained by the study of Cordonnier et al. [11], where the results of Neutron Imaging using water as the material, it was found that a contrast colour (blackish grey) appeared which indicated that the neutrons were absorbed. Octopus 8.8 software was utilized, to determine the material and non-material parts of the image. Some parameters must be met to produce a decent sinogram inserted into

VG Studio Max 3.0 software. These parameters are scale, slice number, step size, central value, arctan, rotation axis, outer slice, lower limit, and upper limit. All these parameters are vital to produce clear sinogram images so that when the images are reconstructed into a 3D image, the image will be clear. After that, using VG Studio Max 3.0 software, we determined the material (matrix) and non-material (pores) of the samples and produced a 3D image that can be horizontally or vertically sliced. The horizontal slice of core X4 can be seen in Figure 5.

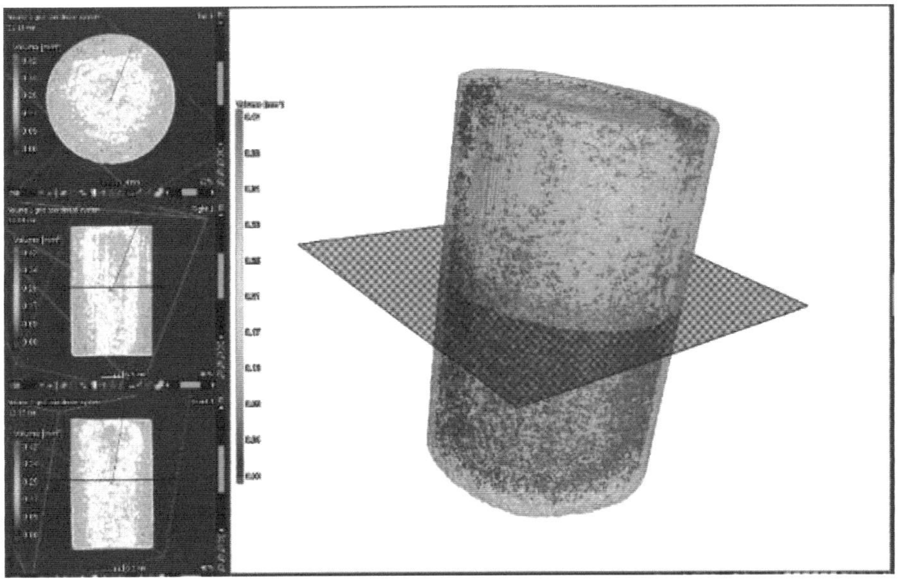

Figure 5. Horizontal slice of core X4.

The same processes were also implemented on the following core, Core X3. The horizontal slice image of Core X3 can be seen in Figure 6.

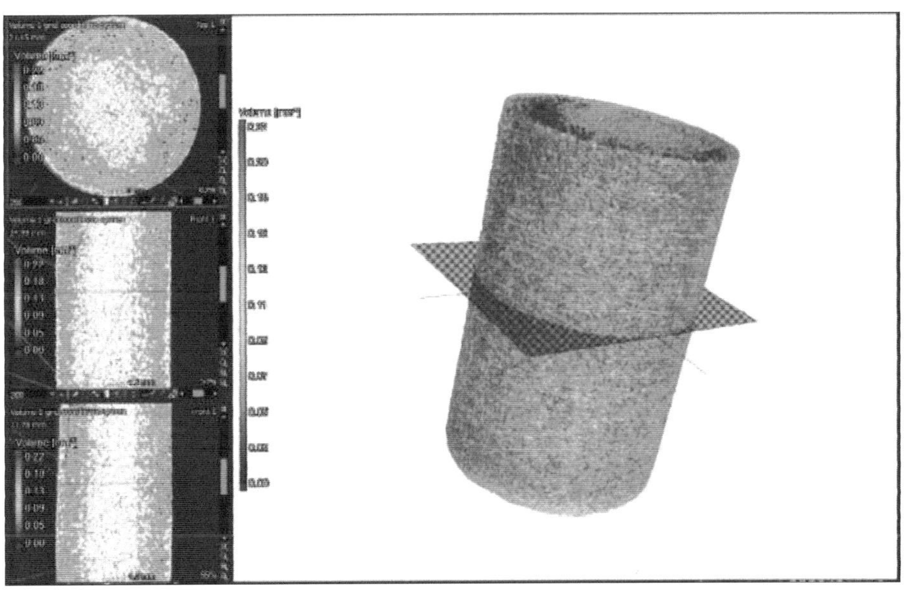

Figure 6. Horizontal slice of core X3.

Similar processes were also implemented on the following core, Core X7. The horizontal slice image of Core X7 can be seen in Figure 7.

Figure 7. Horizontal slice of core X7.

Based on the data shown on the left side of Figures 5, 6, and 7, we can see several concentrated pores in the middle of these three cores. The pores on the centre areas represent effective porosity which consists of interconnected pores only [5]. The effective porosities (interconnected pores) on these three cores are concentrated in the middle, from top to bottom. The pores in Berea sandstone are very small, so the pores outside of the connected pores are determined by the software as not core pores, which means they are "materials" and not pores. Based on the data generated by VG Studio Max 3.0 software, we determined the material and non-material volume and obtained the total porosity values of Core X4, Core X3, and Core X7, which were respectively at 20.04%; 20.11%; and 20.78%.

The results obtained for these three cores are similar because these three cores are derived from one whole core with a length of 12-inch or one foot, which is then cut into three cores (Core X4, Core X3, and Core X7) or commonly referred to as Core Plugs. Therefore, the rock characteristics of these three cores are very similar, which is indicated by the total porosity values of the three cores being almost the same.

4. Conclusion

Using the Non-Destructive Testing (NDT) method, sandstone pore visualisation analysis can be performed using Neutron Tomography by producing 2D and 3D image visualisations from Core X4, Core X3, and Core X7 samples. The core pores that are concentrated in the middle of the three cores are effective porosity. The total porosity values for Core X4, Core X3, and Core X7 are 20.04%; 20.11%; and 20.78%, respectively.

Acknowledgement

The author would like to thank PSTBM – BATAN, Universitas Trisakti, International Conference on Environmental and Sustainability Context (ICoESCo) 2021, and the publishers who have published this paper.

References

[1] Bharoto, Ramadhani B, Akbar A F, Sukaryo, S G, Priyanto, T H, Kurniati, M and Fadhila, F A 2020 *J Phy: Conf Ser* **1436**

[2] Susanto A T, Kurnianto K, Handoyo D, Suryaningsih F 2018 *Proceedings of the Meeting and Scientific Presentation of Basic Research in Nuclear Science and Technology* 73-80

[3] Sutiarso, Akbar F, Indradjaja A 2018 *Proceedings of the 2012 Materials Science and Technology Scientific Meeting* 182

[4] Sutiarso, Bharoto, Setiawan, Juliyani, Fahrurrozi 2010 *Proceedings of the VI National Seminar on Nuclear Technology HR*

[5] Koesoemadinata R 1978 *Geologi Minyak dan Gas Bumi* (Bandung: Penerbit IT)

[6] De Beer, Frikkie C, Middleton M F, and Hilson J 2004 *Appl Radiat Isot* **61** 487-495

[7] De Beer F, Middleton M, and Zadorozhnaya V 2007 *ASEG Extended Abstracts* **1** 1-4

[8] A Tengattini, N Lenoir, E Andò, and G Viggiani 2021 *Geomech Energy Environ* 1-19

[9] Skauge A, Garnes J M, Mørner O J, Torske L 1992 *J Pet Sci Eng*

[10] Kreger W E 1969 *Nucl Appl Technol* **7(2)**

[11] Cordonnier B, Pluymakers A, Tengattini A, Marti S, Kaestner A, Fusseis F, and Renard F 2019 *Front Earth Sci* **7(36)** 1-11

[12] Jones K W, Feng, Lindquist H, Adler W B, Thovert P M, Vekemans J F and Riekel B C 2003 *Geol Soc Spec Publ* **215** 39-49

Abatement cost for selectivity negative emissions technology in power plant Indonesia with aim/end-use model

Z D Nurfajrin[1*], B Satiyawira[1]

[1]Universitas Trisakti- Jl. Kyai Tapa No.1, Grogol Jakarta 11440, Indonesia

zakiahdarajat.zdn@trisakti.ac.id

Abstract. The Indonesian government has followed up the Paris Agreement with Law No. 16 of 2016 by setting an ambitious emission reduction target of 29% by 2030, and this figure could even increase to 41% if supported by international assistance. In line with this, mitigation efforts are carried out in the energy sector. Especially in the energy sector, it can have a significant impact when compared to other sectors due to an increase in energy demand, rapid economic growth, and an increase in living standards that will push the rate of emission growth in the energy sector up to 6. 7% per year. The bottom-up AIM/end-use energy model can select the technologies in the energy sector that are optimal in reducing emissions and costs as a long-term strategy in developing national low-carbon technology. This model can use the Marginal Abatement Cost (MAC) approach to evaluate the potential for GHG emission reductions by adding a certain amount of costs for each selected technology in the target year compared to the reference technology in the baseline scenario. In this study, three scenarios were used as mitigation actions, namely CM1, CM2, CM3. The Abatement Cost Curve tools with an assumed optimum tax value of 100 USD/ton CO_2eq, in the highest GHG emission reduction potential, are in the CM3 scenario, which has the most significant reduction potential, and the mitigation costs are not much different from other scenarios. For example, PLTU – supercritical, which can reduce a significant GHG of 37.39 Mtoe CO_2eq with an emission reduction cost of -23.66 $/Mtoe CO_2eq.

1. Introduction

IPCC (Intergovernmental Panel on Climate Change) in 2007 has reported an increase in global earth temperature of 1.1 – 6.4 °C, encouraging sea level rise of around 16.5 – 53.8 cm in 2100. Of course, this can result in a land loss on the earth's surface. According to estimates, if sea levels rise by about 1 cm, several islands in the South Pacific region and the Indian Ocean will completely disappear. The increase in the earth's temperature makes the climate challenge to predict and is at risk of causing extreme weather. In the 2015 Paris Agreement, countries participating in emission mitigation efforts have agreed to limit the average global temperature increase to less than 2°C and an ambitious target of reaching 1.5°C. The Indonesian government followed up the agreement with Law no. 16 of 2016 concerning the ratification of the Paris Agreement on the United Nations Framework Convention on Climate Change [1,2].

For Indonesia, the increase in the earth's temperature threatens many islands due to rising sea levels. Based on these problems, the Indonesian government has supported efforts to mitigate greenhouse gas emissions in Indonesia. Through its INDC, Indonesia has set an ambitious reduction target of 29% by 2030, and this figure could even increase to 41% if supported by international assistance [3]. The Indonesian government has also stated its commitment to create clean and low-carbon national development and actively contribute to mitigating carbon emissions. The government launched an implementation directive for technical ministries called the RAN (National Action Plan) on GHG as stipulated in Presidential Regulation Number 61 of 2011 [4,5]. The basic principle established is that RAN-GRK does not hinder economic growth, improve people's welfare through sustainable development and protect the poor and vulnerable [6].

In the First Nationally Determined Contribution (NDC), Indonesia in 2016 has shown data on the National GHG emission inventory in 2010, with the forestry sector contributing 49% of total emissions of 110.5 $MtCO_2eq$, while the energy sector occupies the second position with a percentage of contributed emissions of 34 % of the total emission of 647 $MtCO_2eq$ which can be seen in Table 1.

Table 1. Projected BAU and GHG emission reductions from various sectors [3].

No.	**Sector**	**GHG Emission Level 2010***	**GHG Emission Level 2030 ($Mtoe\ CO_2eq$)**			**Annual Average Growth BAU (2010-2030)**
		Mon CO_2eq	**BAU**	**CM1**	**CM2**	
1	Energy*	453.2	1,669	1,355	1,271	4.50
2	Waste	88	296	285	270	4.00
3	IPPU	36	69,6	66,85	66,35	0.10
4	Agriculture	110.5	119,66	110,39	115,86	1.30
5	Forestry**	647	714	217	64	2.70
TOTAL		**1,334**	**2,869**	**2,034**	**1,787**	**3.20**

*Including fugitive; **Including peat fire; CM1 = Counter Measure (Unconditional mitigation scenario); CM2 = Counter Measure (Conditional mitigation scenario)

The IEA in 2019 statistical data has stated that for Indonesia's 2018 GHG emissions in energy use, the first position as a contributor to GHG emissions in Indonesia is occupied by the power generation sector, which is 39.85% of the total emissions in 2018, which is 542 Mton CO_2eq.

Table 2. Share emission by energy sector in 2018 [7].

No.	**Sector**	**Total Emission**	
		Mon CO_2eq	**(%)**
1	Electricity	216	39.85
2	Industry	114	21.03
3	Transportation	154	28.41
4	Residential	23	4.24
5	Agriculture	5	0.92
6	Commercial	2	0.37
7	Final Consumption not elsewhere	1	0.18
8	Other Energy Industries	27	4.98
	TOTAL	**542**	**100**

Viewed from the source of fuel used by the power generation sector, it turns out that coal plays a significant role. That could be the reason why the power generation sector is the largest emitter. In addition, the IPCC states that coal also has the most prominent emission factor compared to other power generation fuel sources with the fossil energy-burning activity in power plants is a significant contributor to GHG emissions in the energy sector, with a cumulative average growth rate for the 2010-2016 period of 7.66%. The Performance Report of the Directorate General of Electricity Indonesia, 2017 discusses the increasing growth rate in the use of fossil fuels as the primary fuel in power plants, which is so significant that it can be caused by the increasing demand for electricity which continues to increase in the context of energy fulfilment and equitable access to national electricity [8]

The increasing national electricity production trend is shown in Table 3, with an average growth rate of 6.63% of electricity production increasing from 169,786 GWh (2010) to 283,801 GWh (2018). This increase was obtained from coal power plants which grew to 58.41% in 2018, increasing from 38% in 2010. Electricity production from oil plants decreased to 5.6% in 2018. Electricity production This oil-fired power plant was dominated by small to medium-scale PLTDs operating in remote areas as a rural electrification improvement program. Meanwhile, the production of NRE power plants - PLTA (Hydroelectric Power Plant) and PLTP (Geothermal Power Plant) is still a priority, and other renewable plants such as PLTS (solar power plant), PLTBn (Biofuel Power Plant), PLTBm (Biomass Powe Plant), and PLTB (Wind Power Plant) - are still in an insignificant portion compared to fossil energy generation [8].

Table 3. Electricity production in Indonesia, 2010-2017 [8].

	2010	**2011**	**2012**	**2013**	**2014**	**2015**	**2016**	**2017**
Electricity Production (TWh)	169.78	183.42	200.32	216.19	228.56	233.98	247.92	254.62

The Indonesian government has issued a policy to reduce dependence on fossil fuels to increase national energy security. Primary energy supply needs to be increased based on new and renewable energy (EBT) to meet energy needs without burdening the trade balance and reducing GHG emissions. Presidential Regulation No. 79 of 2014 concerning KEN (National Energy Policy) states that the Indonesian government prioritizes the use of NRE with a minimum share of 23% in 2025 and increases to 31% in 2050. In addition, a 2019-2038 RUKN suppresses the portion of the fuel energy mix up to 0.1% in 2038 and increases energy efficiency by applying low-carbon technologies.

The biggest challenge today and in the future is how to meet energy demand by providing clean, reliable, and available energy in large quantities. Various poverty alleviation programs and policies will increase people's incomes and enable people to consume various electronic products that require electricity, increasing global energy demand. One thing that can be done in response to this challenge is optimizing alternative resources and increasing energy efficiency in Indonesia. As well as creating an integrated energy system in the context of sustainable energy efficiency improvements by integrating energy efficiency activities in the existing management system to take into account factors costs, environment, energy availability, and business risks [6,9].

Seeing the gap between actual conditions and what is expected, a tool is needed to support the government's efforts in assessing and selecting New and Renewable Energy under social, economic conditions and the availability of technology and resources in Indonesia. In line with this, it can be done by using the bottom-up AIM/end-use energy model based on an optimization model to determine the use of various technologies and energy sources used in the future. The Asian Pacific Integrated Model End-use (AIM/end-use) developed by the National Institute for Environmental Studies (NIES), Japan, is an appropriate energy optimization model used for very detailed technology selection by taking into account the total cost of the system to find a path with the lowest cost [10].

In addition, the AIM/end-use model has a deeper level of technical detail than the top-down energy model. This model is used to assess future energy demand and supply, uses an economic evaluation approach of simulated technology, and usually does not consider the macroeconomic impacts of energy or climate policies or related investments. One of the AIM/End-use energy model outputs that can assess the economic impact of GHG mitigation efforts is MACC (Marginal Abatement Cost Curve). Compared to several other indicators, such as the payback period, MACC is more appropriate to be a reference in making mitigation decisions because it reflects the cost of a mitigation option during the range of mitigation activities. Furthermore, it can guide the cost of reducing per unit of emissions during certain mitigation activities compared to the costs of the activities that become the baseline. Generally, MACC is expressed in units of USD/ton CO_2eq [11,12,13,14,15,16,17,18].

Fabian, 2011 explains the advantages of MAC in the assessment of a GHG mitigation policy:

- the present marginal abatement cost for ant given total reduction amount,
- give the total cost necessary to abate a defined amount of carbon emissions,
- allow the calculation of average abatement cost,
- extensive technological detail,
- the possibility of taking into account technology-specific market distortions,
- straightforward understanding of technology-specific abatement cost,
- MAC, on the bottom-up side, have a model that explicitly maps energy technologies in detail

The fulfilment of Indonesia's Nationally Determined Contribution (NDC) emission reduction target requires substantial funds. Implementing the carbon market must be carried out by considering the various opportunities and risks related to the type of instrument and its scope. To address the opportunities and risks associated with the scope of the carbon market instrument, one of the government policy strategies that can be taken is to consider the potential and costs of the types of mitigation actions carried out. Abatement cost or MAC shows the costs of reducing per-unit emissions throughout the project. moreover, it can make it easier for relevant stakeholders to identify outputs from climate change through the procurement of a budget that has several components such as RAN GRK (National Action Plan for GHG Reduction), RAD GRK (Regional Action Plan for GHG Reduction), Integrated Monitoring, Evaluation and Reporting System. [11,12,19,20,21].

2. Methodology

The AIM/end-use energy model is an energy model developed by Japan's National Institute for Environmental Studies (NIES) for processing data from the model using GAMS (General Algebraic Modeling System) software. The AIM/end-use energy model is a type of model that provides functions that were not present in the previous bottom-up model. The model can calculate changes in energy consumption from technology substitution caused by changes in energy prices in a bottom-up path. Therefore, it is possible to evaluate the efficiency of a single policy and how the effects occur when several policies are combined. By linking the technology selection model with the energy demand model, it is possible to estimate the energy efficiency based on the actual conditions. -respective technology [22,23,24].

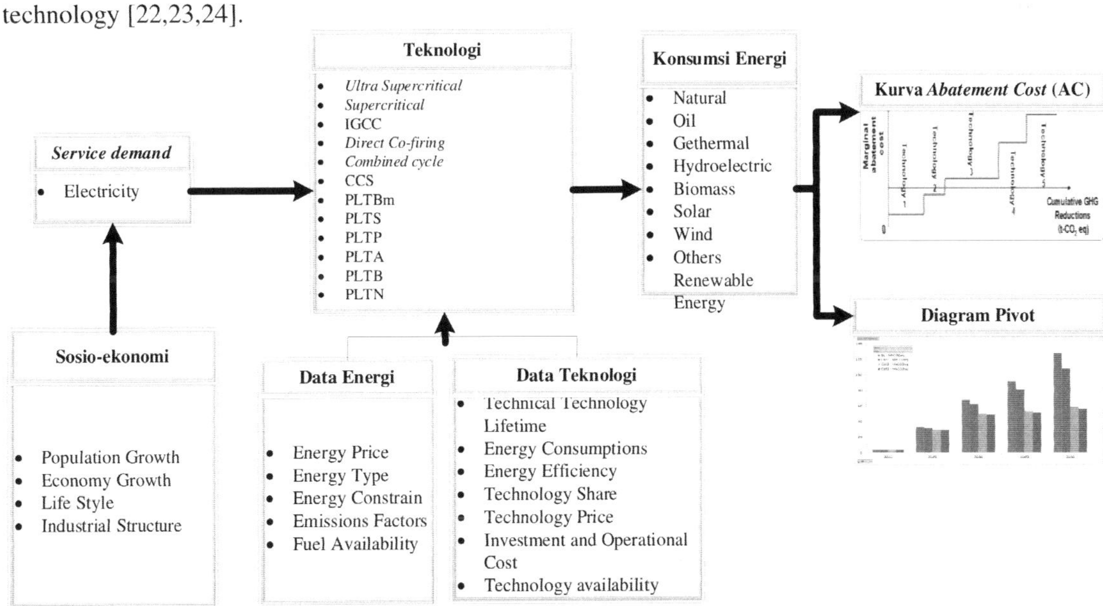

Figure 1. The structure of the AIM/End-use model in the electricity sector.

The structure of the AIM/end-use model for the electricity sector in this study is shown in Figure 1, first determining the service demand for electricity demand up to the simulation target year based on the consideration of socio-economic factors. Furthermore, the AIM/end-use model will determine the type or type of technology to fulfil the service [10]. After determining the type of technology, the energy model will estimate the energy consumption and emissions resulting from the operation of the selected technology. Based on technology selection in this model, a roadmap for low carbon technology can be developed to achieve an emission reduction target [22]. The output of the AIM/end-use model is interpreted in the form of a pivot diagram in the form of the energy mix, emission level, service, stock, and installed technology. From the cost analysis, the output of this model is an abatement cost (AC) curve that estimates the potential for emission reductions against mitigation costs for each selected technology.

The Figure 1 shows an overview of the AIM/end-use framework built to evaluate the potential for energy savings and GHG emission reductions in the Indonesian electricity sector. The AIM/End-use model framework consists of the essential elements of energy, devices, and services. This model uses energy, coal, biomass, natural gas, oil, biofuels, hydropower, geothermal, wind power, solar power, nuclear, and co-firing. Co-firing is a blend of biomass and coal; co-firing is considered very potential to reduce the use of coal in power plants. Biomass feedstock for co-firing can be biomass waste, which is processed in wood pellets or wood chips. The device element consists of various types of technology for each type of fuel used. At the same time, the service is classified into two types, namely internal service and final service. Internal service is the output of one device and, at the same time, is the input of the next device, while the final service is the final service or product produced in a process.

2.1. Marginal Abatement Cost Curve (MACC)

The abatement cost curve is used to evaluate the potential for GHG emission reductions by adding a certain number of costs for each selected technology in the target year compared to the reference technology in the baseline scenario. The abatement cost in each technology is calculated based on the total cost that meets the requirements, and the technology with the abatement cost lower than the selected carbon price will be selected. The basic concept of calculating the cost of reduction is described in the following formula:

$$Abatement\ cost = \frac{\text{Mitigation Cost}}{\text{emission GHG reduction}} \tag{1}$$

GHG emission reduction is defined as the sum of the difference in GHG emission levels from the reference technology in the baseline scenario and the selected low-carbon technology in the mitigation scenario. Meanwhile, mitigation costs are additional costs that include investment costs and O&M costs for the selected technology to the cost of the reference technology. However, if the mitigation cost of the selected technology is lower than the reference technology, then the mitigation action of the selected technology will not only reduce emissions but also reduce the total cost (investment and O&M cost).

2.2. Model assumptions and parameters

Parameters of energy consumption, electricity production for each type of power plant in this study refer to data published by the Directorate General of Electricity (Statistics of Electricity Indonesia). Specifications of Indonesia's existing technology are based on data published by the Indonesian agency for the National Energy Council) [8]. This study sets 2010 as the base year, considering the provisions of international agreements and NDC. Mitigation action planning in the Indonesian power generation sector is developed for a time interval of 10 years to the target year (2050). Parameters used in the AIM/End-use model include energy parameters, including price, fuel, and emission factors resulting from combustion activities at the power plant. The following parameter is the technical technology parameter. Data for these various parameters are collected from relevant literature, as shown in Tables 4 and 5.

2.3. Energy price

Various types of energy are used as energy sources in power plants. Analysis in the AIM/end-use model requires the input of price data for various types of energy from 2010 to projections in 2050. The price data for each type of energy is presented in Table 4.

Table 4. "Actual" and "projected" Energy Prices.

Energy	Price ($/toe)					References
	2010	2018	2020	2025	2030	
Coal	116	90	80	85.6	96.5	[25,26,27]
Natural Gas	180	301	301	301	301	[26,28]
Oil	616	595	595	595	595	[24,25]
Biomass	296	296	296	296.2	296.2	[29]
RDF	59	59	59	58.6	58.6	[30]
Bio-fuel	1180	685	685	685	685	[31]
Geothermal	845	632	632	632	632	[24,25]

2.4. Emissions Factor

The parameters used in the simulation of this AIM/end-use model include technical parameters and emission parameters. Emission parameters consider the emission factor values of various energy sources used. According to the 2006 IPCC guidelines, the main types of GHG emitted from combustion activities at the power plant are CO_2, CH_4, and N_2O, equivalent to CO_2eq units. In addition to factor emission data, other parameters such as calorific value and carbon content of various types of energy are presented in the following Table 5.

Table 5. Calorific value, carbon content, and emission factor of fuel [1].

Fuel	Calorific Value	Carbon Content	Emissions Factor			
	(TJ/Gg)	(%)	(tCO_2/TJ)	(tCH_4/TJ)	(tN_2O/TJ)	(tCO_2e/TJ)
Coal	0.76	42.92-73.82	99.178	0.001	0.0015	997.205
Oil	1.73	85	75.200	0.003	0.0006	752.036
Natural Gas	45.20	71	57.600	0.003	0.0001	576.011
Biofuel	27.00	N/A	-	0.003	0.0006	0.0036
Biomass	0.63	N/A	-	0.030	0.004	0.034

2.5. Data Technology

Low carbon technologies are accounted for and evaluated explicitly in the AIM/End-use model. The goal of implementing low-carbon technology is to improve energy efficiency and reduce GHG emissions. In this research, there are 43 technologies related to the process in power plants that will be used. In detail, the technical parameter data such as investment costs, OM costs, technology lifetime, energy efficiency, capacity factor, and capacity are obtained from various related kinds of literature.

2.6. Projection of Electricity Production and Consumption

Figure 2 shows electricity production and electricity consumption rate (kWh/capita) in 2010 -2050. Actual data on national electricity production from 2010 to 2018 was obtained based on inventory data in the electricity statistics of DJK ESDM, projected data for 2019-2030 was obtained based on RUPTL 2019-2028 and NDC Indonesia, with an average annual growth rate of 6.17%. Then the long-term projection data until 2050 is projected to increase by considering the annual average growth rate of 3.53%. The level of electricity production increased from 169.8 TWh in 2010 to 1126 TWh in 2050. Along with the increase in electricity production, the consumption rate also experienced an increasing trend from 0.77 kWh/capita to 2.91 kWh/capita. The level of electricity production 2010-2050 is applied to all scenarios built to see the mitigation potential of each scenario.

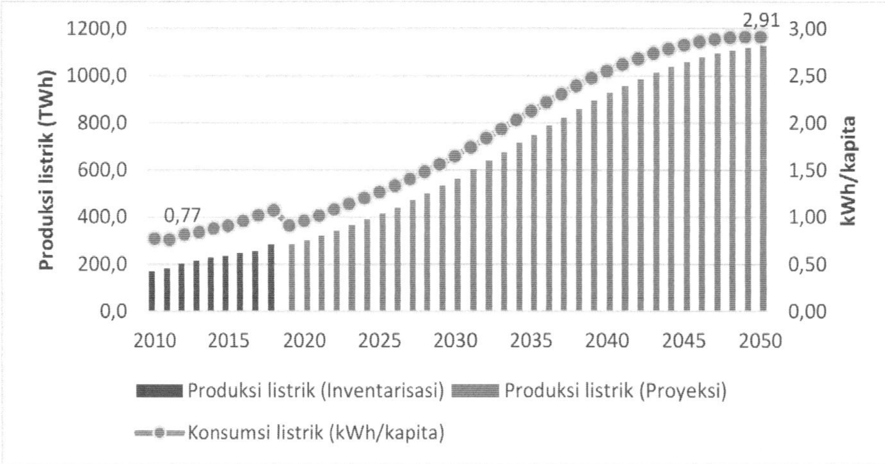

Figure 2. The level of electricity production in the Indonesian electricity sector at the "inventory" and "projected" levels.

2.7. Scenario model

Scenario analysis is an approach to measure the potential for energy savings or emission reductions in a simulation model framework. This study built four scenarios: baseline (BAU) and several alternative mitigation scenarios (CM1, CM2, CM3). Alternative mitigation scenarios are used to carry out more ambitious energy efficiency and GHG emission reductions. The comparison of several scenarios in this study is described in detail in Table 6, as follows:

Table 6. Scenario comparison.

Skenario	**Deskripsi Skenario**
BAU *(Baseline skenario)*	The business as usual (BAU) scenario is a baseline scenario without any intervention efforts to reduce GHG emissions. Baseline scenario assumptions: there is no effort to increase the energy efficiency of conventional power plants, the mix of fuel oil power plants remains the same since the base year. There has been no additional renewable energy generation since the base year. According to the planning scheme for existing plants, the increase in electricity demand is met by constructing conventional coal plants (subcritical), PLTD (Diesel Power Plant), and gas generators.
CM1	Mitigation actions in the CM1 scenario to achieve long-term targets have been aligned with the Indonesia NDC (CM1) follow-up scenario for the electricity sector.
CM2	Mitigation actions in the CM1 scenario to achieve long-term targets have been aligned with the Indonesian NDC (CM2) follow-up scenario for the electricity sector. In this scenario, there is an implementation of IGCC's advanced technology since 2035.
CM3	The utilization of renewable energy sources is more significant than the NDC target, the level of utilization of renewable energy takes into account the availability of fuel and technology readiness, where co-firing has been started since 2030, the use of CCS technology integrated into fossil power plants (coal and gas), and the use of nuclear energy starting in 2040.

Table 7. Detailed scenarios for CM1 and CM2 NDC Indonesia in the energy sector.

No.	Scenario	BAU	Mitigation Scenario CM1	Mitigation Scenario CM2
1.	Efficiency in final energy consumption	In-efficiency in final energy consumption	75%	100%
2.	Implementation of clean coal technology in power plant	0%		
3.	Renewable energy in electricity production	Coal power plant	19.6% (committed 7,4 GW based on RUPTL)	Electricity production of 132,74 TWh
4.	Implementation of biofuel in the transportation sector (Mandatory B30)	0%	90%	100%
5.	Additional gas distribution lines	0%	100%	100%
6.	Additional compressed-natural gas fuel station (SPBG)	0%	100%	100%

3. Result and discussion

The initial step is to validate from 2010 to 2015 to see the reliability of the data entered. For the first five years, there was no increase in GHG emissions due to the absence of the application of renewable technology. The BL (Baseline) scenario can also be referred to as the freezing scenario because there is no mitigation effort in this scenario, assuming the technology used from the beginning to the end of the year is the same, namely conventional direct combustion. Meanwhile, the CM1 and CM2 scenarios have applied technology upgrades according to the predetermined scenario model.

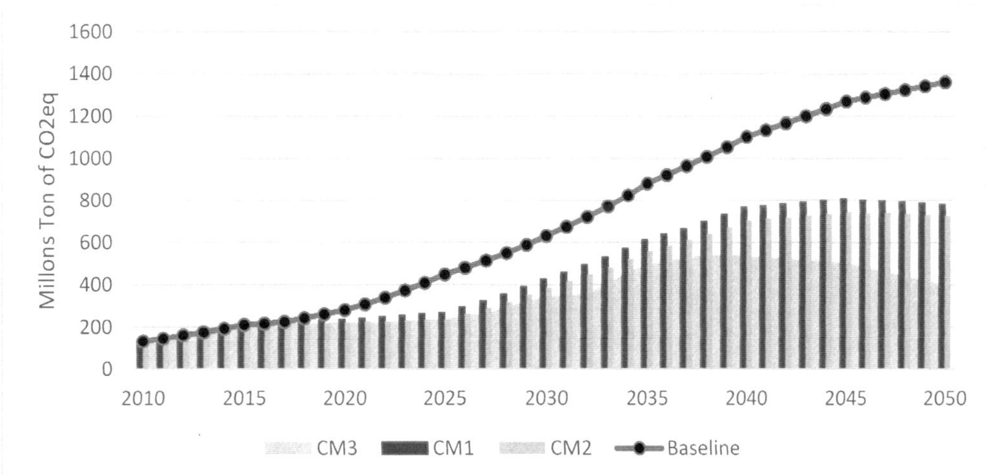

Figure 3. Results of Forecasting GHG Emissions Mitigation Efforts 2010-2050 in various scenarios.

In 2050, CO_2 mitigation is carried out by several mitigation efforts in the CM1 scenario of 578.35 Mtoe CO_2eq or 43%, the CM2 scenario of 632.43 Mtoe CO_2eq or 47%, and the CM3 scenario of 959.92 Mtoe CO_2eq 71% so that it can be proven that applying clean electricity generation technology and upgrading the use of fuel and technology can reduce CO_2 emissions in the atmosphere.

Table 8. Emission calculation result data and percentage of GHG emission reduction against Baseline scenario each year.

YEAR	BL	CM1		CM2		CM3	
	Mtoe CO_2eq	Mtoe oq	Emissions reduction compare with BL scenario	Mtoe CO_2eq	Emissions reduction compare with BL scenario	Mtoe CO_2eq	Emissions reduction compare with BL scenario
2010	133.31	133.31	0%	133.31	0%	133.31	0%
2015	211.59	211.59	0%	211.59	0%	211.59	0%
2020	280.50	236.18	16%	221.10	21%	221.10	21%
2025	448.71	267.88	40%	235.80	47%	235.79	47%
2030	630.33	426.35	32%	387.03	39%	343.76	45%
2035	879.52	611.77	30%	559.25	36%	495.36	44%
2040	1100.49	765.53	30%	702.45	36%	536.02	51%
2045	1269.15	805.38	37%	743.84	41%	500.84	61%
2050	1359.17	780.81	43%	726.74	47%	399.25	71%

Table 8 presents data on GHG emission reductions for the BL scenario and abatement cost values based on the selected technology in 2050. AIM/end-use will select available technologies to be applied in Indonesia based on technology availability, economics, and an environmental perspective, namely emissions, which it produces.

Table 9. GHG emission reduction and abatement cost based on selected technology.

No.	Scenario	Technology	Reduction of GHG Emissions (Mtoe CO_2eq)	Abatement Cost (\$/ton CO_2eq)
1.	CM1	PLTGU – Combined Cycle Power Plant (CCGT)	33.09	-109.75
		PLTG – Open Cycle Gas Turbine (OCGT)	16.21	-98.68
		PLTU – USC (Ultra Supercritical)	8.10	-31.16
		PLTA – Large Hydro	47.49	-24.81
		PLTU – (Supercritical)	37.39	-23.66
		PLTB – Small Onshore	5.21	-22.88
		PLTS – Solar PV Utility-Scale with Li on Battery Storage	1.05	-17.80
		PLTD – Biofuel	7.98	28.68
		PLTBm – Small	2.37	38.67
		PLTP – large	45.08	99.64
2.	CM2	PLTGU – Combined Cycle Power Plant (CCGT)	33.09	-109.75
		PLTG – Open Cycle Gas Turbine (OCGT)	16.21	-98.68
		PLTU – USC (Ultra Supercritical)	8.10	-31.16
		PLTA – Large Hydro	54.39	-24.68
		PLTU – (Supercritical)	37.39	-23.66
		PLTB – Small Onshore	12.11	-21.84
		PLTS – Solar PV Utility-Scale with Li on Battery Storage	4.50	-10.35
		PLTD – Biofuel	7.98	28.68
		PLTBm – Small	2.37	38.67
		PLTP – large	67.15	100.89
3.	CM3	PLTGU – Combined Cycle Power Plant (CCGT)	33.09	-109.75

PLTG – Open Cycle Gas Turbine (OCGT)	16.21	-98.68
PLTU – USC (Ultra Supercritical)	8.10	-31.16
PLTA – Medium Hydro	47.97	-24.06
PLTS – Solar PV Rooftop with Li on Battery Storage	0.00	-24.00
PLTU – SC (Supercritical)	37.39	-23.66
PLTB – Large Onshore	0.00	-22.69
PLTB – Small Onshore	12.11	-21.84
PLTA – Large Hydro	6.41	-12.00
PLTS – Solar PV Utility-Scale with Li on Battery Storage	4.50	-10.35
PTBm – CFB (Circulated Fluidized Bed)	2.17	20.71
Cofiring – Direct Cofiring (Retrofit)	43.26	28.06
PLTBm – BFB (Bubbling Fluidized Bed)	0.00	28.31
PLTD-biodiesel	7.98	28.68
Cofiring – Parallel Cofiring	0.00	29.86
PLTBm – Small	0.22	37.66
PLTBm – Large	67.15	101.88

The emission reduction cost curve or MACC is a graph depicting the MAC and the mitigation potential of all the mitigation activities considered. MACC can be an effective policy-making tool because it can provide a simple means to identify what mitigation activities are the most cost-effective and provide excellent mitigation potential. Visually, mitigation activity is considered carried out up to a specific time or in the target year of 2050.

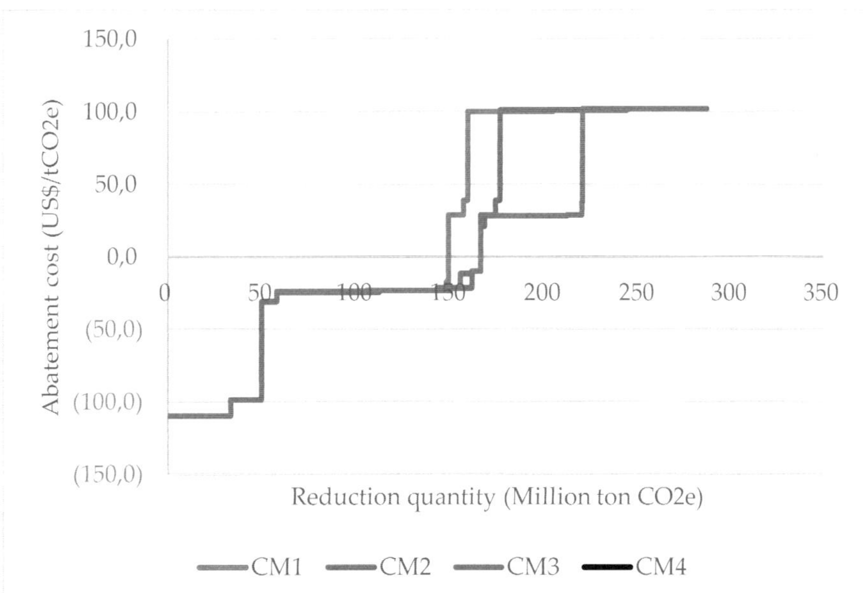

Figure 4. Marginal Abatement Cost Curve in 2050 based on Selected Technology.

This curve is displayed as a line sorted from the most cost-effective to the most expensive to reduce costs. The Y-axis represents the value of the abatement cost, while the X-axis represents the magnitude of the mitigation potential, as shown in Figure 4. Meanwhile, for the line under the X-axis or the abatement cost is negative, the activity can be carried out during the life of the mitigation activity. Provide economic income when compared to activities in the baseline scenario. Based on Table 8 and Figure 4, it is found that several types of technology have added economic value for Indonesia, with low mitigation costs, but have a large enough impact in efforts to reduce GHG emissions. The CM3 scenario

IOP Conf. Series: Earth and Environmental Science **894** (2021) 012011 doi:10.1088/1755-1315/894/1/012011

has the most significant reduction potential, and the mitigation costs are not much different from other scenarios. For example, PLTU – supercritical, which can reduce GHG emissions after significant and direct coal-fired PLTBm, 37.39 Mtoe CO_2eq with an emission reduction cost -23.66 \$/Mtoe CO_2eq.

4. Conclusion

Based on the studies that have been carried out, it can be concluded that Indonesia has enormous potential for renewable energy to be applied to the electricity sector in the long term. Reducing emissions in the CM1 scenario by 578.35 Mtoe CO_2eq or 43%, in the CM2 scenario by 632.43 Mtoe CO_2eq or 47%, and CM3 scenario by 959.92 Mtoe CO_2eq 71%. We have an abatement cost value that can provide economical income for Indonesia. However, in mitigation activities with a relatively high-cost value, the government can implement a policy or program with the target that the value of the mitigation cost can decrease in the long term.

References

[1] IPCC 2007 Climate Change – Impacts, Adaptation, and Vulnerability. Working Group II Contribution to the Fourth Assessment Report of the Intergovernmental Panel on Climate Change
[2] Undang-Undang No. 16 Tahun 2016
[3] First Nationally Determined Contribution Republic of Indonesia 2016
[4] Syamsidar T, Heiner V L 2011 *Guideline for Implementing Green House Gas Emission Reduction Action Plan* (Ministry of National Development Planning Indonesia)
[5] Keputusan Presiden Nomor 61 Tahun 2011
[6] Yoesgiantoro D 2017 *Kebijakan Energi Lingkungan* (Jakarta: LP3S)
[7] IEA 2019 The Future of Hydrogen: Seizing today's opportunities. Report for the G20, Japan.
[8] Gertler P, Shelef O, Wolfram C, Fuchs A 2013 *How Pro-PoorGrowth Affects the Demand for Energy* (Cambridge: National Bureau of Economic Research)
[9] Hibino G, Pandey R, Matsuoka Y and Kainuma M 2006 *Manual Enduse Model/Asia-Pacific Integrated Model (AIM)* (Japan: NIES)
[10] Kainuma M, Matsuoka Y, Morita T 2010 *AIM/end-use Model and Its Application to Forecast Japanese Carbon Dioxide Emissions* (Japan: NIES and Kyoto University)
[11] Andi A, Deden D, Sukandar, Donny W, Mohamad I, Ridwan P, Aldi A, Debi N 2018 *Stategi Implementasi Instrumen Mitigasi Bebasis Pasar Berdasarkan Potensi dan Biaya Mitigasi. Policy Brief #5* (Kementerian Lingkungan Hidup dan Kehutanan)
[12] Western Australian Local Government Association 2014 *Guidelines for Developing a Marginal Abatement Cost Curve (MACC)* (Australia: WALGA)
[13] Wetzelaer B J H W, der Linden N H V, Groenenberg H, de Conick H C 2007 *GHG Marginal Abatement Cost Curves for the Non-Annex I Region* (Enegy Research Centre at the Netherlands)
[14] Miranti A, P Setyanto, M Ardiansyah 2016 *Jurnal Ilmu Lingkungan* **14(1)** 39-48
[15] Eka H P, Dudi R 2019 *Pendanaan Pulik Untuk Pengendalian Perubahan Iklim Indonesia Tahun 2016-2018* (Badan Kebijakan Fiskal-Kementrian Keuangan Repulik Indonesia)
[16] Dicky E H, Andi S, Debi N 2018 *Pengantar Pasar Karbon Untuk Pengendalian Perubahan Iklim.* (Jakarta: PMR (Partnership for Market Readiness Indonesia))
[17] Fabian K *2011 33rd IAEE International Conference, Rio de Janeiro Brazil*
[18] Nadine I and Christoper K 2016 *Energies* **9** 227
[19] Addams L, G Boccaletti, M Kerlin, M Stuchtey 2009 *Charting Our Water Future – Economic Frameworks to Inform Decision-making* (New York: McKinsey & Company)
[20] Agus S, Sizanty S, Bowo W S, Martha M, Fathurrahman, Melany T 2009 *National Economic, Environmental and Development Study (NEEDS) for Climate Change* (Jakarta: National Council on the Climate Change Republic Indonesia)
[21] Kenneth G, James H S 2018 *Journal of Economic Perspectives*

[22] BAPPENAS 2013 *Kajian Pengembangan Model Dalam Mendukung Perencanaan Energi* (Jakarta: Direktorat Sumber Daya Energi Mineral dan Pertambangan)
[23] Jin-Hua X, Bo-Wen Y and Ying F 2016 *International Journal of Greenhouse Gas Control* **44** 199-216
[24] PLN 2018 *Buku Statistik PLN 2018*
[25] PLN 2018 *Laporan Keuangan Konsolidasian Intern PT. Perusahan Listrik Negara*
[26] PGN 2010 *Laporan Tahunan PGN 2010*
[27] EIA 2018 *AEO2018 National Energy Modeling System run*
[28] PGN 2018 *Laporan Tahunan PGN 2018*
[29] Rosmiati Xxx *Perhitungan Kebutuhan Bahan Bakar Cangkang Sawit Dan Tongkol Jagung Pada Unit Thermal Oil Heater di PT Shamrock Manufacturing Corpora*
[30] Paramita 2018 *Sustainability of Refuse Derived Fuel Potential from Municipal Solid Waste for Cement's Alternative Fuel in Indonesia* (A Case at Jeruklegi Landfill, in Cilacap)
[31] http://www.palmoilanalytics.com, accessed at 11st July 2021

Assessing implementation of low-carbon technologies in Thekelan Hamlet, Indonesia using participatory rural appraisal method

A Rezagama[1], M A Budihardjo[1], B Zaman[1], E Yohana[2]

[1]Department of Environmental Engineering, Diponegoro University, Semarang Indonesia 50275

[2]Department of Mechanical Engineering, Diponegoro University, Semarang Indonesia 50275

aryarezagama@lecturer.undip.ac.id

Abstract. Thekelan hamlet is located at the foot of Mount Merbabu makes the electricity that reaches this village experience a reasonably high loss percentage. By utilizing the sunlight, the community can make a power plant with solar cell technology. In addition, the majority of residents in Thekelan Hamlet still use wood-fired stoves for cooking. This condition is, of course, not environmentally friendly because the combustion releases high carbon into the atmosphere. Therefore, the prospect of renewable energy has not yet been fully exploited. This situation is due to the low level of knowledge, education, and capital owned by residents. The Participatory Rural Appraisal (PRA) approach were used to develop low-carbon society in Thekelan Hamlet. This approach comprises three stage including workshop, action, and reflection stage. Result shows that the community in Thekelan District were able to implement low-carbon society. However, the capital price of low-carbon technology such as solar cells and dew catcher is still high with respect to the output that they can generate. Therefore, the community is agree to contribute and join to the development of this program in the future since they realize that low-carbon society can induce economic development of the hamlet.

1. Introduction

Thekelan hamlet is an enclosed area or an area surrounded by the Mount Merbabu National Park area, which is located in Batur Village, Kopeng District, Semarang Regency. Mount Merbabu National Park is a conservation forest area of 5,725 hectares [1]. The number of residents living in Thekelan Hamlet is 280 families, with 720 residents employed in the livestock and agriculture sectors. They are active in the Mount Merbabu National Park area to take grass as a source of animal feed and firewood for cooking. The utilization is increasing in line with the community's increasing needs due to the low level of community income and the increase in population. The number of dairy cows in this village reaches 500 cows which the majority coming from Holstein Fresian breeds. Each house has an average of 3 cows, of which two of them are dairy cows [2,3].

Besides, Thekelan Hamlet has another energy potential that can be developed as a ecotourism spots through water energy and sunlight. There are also several small rivers as a source of water for irrigating the rice fields of the Thekelan Hamlet community and have the potential to be a location for the construction of micro-hydro turbines. Technically, the opportunity to build micro-hydro electricity around forest areas is tremendous and can be a solution to meet the needs of cheap electricity in remote areas [4]. Thekelan hamlet is also exposed to sufficient sunlight so that it is possible to use sunlight as a solar power plant. Thekelan hamlet has also attract many tourist since this area are considered as starting point for Mount Merbabu climbers. The development of tourism village in this area are intriguing and also becoming one of local government target [5]. Therefore, sustainable ecotourism can be another path that want to be achieve by the communities to increase the economic status of the society.

The target for developing mixed energy from renewable sources in Indonesia in 2025 is 23% [6]. Therefore, introducing renewable energy technology according to a local potential such as solar sources, biogas, and hydropower is needed to achieve the target. This system will be developed continuously into a grid system between houses in the villagers. The development of Thekelan Hamlet is interesting as a pilot village with a complete concept of using renewable energy to gain additional knowledge. In this study, we assess the technological implementation of dew nets and solar cells as part of low-carbon society development in Thekelan hamlets by using participatory rural appraisal (PRA) approach. The use of micro-hydro turbines is still on feasibility stage which cannot be explained in this action research. The potential of biomass utilization is already discussed and published elsewhere [2].

2. Methodology

In this study, a participatory rural appraisal (PRA) approach was used as part of action research (AR) which aims to improve the living condition of the community as the object. This approach is divided into 5 phases: identification, data collection, data interpretation, action plan development, and reflection. Most PRA is done by non-government organizations (NGOs) to solve problems and find solutions by working together with the community [2,7]. People will be interested in implementing technology through a continuous interdisciplinary learning process. The community will be invited to discuss through workshops and produce appropriate action plans to realize a low-carbon society in Thekelan Hamlet.

The lack of public knowledge about renewable energy utilization technology makes the mentoring process take a long time. So, this mentoring strategy is carried out simultaneously with economic development based on the potential of the surrounding nature. The PRA approach included in this study are considered as follows:

a. The socialization and workshop stage aims to increase public awareness about the use of renewable energy and identify all problems in implementing low-carbon technology. Data interpretation which seeks to map the community strengths and weaknesses, is also included in this stage.
b. The action stage aims to elaborate and educate the community with low-carbon technology and perform routine maintenance. This stage may be essential to increase public awareness of protecting the environment, such as managing waste, wastewater and reducing carbon by switching from wood-burning stoves.
c. The reflection stage is a monitoring program to ensure the sustainability of the action plan

3. Result and discussion

3.1. Workshop stage

In this activity, the community was expected to aware of and understands the vitality of a local leader. The scope of discussion in this event includes several important points as follows:

a. The proper management of community and stakeholder relations in the establishment of low carbon society
b. The obstacles and challenges in the establishment and operation of the low carbon society
c. The vital role of local leaders
d. The potential and case studies of the development of Thekelan Hamlet, along with recommendations

Thekelan has potential as a place for the construction of solar cells. The air temperature in Thekelan Hamlet tends not to increase because of its location significantly at the foot of the mountain. In addition, it is located in a tropical climate zone, so that the solar radiation received is quite adequate in the utilization of solar energy through solar cells. Although fog often occurs in Thekelan Hamlet, the mist only lasts in the morning, and the area is relatively clean from air pollution to support maximum solar cell yields. Thekelan hamlet has a large area of land that can be used as a solar cell arrangement. The amount of solar cell energy produced is directly proportional to the square meter occupied by the solar cell [8].

A dew catcher is a tool used to harvest mist with high moisture content to produce water droplets. More water droplets will collect and can be used for agricultural irrigation needs [9]. Thekelan Hamlet has drought problems during the dry season where the available water sources are only sufficient for household needs such as drinking water. Meanwhile, vegetable commodities grown in Thekelan Hamlet require quite a lot of water. As a result, in the dry season, Thekelan Hamlet can only plant drought-resistant crops. This program is implemented as a solution to solving these problems. The prototype of this dew catcher is expected to be a new source of water that can support the running of agricultural activities in Thekelan Hamlet to increase agricultural yields to achieve a prosperous life.

3.2. Action stage

The prototype of this dew catcher is a miniature version of the conventional dew catcher carried out in several places even outside Indonesia. This tool uses several materials such as plastic nets as the primary material, bamboo, ropes, gutters, and water barrels. This prototype is in the form of a trap stretched between 2 bamboos by tying the net using a string. The nets that have been laid out are then given gutters as water reservoirs to then flow into water barrels (See Figure 1). Farmers can use the water collected in the barrels to water the crops in the fields.

Figure 1. The installation of dew catcher.

Thekelan hamlet is located at the foot of Mount Merbabu, so the interconnection of electricity from the city to this hamlet experiences a reasonably high loss percentage. Solar energy can be utilized as an alternative energy source for the Thekelan Hamlet power plant to overcome these problems. Solar cells use solar energy directly into electrical energy [4]. Solar cells are an environmentally friendly technology. In operation, according to Yuliananda et al., for solar cells to get the maximum value, the factors that need to be considered are ambient air temperature, solar radiation, blowing wind speed, the orientation of solar panels, and position of solar cells to the sun [10].

The inverter installation program aims to utilize new renewable energy to the maximum. With the installation of an inverter in the solar system, the electricity generated by solar panels can supply the power needs of various electronic devices and lights in the basecamp to meet the basecamp's energy needs as a whole and reduce the use of electricity. The existence of an inverter device can convert the solar panel output current, which is direct current (DC), into alternating current (AC) with a frequency of 50 hertz, which is per Indonesian electricity frequency standards to achieve optimal utilization of new renewable energy (See Figure 2).

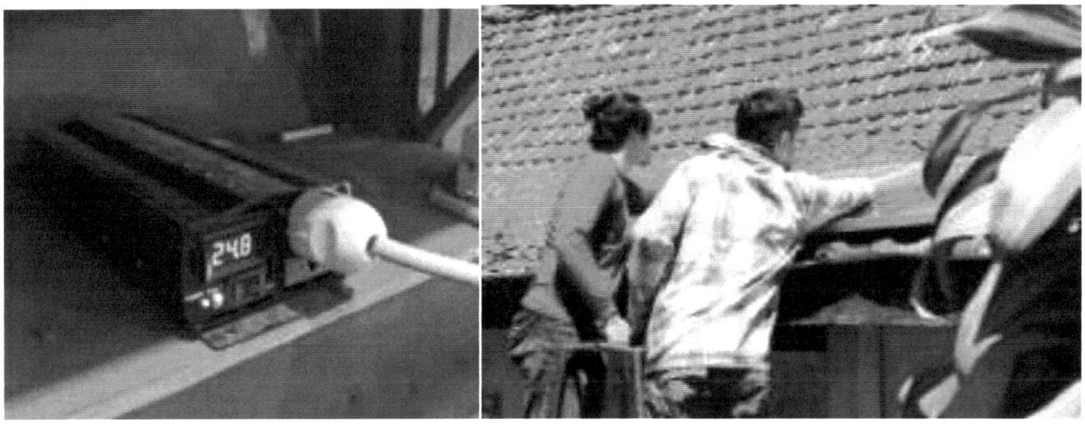

Figure 2. The installation of solar panel in Mount Merbabu climber's basecamp.

3.3. Reflection stage
The use of solar panels at the tourist base camp can be used to replace conventional electricity for one house, which reduces electricity use by 70-100 thousand rupiahs per month. This value is relatively smaller than the investment cost. Still, the existence of solar panels is an alternative where there are frequent power outages and the cost of replacing government subsidies. In addition, there is also a dew-catching technology that is expected to help farmers during the dry season. Seeing so much potential in Thekelan Hamlet to be brought to the Tourism Hamlet, it is necessary to have sustainable planning and assistance so that the quality of tourism management in Thekelan Hamlet is getting better. The village has the opportunity to develop further its existing low-carbon society potential. The capital price of solar cells and dew catcher are still high compare to the generated output of the technology. The community were agree to enhance the efficiency of the technology, so continuous improvement will be needed to boost the implementation of those technologies in the society.

4. Conclusion
The uniqueness of Thekelan Hamlet, which is difficult to find elsewhere in Indonesia, is thick with inter-religious harmony amid the heterogeneity of beliefs held by the people in the village. This situation makes Thekelan Hamlet one of the well-known tourist destinations in various parts of Indonesia. However, the use of firewood and electricity that is quite large can increase the potential for releasing large amounts of carbon into the atmosphere. Therefore, several alternatives such as installing solar panels and dew catchers are expected to reduce the possibility of releasing carbon emissions into the atmosphere. The action research shows that the two technologies still have weaknesses where the output produced is not as large as expected. Efforts to reduce the use of electrical energy from the state electrical company have not been achieved maximum results. However, as part of continuous improvement, the community is willing to follow the development of technology implementation to realize a low-carbon society can run well.

Acknowledgment
This research was funded by Universitas Diponegoro under SAPBN 2021 (IDBU scheme) number 330-04/UN7.6.1/PM/2021

References
[1] Andarani P, Lestari D F, Rezagama A and Sariffuddin S 2018 *E3S Web Conf* **73** 02019

IOP Conf. Series: Earth and Environmental Science **894** (2021) 012012 doi:10.1088/1755-1315/894/1/012012

[2] Budihardjo M A, Zaman B, Rezagama A, Kurnianto N T, Aulia T, Ramadan B S, Fitria S, Muhammad F I 2020 *IOP Conf Series: Earth and Env Sci* **448** 012110
[3] Susana I G B and Suartika I M 2017 *Logic : Jurnal Rancang Bangun dan Teknologi* **17(3)** 163-165
[4] Anwar 2013 *Politeknologi* **12(2)** 153-159
[5] Rezagama A, Budihardjo M A, Zaman B, Yohana E, Ramadan B S and Safitri R P 2021 *IOP Conf Series: Earth and Env Sci* **623** 012029
[6] Setyowati A B 2020 *Sustainability* **12(4)** 1603
[7] Lara C S, Crispin A F, Tellez M C L 2018 *J Clean Prod* **172** 4254-4262
[8] Effendi A 2012 *Jurnal Teknik Elektro* **1(1)**
[9] Jarimi H, Powell R and Riffat S 2020 *Int J Low-Carbon Technol* **15(2)** 253-276
[10] Yuliananda S, Sarya G and Hastijanti R A R 2015 *Jurnal pengabdian LPPM Untag Surabaya* **1(2)** 193-202

A review of anaerobic landfill bioreactor using leachate recirculation to increase methane gas recovery

M A Budihardjo[1*], B S Ramadan[1], E Yohana[2], Syafrudin, F Rahmawati[1], R Ardiana[1], D B Susilo[1], N Ikhlas[3], A Karmilia[1]

[1]Department of Environmental Engineering, Diponegoro University, Semarang Indonesia 50275
[2]Department of Mechanical Engineering, Diponegoro University, Semarang Indonesia 50275
[3]Department of Environmental Engineering, Institut Teknologi Pembangunan Surabaya, Surabaya Indonesia 60188

m.budihardjo@ft.undip.ac.id

Abstract. Municipal Solid Waste (MSW) treatment with anaerobic landfill bioreactor utilizes landfill as a place of biodegradation and produces methane gas which can be used as renewable alternative energy source. Anaerobic landfill bioreactor technology is a landfill development method that can increase waste degradation and increase biogas production. The increase of biogas and the removal of pollutants from leachate needs to pay attention to the factors that influence the success of anaerobic landfill bioreactor, including pH value, temperature, water content, and COD concentration after recirculation, and methane production. The relationship between these factors was discussed in depth in this paper. The method used is a narrative review where metadata is obtained from Google Scholar and Web of Science. This study explains the development of an anaerobic landfill bioreactor and conducts a synthesis for future research development plans by leachate recirculation.

1. Introduction

Municipal solid waste (MSW) in developing countries generally uses the landfilling method for waste disposal. Cost-effectiveness, easier maintenance, large storage capacity, and the ability to process organic and inorganic waste are the main factors in the use of landfilling [1, 2, 3]. The use of the landfilling method cannot be separated from the negative impacts caused by the operation of landfills in the form of air, water, and soil pollution. These impacts need to be given more attention because they affect human health and the environment, so long-term sustainable management is needed [4, 5]. Air pollution comes from greenhouse gas emissions, while the decrease in soil and water quality is caused by leachate. Both of these pollutants are produced due to the biodegradation process of organic waste material [6, 4].

Because of the high composition of organic waste, greenhouse gases (GHG) and leachate will be emitted to the environment. Moreover, the potential increase in the amount of waste makes management even more complicated. Therefore, mechanical biological treatment (MBT) using anaerobic landfill bioreactors is recognized as technology that can reduce the number of GHG and leachate pollute the environment. Anaerobic landfill bioreactor technology utilizes leachate recirculation to accelerate the process of stabilizing urban waste, minimizing leachate content, and recovery from methane gas can be achieved [7, 8]. Waste stabilization can be achieved faster, and the methane gas produced can be used economically using this technology [9].

Several researchers have carried out a study of anaerobic landfill bioreactors with leachate recirculation to improve leachate quality. An anaerobic landfill bioreactor produces more gas methane than the aerobic phase because it reaches the optimum conditions for methane gas production [10]. In addition, in the more extended thermophilic phase, the chemical oxygen demand (COD) value tends to decrease lower. Another study also revealed that reactors applying leachate recirculation would increase the decrease in COD concentrations higher than reactors without leachate recirculation [11]. Utilization of leachate with high COD was also carried out by Feng et al., where the study reduced the COD concentration in leachate by 90.7% [12]. This situation can happen because hydrolytic acidification is increased and reduces the pH during acidification. In addition, an anaerobic bioreactor with leachate recirculation and addition of water will reach peak temperatures (60°C – 61°C, respectively) which will increase bacterial metabolic activity so that efficient biodegradation occurs. This condition will increase methane production [13, 14]. Various studies have shown that the anaerobic landfill bioreactor shows satisfactory results when applied on a larger scale. However, no paper has been found that describes these findings in a review article. The relationship between the factors that influence the success of the anaerobic landfill bioreactor process is well understood and can be used as a determinant and decision making how to apply the anaerobic landfill bioreactor. This paper aims to explain the development of an anaerobic landfill bioreactor, describe the factors that influence the success of an anaerobic landfill bioreactor using a narrative method from related references, and carry out a synthesis for further research development plans by combining leachate recirculation with heating elements or using hybrid technology.

2. Literature review

2.1. Current anaerobic landfill bioreactor practices

Many studies on anaerobic landfill bioreactors have been carried out both on laboratory and pilot-scale applications. These studies can be seen in Table 1.

Table 1. Studies on anaerobic landfill bioreactor.

References	Practices
Gunarathna et al. (2011)	Rainfall affects the formation of leachate and biological degradation process.
Ayuningtias et al. (2013)	The concentration of organic matter in lysimeter with recirculation tend to be higher than without recirculation.
Zhang et al. (2017)	The effect of semi-aeration and recirculation can reduce methyl mercaptan concentration in landfills reaching 87.4 – 94.9%, indicating low COD concentration.
Kahar et al. (2018)	Comparison for internal pH (in bioreactor) 7.2 and external pH 8 for mass transfer of leachate and dissolved materials
Rasapoor et al. (2020)	The formation of LFG can be increased significantly with a moisture content of 45%.

From the several studies described above, the development of each bioreactor has different specifications because differences influence the characteristics of the landfill in each research area, so the results obtained are also varied. It is necessary to compare the factors that influence bioreactors' success to determine the relationship between them.

3. Result and discussion

3.1. Factor Affecting Anaerobic Landfill Bioreactor Performance

Anaerobic landfill bioreactors with leachate recirculation shows that the success of the process is influenced by several factors such as pH, temperature, water content and initial COD concentration.

3.1.1. pH value. Optimal pH value for bacteria to metabolize is 6.5-8.0 [15]. The decrease in pH value in the early stages is caused by the accumulation of acid due to the hydrolysis process because microbes require an environment with different acidity levels to carry out their activities [10]. Biogas production will achieve optimum results at the end of the hydrolysis stage if the pH value is in the range of 6.8-7.2 [16].

3.1.2. Temperature. The hydrolysis process causes the temperature to rise and fall again due to bacterial activity [14]. The initial temperature in the bioreactor is carried out between 28°C-30°C [10]. The temperature will increase and reach the thermophilic phase (45°C-60°C), which is the phase where bacterial metabolic activity occurs optimally and biodegradation occurs more quickly [13]. After reaching the thermophilic phase, the temperature will gradually decrease until it reaches a temperature of 28°C-30°C, which indicates that bacterial activity is decreasing [14].

3.1.3. Water content. The water content will encourage microbial activity to increase biogas formation in the range of 85-90% [17]. The water content is influenced by the pressure of the overburden and the level of degradation of municipal solid waste or MSW [18]. Deeper landfills have higher moisture content, high-density values, and lower porosity [19]. The optimal water content in gas formation is in the range of 40%-60% [20]. The conventional landfill has a high-density value but low water content. For that, it is necessary to adjust the humidity to increase biogas production [21].

3.1.4. Initial COD (chemical oxygen demand) concentration. COD growth was relatively slower before day 10 because the lag phase represented a delayed response of microbes to environmental changes [22, 12]. COD concentration increased rapidly in the next 10 days, as leachate recirculation increased the moisture in the waste [12]. Recirculating leachate increases the concentration of organic substrates in the waste pile, which is decomposed by microbes so that the concentration increases and the COD degradation rate decreases at the final stage [23]. The reactor with leachate recirculation had a higher COD removal performance than the reactor without recirculation [14]. The concentration given is 7.406 mg/L with an efficiency of removing pollutants of 68%. The COD value decreased since the beginning, and after entering the thermophilic phase, the decrease would be more significant and would continue to decline until the end [10]. Methane gas production will continue to increase until the bioreactor temperature reaches its peak, after which the temperature in the bioreactor decreases, causing methane production to decrease [10]. More methane gas is produced in an anaerobic landfill than in anaerobic landfill because stable conditions for methane production are achieved [8, 12]. The COD reactor of 6,000 mg/L showed the highest volume of methane at 89%, while the control reactor was 78% and the COD reactor was 9,000 mg/L 71%. This difference is caused by the proportion of methanobacterium that utilizes H_2/CO_2 to form methane. Therefore, the correct concentration of COD during recirculation will increase methane production.

3.2. Future development of anaerobic landfill bioreactor

In addition to leachate recirculation, the development of a bioreactor can be done by the addition of temperature controller to obtain an ideal bioprocess condition and endorse a rapid organic matter decomposition by the microorganism [24]. Another technology that can be developed from bioreactors is hybrid technology. Aerobic bioreactors produce less gas than hybrid bioreactors, with 80.5 $\mu g/m^3$ for leaf litter and 44.3 $\mu g/m^3$ for vegetable waste. Due to more prolonged aeration, there is more oxidized organic material compared to hybrid bioreactors. Bacterial metabolic activity that occurs does not produce methane [25, 26]. On the other hand, anaerobic bioreactors only produce methane gas of 30.2 μg/m^3 and 54.4 μg/m^3 for leaf waste and vegetable waste, respectively. This low yield can be caused by not achieving stable conditions for the methane gas production process during the processing.

4. Conclusion

This paper shows that an anaerobic landfill bioreactor with leachate recirculation effectively reduces pollutants and increases methane gas production from landfills. Factors that affect the anaerobic landfill bioreactor process are pH, temperature, water content, COD concentration, and methane gas production. Some of these factors are very helpful in deciding the design of bioreactor development to optimize the desired results. Future research development should employ temperature controller to obtain an ideal condition.

Acknowledgement

This research was supported by SAPBN Universitas Diponegoro 2021 Number 329-119/UN7.6.1/PP/2021.

References

[1] Kumar S, Chiemchaisri C and Mudhoo A 2011 *Critical Rev Biotech* **31(1)** 77–97
[2] Li W, Sun Y, Wang H and Wang Y N 2018 *Chemosphere* **209** 839–847
[3] Wu X, Yue B, Huang Q, Wang Q, Lin Y, Zhang W and Yan Z 2017 *J Environ Sci (China)* **54** 288–297
[4] Pellera, Maria F, Pasparakis E and Gidarakos E 2016 *Waste Manag* **56** 181–189
[5] van Turnhout, A G, Brandstätter C, Kleerebezem R, Fellner J and Heimovaara T J 2018 *Waste Manag* **71** 246–254
[6] Pariatamby, Agamuthu and Tanaka M 2014 *Munic Sol Waste Manag in Asia and the Pacific Island*
[7] Reinhart D R and Townsend T G 1998
[8] Xu Q, Jin X, Ma Z, Tao H and Ko J H 2014 *Biores Tech* **168** 92–96
[9] Reza B and Sattler M 2016 *J Clean Prod* **112** 2664–2673
[10] Budihardjo M A, Rizaldianto A R and Muhammad F I 2019 *IOP Conf Ser: Mat Sci Eng* **669** 012021
[11] Priyambada I B and Oktiawan W 2020 *IOP Conf Ser: Earth Environ Sci* **448** 012127
[12] Feng S, Hong X, Wang T, Huang X, Tong Y and Yang H 2019 *Biores Tech*
[13] Kumar S, Chiemchaisri C and Mudhoo A 2011 *Critic Rev Biotech*
[14] Oktiawan W, Wardhana I W, Sutrisno E, Gorat D and Rizaldianto A R 2019 *E3S Web of Conf* **125** 07002
[15] Sekman E, Top S, Varank G and Bilgili M S 2011 *Fres Environ Bull* **20(7)**
[16] Yuan H and Zhu N 2016 *Renew Sustain Energy Rev* **58** 429-438
[17] Panigrahi S and Dubey B K 2019 *Renew Energy* **149** 779-797
[18] Zhan L T, Xu H, Chen Y M, Lan J W, Lin W A, Xu X B and He P J 2017 *Waste Manag* **68** 307-318
[19] Hu J, Wu X W, Ke H, Xu X B, Lan J W and Zhan L T 2019 *Eng Geo* **254** 1-12
[20] Caicedo L M, Wang H, Lu W, De Clercq D, Liu Y, Xu S and Ni Z 2017 *Biores Tech* **233** 332-341
[21] Rasapoor M, Young B, Brar R and Baroutian S 2020 *Energy Convers and Manag* **216** 112947
[22] Mao C, Wang X, Xi J, Feng Y and Ren G 2017 *Energy* **135** 352-360
[23] Jiang Y, Dennehy C, Lawlor P G, Hu Z, McCabe M, Cormican P, Zhan X and Gardiner G E 2018 *Waste Manag* **114** 96-106
[24] Damsir and Ansyori 2017 *J Ind Serv* **3 (1)**
[25] Berge N D, Reinhart D R and Townsend T G 2005 *Critic Rev in Environ Sci Tech* **35(4)** 365–399
[26] Cossu R, Morello L, Raga R and Cerminara G 2016 *Waste Manag* **55** 83–92

IOP Conf. Series: Earth and Environmental Science **894** (2021) 012014 doi:10.1088/1755-1315/894/1/012014

Determination of load capacity of BOD pollutant in Cisadane River with Qual2kw model & its effect on DO parameters

I Rahmandani[1], D Hendrawan[1*], W Astono[1]

Department of Environmental Engineering, Faculty of Landscape Architecture and Environmental Technology, Universitas Trisakti, Jakarta, Indonesia.

dianahendrawan@trisakti.ac.id

Abstract. The research to determine the carrying capacity of pollutant loads in the middle to downstream Cisadane River segment started from Jl. Teuku Umar, Tangerang and ends at Muara Tanjung Burung, Tangerang Regency with a research time starting from March – June 2021. Increased land-use change due to human and industrial activities so that population growth and limited residential land make watersheds a target for land conversion. The purpose of this study is to identify sources of pollutants that have the potential to contaminate the Cisadane River in the middle to downstream segments, analyze water quality and pollutant load capacity, the Cisadane River in the middle to downstream segments along 34 km. The method used for identification of pollutant sources is carried out by conducting field surveys, water quality analysis is carried out by SNI and compared with quality standards according to Peraturan Pemerintah No. 22 of 2021, as well as analysis of the Pollutant Load Capacity (PLC) BOD and DO using the QUAL2Kw model. The results of identifying potential pollutant sources in the middle to a downstream segment of the Cisadane River are dominated by settlements, agriculture, industry, and other domestic activities, namely restaurants and stalls. The results of water quality analysis for BOD parameters ranged from 2.51 mg/L - 5.1 mg/L, and DO range from 5.4 mg/L – 7,2 mg/L. BOD parameters from all points 1-2 meet the quality standard class 2 (3 mg/L) & points 4-6 do not meet the quality standard. The DO parameter is still in good condition because the more significant the DO value, the waters are in good condition. The high levels of BOD in the Cisadane River are influenced by domestic waste that enters the river, and the land is dominated by built-up spaces as settlements, schools, hotels and malls. Land use around the river is very influential on the load of pollutants that enter the river. The lowest DTBP of the Cisadane River for BOD is 22497,9 kg/day, and the highest is 33201,6 kg/day. The decrease in the concentration of BOD load must be lowered in the Cisadane River by 29% - 42%.

1. Introduction

Cisadane River crosses West Java and Banten provinces. The river's upstream reaches are located in Mount Pangrango, bypassing on the west side of Bogor Regency, towards Tangerang regency and downstream to Tanjung Burung. Cisadane River water quality the lower the decrease with the higher the pollution level. Source pollution from various activities in the Cisadane watershed from households, agriculture and industry [1].

Cisadane River is a body of water that can recover and perform self-cleaning within certain limits. This capability occurs in conditions where the pollution does not exceed the threshold or environmental carrying capacity. The high pollutant load that enters the river body can interfere with the river's ability

IOP Conf. Series: Earth and Environmental Science **894** (2021) 012014 doi:10.1088/1755-1315/894/1/012014

to recover itself self-purification and reduce the carrying capacity of the river's pollutant load. The water quality of the Cisadane River is decreasing downstream with higher levels of pollution. Sources of pollution from various activities in the Cisadane watershed are households, agriculture and industry. The results of human activities that are not utilized are discharged into the Cisadane River, and Cisadane River tributaries and the level of pollution is most severe in rivers that pass through big cities [2].

Therefore, in this study, the determination of the BOD Pollutant Load Capacity of the Cisadane River use the Qual2kw Model, and its effect on the DO Parameter was carried out. In designing water quality programs, the application of the Qual2Kw method can be used to see the pattern of pollutant behaviour trends along the river. The water quality modelling aimed to 1) Identify sources of pollutants that can contaminate the Cisadane River in the middle to downstream segments. 2) Analyzing the pollutant load capacity in the middle to downstream Cisadane River segment using the Qual2Kw model. 3) Provide alternative scenarios in reducing pollutants in the Cisadane River in the middle to downstream segments.

2. Methodhology

2.1. Study area

This research was conducted from February-June 2021 in the Cisadane River Kota Tangerang starts from the Teuku Umar Bridge to the Tanjung Burung estuary 34 km long. The scope of the study area is along the Cisadane Rivers, and Exsitu testing was carried out at the Trisakti University Environmental Engineering Laboratory. Point
sampling is in Table 1.

Table 1. Cisadane river segmentation.

Reach	Location	Coordinat		Distance (Km)
		BT	LS	
1	Teuku Umar street	106°37'33.9"E	6°11'44.8"S	0
2	Karel Satsuit Tubun bridge	106°37'59.1"E	6°10'06.6"S	4,60
3	Marsekal Surya Darma street	106°36'50.5"E	6°07'40.6"S	6,4
4	Kedaung Barat street	106°38'07.8"E	6°06'24.6"S	3,41
5	Kali Baru street	106°37'59.6"E	6°03'42.5"S	6,47
6	Tanjung burung estuary	106°38'40.82"E	6°0'20.66"S	7,39

2.2. Method sampling

The determination of the sampling location was carried out based on previous field surveys in the study area. Namely, the location was quite representative of input from the surrounding area that affected the water quality of the Cisadane River, such as residential activities, land use and industrial activities while still considering the ease of access and cost. River sample taking was determined from river debit, and it was knowns that the debit of Cisadane River was ranged 86,79- 128,09 m^3/s, and the sample was taken from two points ½-¾ of river width from half the depth of the river or using an integrated sampler tool to obtain even sample from surface to bottom and then to be mixed [3].

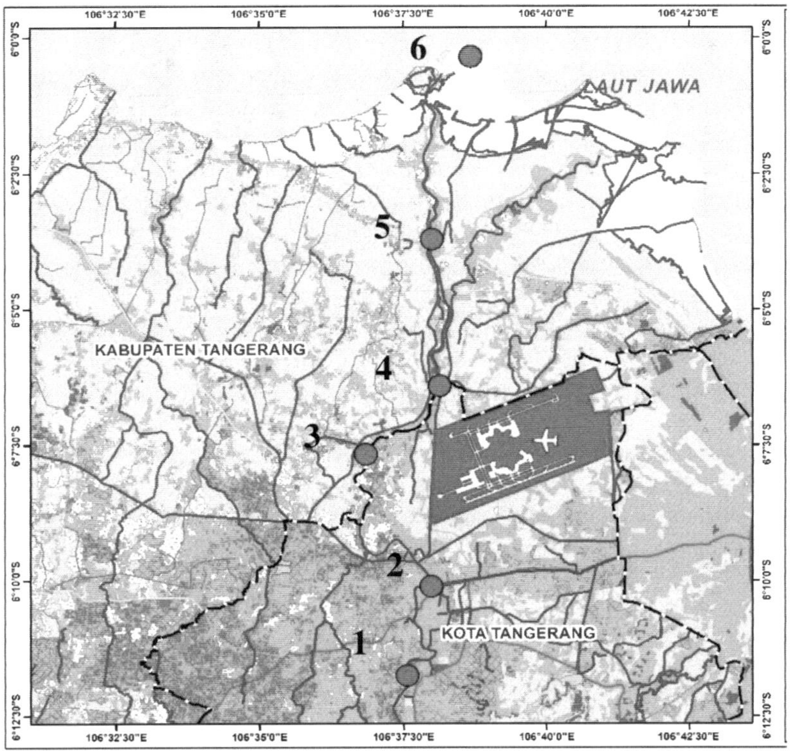

Figure 1. Map of study area and segmentation Cisadane River (Red dot represent the sampling location).

2.3. Qual2Kw model

In this study, the analysis of the calculation of the pollutant load capacity using the Qual2Kw method. This analysis determines the Pollutant Load Carrying Capacity (PLCC) referring to the regulation of the Peraturan Menteri Negara Lingkungan Hidup No.110 Tahun 2010 concerning Water Pollution Control Management [4]. The following are the main steps in the calculation/analysis of the water pollution load capacity using the Qual2Kw method, as follows:

1. Input Data
   The modelling input consisted of river segment borders, hydrological data and river morphometry, climatological data, water quality data, and pollutant potentials.
2. Model Running
   Model running was performed after data entry, and the Qual2Kw program was executed through qual2kw.exe.
3. Calibration and Reliability Test
   The result of the Qual2Kw model running was calibrate using trial and error and adding or reducing the pollutant load to produce the closest estimated value to the observed pollutant load [5]. Afterwards, the result of the trial and error was validated using this formula:

$$\%\ \textbf{Error} = \frac{\textit{BOD Data (mg/L)} - \textit{BOD Model (mg/L)}}{\textbf{BOD Data (mg/L)}}\ x\ \mathbf{100\%} \qquad (2.1)$$

A validation value equal to 50% indicates that the model's suitability with field data is very poor. If the validation value is below 20%, it can state that the model is acceptable. If the validation value on the model is wrong, it is necessary to re-calibrate until the model matches or is close to the data in the field.

2.4. Simulation techniques

Scenarios of reducing the pollutant load can be done by simulation based on the target to be achieved, namely by water quality standards according to Peraturan Pemerintah No. 22 Tahun 2021 Kelas II on water quality management and water pollution control. The scenario of reducing the pollutant load is carried out so that the pollutant load does not exceed the capacity of the Cisadane River [6].

As for the determination of the pollutant load reduction scenario that can be done, one of the potential sources of pollution in the middle-downstream segment of the Cisadane River is domestic waste from settlements. With the WWTP communal for each section that exceeds the quality standards that have been set, it can reduce the pollutant load that enters the river body.

3. Results and discussion

3.1. Calibration and validation

The model calibration result for the water quality data at six locations is shown in Figure 2. The model calibration results are conformed to be measured data because model data has been similar to the input data (black box). The calibrated parameters values in the model are presented in Table 2. The relative mean error between the simulated and observed for BOD and DO are 9% and 8% when the result is below 20% error value. It can be accepted.

Table 2. BOD model simulation verification.

	BOD (mg/L)		
Reach	**Model**	**Data**	**% Deviation**
1	2.63	2.51	4.6%
2	2.23	2.09	6.3%
3	4	3.84	4%
4	4.75	5.05	5.9%
5	5.76	5.03	12.7%
6	3.94	3.23	18%
Average	**3.89**	**3.63**	**9%**

Table 3. DO model simulation verification.

	DO (mg/L)		
Reach	**Model**	**Data**	**% Deviation**
1	5.30	5.74	7.7%
2	5.76	6.19	6.9%
3	5.66	6.07	6.8%
4	5.63	6.76	16.7%
5	5.97	5.97	0%
6	5.35	5.84	8.4%
Average	**5.61**	**6.10**	**8%**

The results of the DO model simulation compared with the field analysis data will produce the percentage deviation, which can be seen in Table 3. The average deviation percentage from the DO model simulation results is 8% so that the model simulation results can be accepted and are considered to be able to describe DO parameters in the middle to a downstream segment of the Cisadane River.

3.2. Simulation result

The model that has been run will produce a simulation graph of the modelled parameters of the Cisadane River in the middle to downstream segments. In the model simulation results graph, the black box is the data from the field analysis, while the black-red straight line is the result of the model simulation.

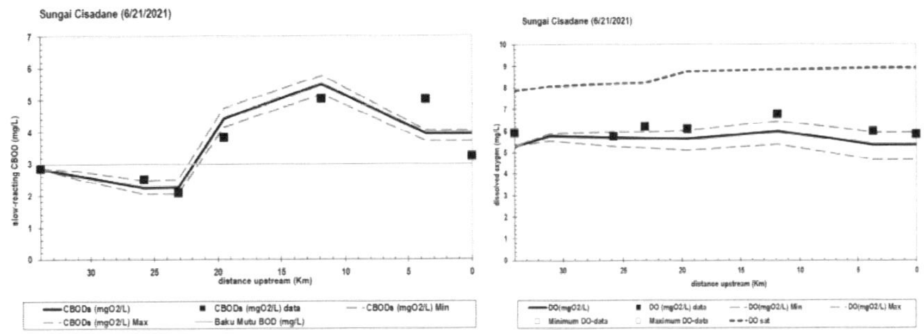

(a) (b)

Figure 2. (a) The Simulation results of the Cisadane River BOD, (b) DO model for the middle-downstream segment.

From Figure 2a, the BOD content in the Cisadane River in the middle to downstream segment increases at km 20. It is indicated by the trend line (model), which seems to have followed the trend of the black box point (data). The greater the BOD content in the river, the smaller the DO content because BOD and DO are inversely proportional to waters. Even though there was a difference between the measurements and the simulation model, the analyzed BOD concentrations remained in the same 2-5.1 mg/L range. When compared with the quality standard with model simulation results and existing field data, the water quality of the Cisadane River for the BOD parameter exceeds the quality standard for points 3-6.

From the model simulation results and measurements, it can be seen that the BOD quality in the middle to a downstream segment of the Cisadane River has exceeded the river water quality standard of class II according to PP No. 22 of 2021, so that a pollutant load reduction scenario is needed for these parameters. Before determining the pollutant load reduction scenario, the calculation of the pollutant load from the simulation model results is carried out. The calculation of the pollutant load from the simulation model can be seen in Table 4 below:

Table 4. Overload of BOD pollutants in the Cisadane River middle- downstream segment.

Reach	**Location**	**Upstream**	**Downstream**	**Debit**	**Model BOD Load**	**BOD Capacity**	**Difference**	**% BOD to be lowered**
		(km)	**(km)**	**m³/s**	**kg/day**	**kg/day**		
0	Jl.BSD Boulevard		34	86,80	21297.99	22497,88	1199.89	0%
1	Jl. Teuku Umar	34	28	96,54	21936.64	25022,79	3086.14	0%
2	Jembatan Karel Satsuit Tubun	28	23,6	104,45	20124.23	27072,96	6948.73	0%
3	Jl. Marsekal Surya Darma	23,6	22,6	112,69	38944.49	29208,37	-9736.12	25%
4	Jl. Kedaung Barat	22,6	16,45	117,42	48190.15	30435,88	-17754.26	37%
5	Jl. Raya Kali Baru	16,45	7,25	121,79	60608.67	31567,02	-29041.66	48%

Reach	Location	Upstream	Downstream	Debit	Model BOD Load	BOD Capacity	Difference	% BOD to be lowered
		(km)	(km)	m^3/s	kg/day	kg/day		
6	Muara Tanjung Burung	7,25	0	128,09	43604.75	33201,58	-10403.16	24%

Description: ▬▬ = negative BOD difference

From Table 4, it can be seen that the difference between the capacity and the pollutant load has a negative value. This negative capacity value means that the Cisadane River in the middle to the downstream segment has no pollution load capacity for the BOD point because the BOD pollution load that enters the river exceeds the allowable pollution load by class II water quality standards. Therefore, the BOD pollutant load along the middle to a downstream segment of the Cisadane River must be reduced, especially on sections 4, 5, 6, which are the highest pollutant loads. It is caused by dense residential activities that dump organic waste into the Cisadane River.

3.3. Pollutant load reduction scenario

Scenarios of reducing the pollutant load in the middle to downstream Cisadane River segments need to be carried out so that the river does not exceed the maximum limit of incoming pollution loads. From the data analysis of the pollutant load capacity of the Cisadane River, it is seen that points 4-6 have an overload for the BOD parameter. The results of the pollutant's load reduction scenario model with the installation of WWTP communal can be seen in Table 5 as follows:

Table 5. BOD pollutant load calculation results with communal WWTP application scenarios.

Reach	Location	Upstream	Downstream	BOD Scenarios	Model BOD Load	BOD Capacity	Difference	% BOD to be lowered
		(km)	(km)	mg/L	kg/day	kg/day		
0	Jl.BSD Boulevard		34	2.84	21297.99	22497.88	1199.89	0%
1	Jl. Teuku Umar	34	28	2.63	21936.64	25022.79	3086.14	0%
2	Jembatan Karel Satsuit Tubun	28	23,6	2.23	20124.23	27072.96	6948.73	0%
3	Jl. Marsekal Surya Darma	23,6	22,6	2.80	27222.92	29208.37	1985.45	0%
4	Jl. Kedaung Barat	22,6	16,45	2.90	29451.46	30435.88	984.42	0%
5	Jl. Raya Kali Baru	16,45	7,25	2.62	27531.72	31567.02	4035.30	0%
6	Muara Tanjung burung	7,25	0	2.62	28957.34	33201.58	4244.25	0%

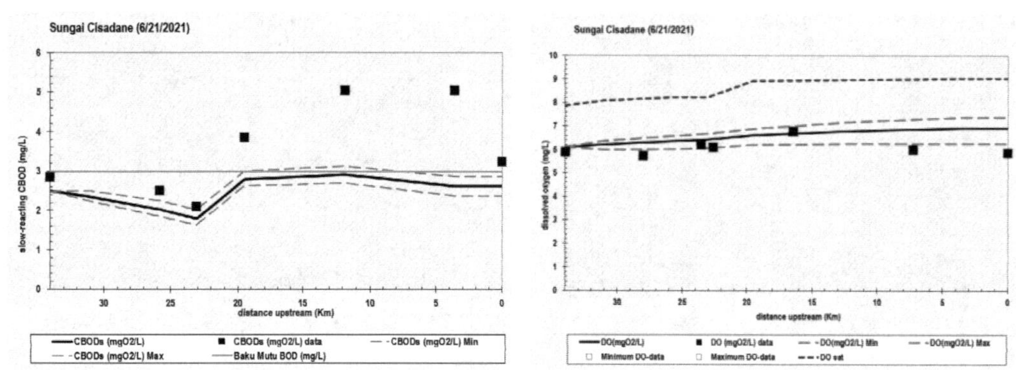

(a) (b)

Figure 3. (a) Decrease in BOD in the Cisadane River middle-downstream segment result of scenario with implementation of communal WWTP.

From Figure 3, it can be seen after the reduction scenario with the application of communal WWTP with BOD concentration at points 3-6 in the Cisadane River, the middle to downstream segment, has met the quality standard of PP No. 22 the Year 2021 class II. The results of the calculation of the BOD pollutant load obtained the lowest BOD load value of 20124.23 kg/day and the highest of 29451.46 kg/day with the largest BOD load capacity of 33201.58 kg/day so that the average BOD percentage that must be reduced is there is no more. Therefore, when the concentration of BOD decreases, the DO value in the middle to a downstream segment of the Cisadane River increases because the pollutant load entering the river body is processed first with communal WWTPs.

4. Conclusion

Potential sources of pollution in the middle to downstream Cisadane River segments at points 1-3 are dominated by settlements, industry, offices, schools and shopping centers. Meanwhile, points 4-6 are dominated by settlements, agriculture, rice fields and restaurants. In addition, the Cisadane River has irrigation canals that can affect water quality. The pollutant's load reduction scenario results by applying WWTP communal use of anaerobic-aerobic biofilter has the lowest BOD load of 20124.23 kg/day, and the highest was 29451.46 kg/day. After the done scenario with the installation of the Cisadane River communal WWTP in the middle to downstream segments in good condition for the pollutant load capacity of the BOD parameter.

References

[1] Rosarina D and Laksanawat E 2018 *Jurnal Redoks* **3** 2
[2] Siahaan R 2011 *Jurnal Ilmiah Sains* **11**(2)
[3] SNI 6 2008 *Air dan Air Limbah-Bagian 59: Metode Pengambilan Contoh Air Limbah* (Bandung: Badan Standarisasi Nasional)
[4] KEMENLH 2010 *Keputusan Menterin Negara Lingkungan Hidup Nomor 110 Pedoman Penetapan Daya Tampung Beban Pencemaran Air pada Sumber Air* (Jakarta: Kementrian Lingkungan Hidup)
[5] Darajati A 2018 *Indonesian Journal of Geography*
[6] Peraturan Pemerintah Republik Indonesia Nomor 22 Tahun 2021 *Tentang Penyelenggaran Perlindungan dan Pengelolaan Lingkungan Hidup.*

The use of subsurface constructed wetland for the treatment of tofu industrial wastewater in Semanan, Jakarta Barat

M R Raivaldi[1], R Hadisoebroto[1] and D I Hendrawan[1]*

[1]Department of Environmental Engineering, Faculty of Landscape Architecture and Environmental Technology, Universitas Trisakti, Jakarta, Indonesia

dianahendrawan@trisakti.ac.id

Abstract. The tofu preparing industry, typically done by the little businesses, can be found in almost every city in Indonesia and is dominated by industrial house stairways. The largest industry does not have a unit of processing wastewater. The wastewater focus is as yet over the edge quality norm, which messed contamination up. Constructed wetlands are affordable and reliable green technologies to diminish water contamination by utilizing plants. This study aims to know the quality of tofu industrial wastewater in Semanan, Jakarta Barat and analyze the removal of pollutants. The performance of the horizontal flow type Sub-Surface Flow constructed wetland with *Typha latifolia* plant. The media used were soil, sand, and gravel with a *Typha latifolia* plant. Testing the quality of tofu wastewater in Constructed wetland every 7 days at inlet and outlet, the sample was taken for TSS, BOD, and COD concentration. This research showed that the reactor could decrease TSS by 91.17% on day 36, a decrease of BOD 89.30% on day 15, and COD 87.86% on day 36. Those plants can produce growth fertile. These results fulfil the requirement of the quality standards of the Decree of the Minister of Environment No. 05 of 2014.

1. Introduction

The rapid growth of the population and the industrial sector in Indonesia, especially in urban areas, has a severe impact on the environment's carrying capacity. These impacts must be addressed and handled appropriately, especially in managing clean water and industrial and household waste. The increase in population and growth in the industrial sector increases the consumption of clean water that can cause an increasing amount of wastewater. Disposal of wastewater without undergoing a treatment process can cause environmental pollution, such as pollution in water bodies, incredibly raw water sources for drinking water that cause both surface water and groundwater.

Tofu is one of the favourite food that many people can consume, from low to high class. Tofu itself is known as healthy food with a low price. Almost every town in Indonesia has a tofu industry. Generally, the tofu industry is a small industry managed by the people, and some of them are included in the Tofu Entrepreneurs Cooperative (KOPTI), such as the one in Semanan, West Jakarta.

Tofu industrial wastewater that is disposed of without further processing can cause several problems for the environment. Usually, the problems that arise are in the form of deposition and decomposition of organic matter in water bodies receiving waste which ultimately worsens water quality [1]. In general, water pollution by wastewater from the tofu industry can diminish the dissolved oxygen content in the water (due to the decomposition of organic matter) and create an anaerobic atmosphere that disrupts ecosystems and life in the waters. According to the Regulation of the Minister of the Environment No. 5 of 2014 concerning Quality Standards for Tofu Industrial Wastewater, it is a form of regulation as a guide to the quality of tofu wastewater effluent standards which have a value of BOD (150 mg/L), COD (300 mg/L), TSS (200 mg/L) and pH (6-9) [2].

Efforts with simple, cheap, easy, efficient, and economical technology include operations and maintenance that do not require a particular person to be applied by the community, such as an artificial wetland system (Constructed Wetland). The use of aquatic plants in the Constructed Wetland system as an alternative means of treating wastewater has been widely used in several countries. However, in Indonesia, its development has not been so popular because studies and publications on the ability of these aquatic plants are still lacking. Based on the morphology of the Cattail plant (*Typha latifolia*), it is very suitable for processing with the Constructed Wetland system. Cattail plants have a root system that can absorb organic substances in the water chart. Meanwhile, Cattail plants are very numerous and thrive in Indonesia [3].

Constructed wetland is a planned or controlled processing system designed and constructed using natural processes. The Advantages of Constructed wetland that compared to sewage treatment facilities conventional investment, operation and maintenance costs is cheaper. [4]. Constructed wetlands can adequately dispense with suspended solids, natural poisons, and supplements from wastewater [5]. The constructed wetland framework has two sorts, in particular the kind of stream above Free Water Surface Flow (surface) and Sub-Surface Flow (subsurface flow) [6]. Notwithstanding, because the Constructed wetland framework in a traditional manner requires huge land, successive blockages, slow mass exchange, and helpless root entrance into layered soil sections, innovation is expected to defeat this issue.

Another methodology in constructed wetland frameworks expands proficiency in wastewater treatment by consolidating the advantages of phytoremediation and connected development frameworks based on bio-rack technique (engineering procedures) [7]. This bio-rack technique can make preferable conditions to plant growth, the influence of microorganisms in plants, better oxygen diffusion, and detect changes in the environment for the object of study.

Because tofu industrial wastewater can cause environmental pollution, a cheap and efficient treatment effort can reduce the negative impact of the wastewater with constructed wetlands. The objective of this study was to know the quality of tofu industrial wastewater in Semanan, West Jakarta and analyze the removal of pollutants. The performance of the horizontal flow type Sub-Surface Flow constructed wetland with *Typha latifolia* plant.

2. Methodology

2.1. Location

The study was conducted in Tofu Industry Semanan Village, Kalideres Sub-district, West Jakarta. The area of Semanan Village is 598 Ha with a population of more or less 79.306 people spread in 12 hamlets with a population density of 13.262 people/ km^2.

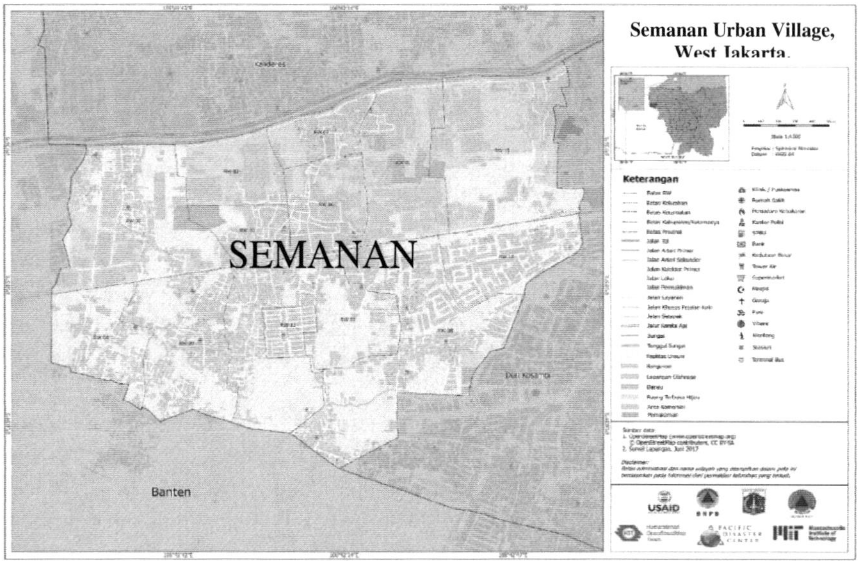

Figure 1. Research location.

2.2. Tofu industrial wastewater
The wastewater used is tofu industrial waste Semanan Urban Village, West Jakarta, from the sewer. The tofu wastewater is directly disposed of through the sewer before being channelled into a ditch or drainage channel.

2.3. The constructed wetland
This research was implemented in Semanan Village, West Jakarta, and for the constructed wetland at the researcher's own house, which is located in Tangerang. This Wetland Construction measures 400cm x 100cm x 100cm. This wetland system uses a type of sub-surface flow (Sub-Surface Flow). The Constructed Wetland system was constructed for the tofu industry. It requires a large area, and primary wastewater treatment is needed because the wetland system is constructed as secondary wastewater treatment. New technology is expected to beat this issue.

2.4. Media and plant
Media containing gravel, sand, topsoil mixed with 10% compost. The plant used *Typha latifolia*, which can retention pollutants as nutrients required [8].

2.5. Data analysis
Utilizing + formulas to the calculation of the effectiveness of pollutant removal in constructed wetlands.

$$Efficiency\,(\%) = \frac{Ci - Co}{Ci}\, x\ 100\% \qquad (2.1)$$

where:

Ci: Value of tofu wastewater parameters before treatment.
Co: Value of parameters for tofu wastewater after treatment.

2.6. Quality standard
Tofu industrial wastewater is the wastewater used from this research that quality is not following the quality standard necessities to release into the environment. Through table 1, the quality of tofu industrial wastewater before processing can be shown.

Table 1. Quality of tofu industrial wastewater in Semanan.

Parameter	Unit	Quality of wastewater	Quality Standard*
pH	-	4.6 – 4.87	6-9
TSS	mg/l	442 – 840	200
BOD	mg/l	3,072 – 3,649	150
COD	mg/l	5,190 – 6,192	300
T-N	mg/l	21.5 – 52.7	-
T-P	mg/l	51.9 – 80.2	-

*Quality Standards: Ministry of Environment and Forestry Indonesia No.05/2014.

3. Results and discussion

The highest TSS concentration from figure 2 shows that after passing through the constructed wetland is 396 mg/l, and the lowest is 17.3 mg/L. On day 36, the highest efficiency occurred with 91.17%, while the lowest concentration occurred on day 8 with a removal efficiency of 78.16%. After the 15th day, because the ability to reduce TSS content increases, the efficiency of removing wastewater content depends on the concentration and length of detention time in a constructed wetland. Moreover, this result has been fulfilled compare to the quality standard by the Minister of Environment Regulation No. 04 of 2014, where the TSS concentration value is 200 mg/L.

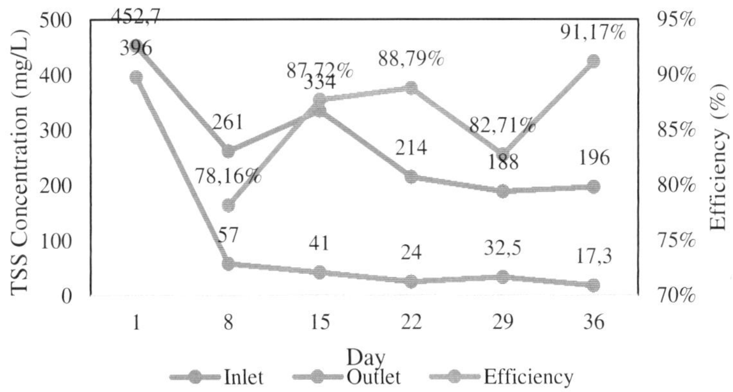

Figure 2. TSS concentration.

The absorption process by plants, the decomposition of dissolved organic matter and the deposition of the decomposition products of organic matter causes a decrease in TSS levels after processing using plants and a significant decrease from TSS focus because of calm water conditions making particles or colloids effectively enter and settle at the lower part of the compartment. [9]. The system, through filtration and sedimentation in the Constructed Wetland, TSS were reduced [10].

In Figure 3, it can be seen that the COD concentration at the outlet has a COD highest value was 2,592 mg/L and had a COD concentration lowest value was 102 mg/L. This result fulfils the quality standard by the Minister of Environment Regulation No. 04 of 2014, where the COD concentration value is 300 mg/L. The high amount of ingredients organic matter contained in tofu industrial wastewater causes an increase or decrease in concentration. Here is a graph of COD parameter removal efficiency in a constructed wetland that can be seen in Figure 3.

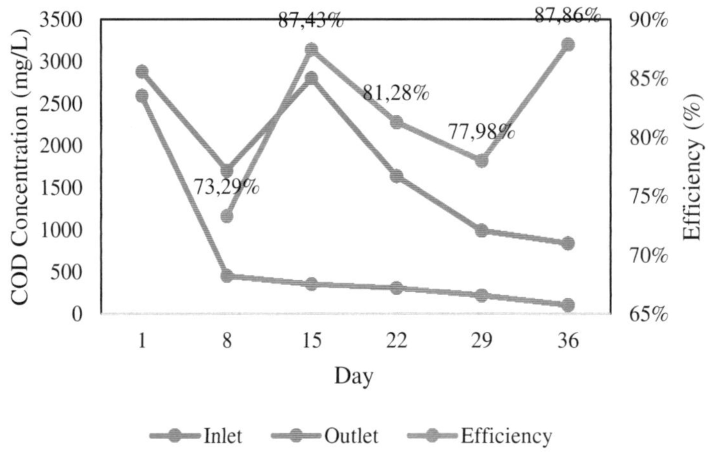

Figure 3. COD concentration.

Figure 3 shows that the removal of COD concentration in constructed wetland has the highest removal that is 87.86%, on the 36th day, and the lowest removal of COD concentration was 73.29% on the 8th day. It shows that the high and low are caused by the performance of microorganisms and pollutant concentrations in a constructed wetland.

The increase in COD concentration was caused by the water hyacinth plants having passed the immersion point. The ability of plants to assimilate natural materials was reduced, and even the convergence of natural materials in wastewater expanded because plants were considered to provide natural ingredients that had been preserved once again. [8]. The process of reducing COD in wastewater can occur due to the role of aquatic plants, carbon adsorption, and organic materials and microorganisms present in artificial wetlands. The cycle that happens because of natural substances contained in wastewater is used by plants as nutrients and for the course of photosynthesis [11].

In Figure 4, it can be seen that the removal of BOD concentration in constructed wetland has the lowest removal that is 74.37%, on the 8th day and the highest removal was 89.30% on the 15th day. There are many organic substances in the tofu industrial wastewater, which is shown from the highest BOD_5 value so that microorganisms require a great deal of oxygen to break down natural substances. Here is a graph of BOD parameter removal efficiency in a constructed wetland that can be seen in Figure 4.

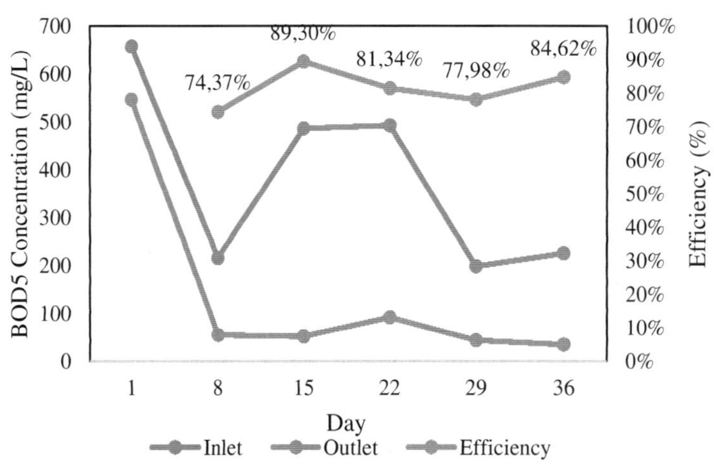

Figure 4. BOD_5 concentration.

The highest BOD_5 showed from Figure 4 the value at the outlet occurred on the 22nd day as for the 36th day, the lowest concentration occurred. The impact of maintenance time additionally assumed a part during the time spent decreasing waste content. The decrease in wastewater treatment is due to bacteria and microbes found in soil and plant roots working on setting aside BOD parameters. The decrease in BOD levels which is getting higher with lower water levels is brought about by a large portion of the natural matter is as wastewater that enters the Constructed wetland framework through adsorption cycles. Disintegration measures on the substrate and the lower the water level, the contact time, and the decay interaction between the wastewater and the substrate contained in the Constructed wetlands are more viable. Hence, the decrease in proficiency is higher [12]. The high concentration of BOD in liquid tofu waste is caused by Tofu wastewater that contains high organic matter, so BOD and COD levels are also relatively high. If directly discharged into water bodies, it will reduce the environmental carrying capacity [13].

Plants help bacteria debase natural mixtures through their underlying foundations delivering exudate into the rhizosphere zone, and the exudate triggers the corruption interaction by bacteria [14]. The availability of oxygen for biological processes is one of the factors. Wherefrom, the course of digestion of microorganisms, created CO_2, which is then utilized by plants to do photosynthesis, from the photosynthesis interaction delivered glucose/carbs which will become supplements for plant notwithstanding oxygen delivered which will be reused by vigorous microorganisms to decay the natural substance in tofu wastewater [8]. The opportunity to deposit solid particles will be given when the wastewater passes through the soil particles at a specific detention time. Through this deposition process, the need for oxygen will be reduced for further biological treatment.

The BOD value is additionally influenced by the presence of plants that cover wastewater surfaces. The presence of these plants can assimilate natural substances contained in wastewater—the more plants, the more adsorbed natural matter and natural matter that fewer microorganisms should corrupt. The less natural matter we need to debase by microorganisms, the oxygen content in the wastewater increments. The lessening in wastewater treatment is because of microorganisms and organisms found in soil and plant attaches attempting to save BOD parameters [15]. Mainly living bodies in profoundly high-impact soil need oxygen to help the action of outlining organic matter. Generally, an effective wastewater treatment system equipped artificial wetlands with proven plant growth high enough.

4. Conclusion

Sourced on the outcome analysis of tofu industrial wastewater parameters of TSS ranged from 442–840 mg/L, COD ranged from 5,190–6,192 mg/L, and BOD_5 ranged from 3,072–3,649 mg/L while Tofu wastewater quality standards according to Minister of Environment Regulation No. 5 of 2014 for BOD 300 mg/L, COD 150 mg/L and TSS 100 mg/L. The treatment with constructed wetlands by *Typha latifolia* plant has a good performance in tofu industrial wastewater treatment in a constructed wetland system. The planting pattern is made in a zig-zag way to adjust the flow pattern, and while testing the quality of tofu industrial wastewater in effluent constructed wetlands every 7 days, the average TSS removal is 85.71%, the average COD removal is 81.57%, and the average BOD_5 removal is 81.52%.

References

[1] Widyastuti D and Suprayitno D 2020 (Malang: Institut Pertanian Malang)
[2] Indonesia Ministry of Environment, Ministerial Regulation No. 05/2014 on regarding business and/or soybean processing activities 2014
[3] Hidayah and Aditya 2010 *Envirotek: Jurnal Ilmiah Teknik Lingkungan* **2(2)** p 11-18 ISSN 2085-501-X
[4] Husnabilah A and Tangahu B V 201 (Surabaya: ITS Environmental Engineering Department)
[5] Va V, Setiyawan A S, Soewondo P and Putri D W 2018 *indones.j.urban.environ.technol.* **1(2)** 199 – 214
[6] Prawira J 2015 (Tanjungpinang: Department of Aquatic Resource Management Faculty of Marine and Fisheries, Maritime University Raja Ali Haji)

[7] Valipour A, Raman V K, Ahn Y-H 2015 *Water* **7(1)** 329-347
[8] Qomariah S 2017 *Journal of Civil Engineering Research* 25-32
[9] Oktorina A, Achmad Z and Mary S 2019 *Journal of Physics: Conference Series* 1341
[10] Choi H, Geronimo F K F, Jeon M, Kim L-H 2021 *Water* **13** 1242
[11] Setiyanto R A, Darundiati Y H, Joko T 2016 *Public Health Journal (e-Journal)* **4(1)** Januari 2016 ISSN 2356-3346
[12] De Rozari P, Ledoh S M 2018 *Al-Kimia* 6/2/2018 117-124
[13] Sayow F, Tilaar B V and Augustine K D 2020 *Agri-Sosio Ekonomi* **16** 245–252
[14] Wimbaningrum R, Arianti I, Sulistiyowati H 2020 *Berkala Sainstek 2020* **VIII (1)** 25-28
[15] Haryani M F 2020 *International Journal of Scientific & Technology Research* **9(01)**

Wastewater treatment for tofu home industries in Semanan, West Jakarta using electrocoagulation method with electrode al-stainless steel

A Wijaya[1], B Iswanto[1], M M Sintorini[1*]

Department of Environmental Engineering, Universitas Trisakti, Jakarta, Indonesia

Sintorini@trisakti.ac.id

Abstract. Tofu wastewater is dangerous for the aquatic environment because it contains high BOD, COD, and TSS, which will damage life in the river. Tofu home industries in Semanan have not used a proper wastewater treatment system, and this study was conducted to test the electrocoagulation system to reduce the levels of BOD, COD, and TSS. According to several studies, the electrocoagulation system can reduce BOD, COD, and TSS. It is hoped that this electrocoagulation system can solve environmental problems caused by tofu wastewater. From the research that has been done, namely with the stirring variable at a speed of 150 RPM and at a detention time of 10 minutes with a discharge of 0.6 liter/minute, the percentage of COD removal is 36%, the percentage of BOD removal is 27%, and the percentage of TSS removal is 51%. Then on the stirring variable at a speed of 100 RPM and at a detention time of 10 minutes with a discharge of 0.6 liter/minute, the percentage of COD removal is 32%, the percentage of BOD removal is 15%, and the percentage of TSS removal is 58%.

1. Introduction

Tofu waste is waste generated in making tofu in the form of solid and liquid waste. So far, solid waste can be reused as animal feed, but for liquid waste, the impact will result in a foul smell and discolouration of river water if it is directly disposed of without treatment [1]. In the tofu industry, the wastewater produced generally contains high organic compounds. The organic compounds are usually COD, BOD and TSS. If the industrial tofu waste is not treated correctly, it can cause water pollution, a source of disease, and a disturbing odour, so a process for handling the tofu wastewater is needed [2].

In the electrocoagulation process, active coagulants are released in the form of metal ions at the anode (usually aluminium or iron) into the solution. At the same time, an electrolysis reaction occurs in the form of the release of hydrogen ions at the cathode, the process of releasing active coagulants and the release of hydrogen ions in the electrocoagulation process is influenced by several things, including; electric current, voltage, contact time, temperature, pH and conductivity [3]. So it is hoped that the process of releasing active coagulants and hydrogen ions can reduce the pollutant content of COD, BOD, and TSS. The electrocoagulation process is eco-friendly cause it saves chemical use, energy-saving, low operating cost, and saving space in use [4].

2. Methodology

2.1. Time and location
This research was conducted from March 2021 to August 2021, and this research was carried out in the Environmental Laboratory of the Department of Environmental Engineering, Universitas Trisakti, West Jakarta and the testing and analysis of research results was carried out at the Environmental Laboratory of the Department of Environmental Engineering, Universitas Trisakti. Wastewater samples were taken from the sewerage outlet of the Semanan tofu industrial complex, to be precise, the tofu factory owned by Mr. Haji Cahyo, West Jakarta. The reason for choosing the location was because the location specifically produces tofu, so it is hoped that the wastewater content will not be mixed with other wastewater so that the results obtained are representative.

2.2. Research stages
The treatment system used is continuous electrocoagulation with aluminium-stainless steel electrodes, and the preliminary research is oxidation. The following is a chart that explains the stages of the research.

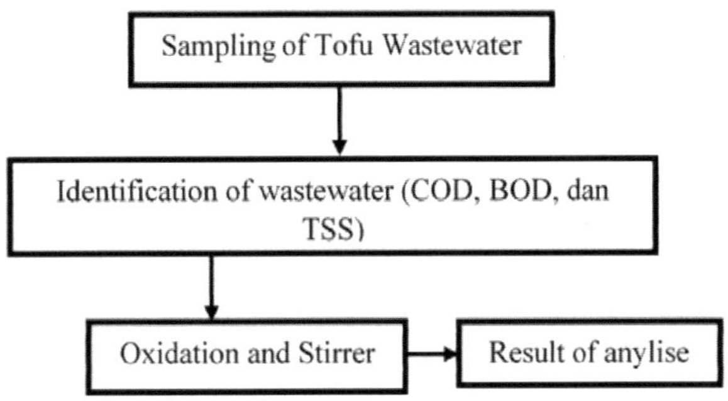

Figure 1. Stage of preliminary research.

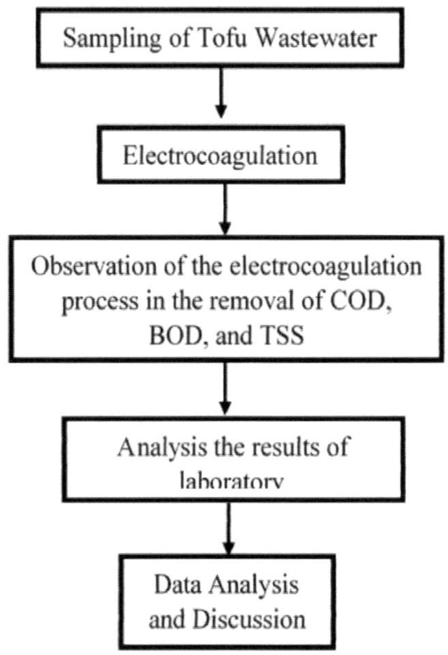

Figure 2. Stage of the primary research.

2.3. A sampling of tofu wastewater
Sampling was carried out at the Semanan tofu factory, where the sampling point was at the tofu wastewater disposal outlet at the Haji Cahyo tofu factory located in the Semanan Tofu Tempe Indonesia Cooperative Area (KOPTI), West Jakarta.

2.4. Characteristic identification of tofu wastewater
In the Table 1, some pollutant parameters will be tested and compared with DKI Jakarta Governor Regulation No. 69 of 2013 and Minister of Environment Regulation No. 5 of 2014.

IOP Conf. Series: Earth and Environmental Science **894** (2021) 012016 doi:10.1088/1755-1315/894/1/012016

Tabel 1. Regulation of tofu wastewater.

Test Parameters	Standard (mg/L)	
	PerGub DKI No.69 of 2013 (mg/L)	Perrin LH No. 5 of 2014 (mg/L)
COD	75	150
BOD_5	100	300
TSS	100	200
Nitrate	10	20

2.5. *Preliminary research*
Preliminary research conducted is to test the process of oxidation and stirring of tofu wastewater samples.

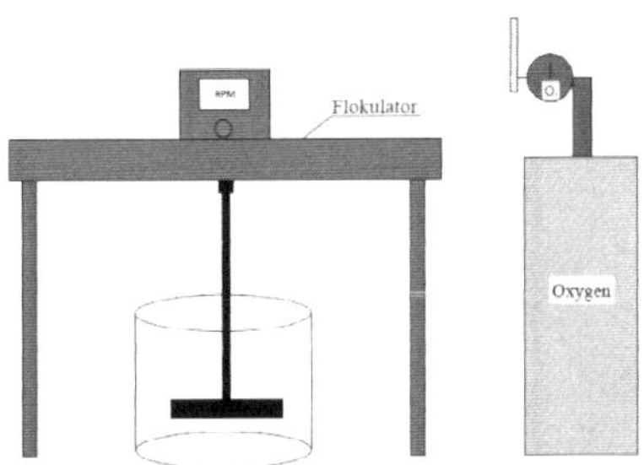

Figure 3. Sketch of oxidation and stirrer reactor.

Figure 4. The operation condition of oxidations.

In Figure 3 and Figure 4, the process that occurs is that oxygen will enter the wastewater to increase the oxygen level in the wastewater, then the function of stirring is for mixed the oxygen with the wastewater so that will make maximum distribution of oxygen.

2.6. *The primary research of electrocoagulation*
The reactor used has a stainless-steel anode that also serves as a stirring rod and uses aluminium as a cathode placed on the base of the reactor. The residence time is setting 20 minutes with a flow rate of 0.3 L/min, a residence time of 10 minutes with a flow rate of 0.6 L/min, and a residence time of 30 seconds with a flow rate of 11 L/min. As well as setting the rotation speed of the stirrer set at 150 rpm, 100 rpm and 50 rpm, and with fixed voltage electricity sourced from 12V 65Ah battery capacity. The following is contained in the Table 2 the variation of the research variables carried out.

Table 2. Variable variation of research.

Rotation Speed (rpm)	Detention Time	Flow Rate (L/Minutes)	Inlet	Outlet 1	Outlet 2	Outlet 3
	20 Minutes	0.3		I	II	III
150	10 Minutes	0.6	A	IV	V	VI
	30 Second	11		VII	VIII	IX
	20 Minutes	0.3		I	II	III
100	10 Minutes	0.6	B	IV	V	VI
	30 Second	11		VII	VIII	IX
	20 Minutes	0.3		I	II	III
50	10 Minutes	0.6	C	IV	V	VI
	30 Second	11		VII	VIII	IX

A reactor of electrocoagulation was used during the study, as shown in Figure 5 below.

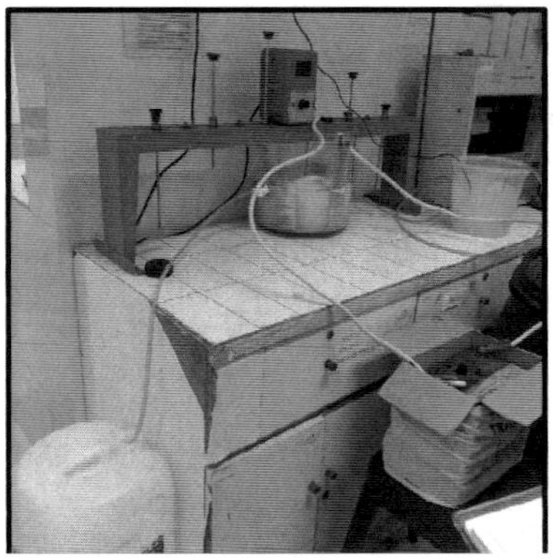

Figure 5. Electrocoagulation reactor.

2.7. Laboratory analysis
A sampling of tofu wastewater was carried out at the wastewater outlet of the tofu industry. The pollutant parameters that will be tested in this study are COD, BOD, TSS and Nitrate. The results of the analysis of the electrocoagulation process will be compared with the Governor's regulation No. 69 of 2013 concerning Wastewater Quality Standards for Activities and/or Businesses and compared with the Regulation of the Minister of the Environment No. 5 of 2014 concerning Wastewater Quality Standards. There is an analytical method used to measure the parameters studied in Table 3.

Table 3. Analysis method.

Parameter	Methods	Analysis Methods
COD	Closed Refluks with titrimetric	SNI 6998.2:2009
BOD	Winkler	SNI 6989.72:2009
TSS	Gravimetric	SNI 6989.3:2004
Nitrate	Spektrofotometri	SNI 6989.9:2004

2.8. Equation and mathematics
To get the efficiency of reducing the pollutant content, the calculation is carried out with the following formula:

$$\text{Efficiency} = \frac{a-b}{a} \times 100\% \quad (2.1)$$

Description:
a: value of sample concentration before electrocoagulation
b: the value of sample concentration after electrocoagulation

3. Results and discussion
By the planned research stages, the initial stage after sampling is to check the characteristics of the tofu wastewater. There are in Table 4 regarding some of the characteristics of the pollutant obtained, namely as follows:

Table 4. Characteristic of tofu wastewater.

Test Parameters	Standard (mg/L)			Description
	Tofu Wastewater (mg/L)	PerGub DKI No.69 of 2013	PerMen LH No. 5 of 2014	
COD	6120	75	150	Does Not Qualify

Test Parameters	Standard (mg/L)			Description
	Tofu Wastewater (mg/L)	PerGub DKI No.69 of 2013	PerMen LH No. 5 of 2014	
BOD_5	4898	100	300	Does Not Qualify
TSS	570	100	200	Does Not Qualify
Nitrate	4	10	20	Qualified

Testing the characteristics of wastewater is used to determine how much pollutant content is in the wastewater and how much removal efficiency is obtained from the electrocoagulation process. Table 5 shows the % removal of COD (chemical oxygen demand) obtained from the electrocoagulation process.

Table 5. Result of COD analysis.

Rotation Speed (rpm)		COD (Mg/L)	%Efficiency	Standard		Description
				PerGub DKI No.69/2013	PerMen LH No. 5/2014	
150 rpm	Inlet	6120		150 mg/L	150 mg/L	Does Not Qualify
	30 Second	4896	20%			Does Not Qualify
	10 Minutes	3916.8	36%			Does Not Qualify
	20 Minutes	4406	28%			Does Not Qualify
100 rpm	Inlet	6854.4				Does Not Qualify
	30 Second	4896	29%			Does Not Qualify
	10 Minutes	4651.2	32%			Does Not Qualify
	20 Minutes	4896	29%			Does Not Qualify
50 rpm	Inlet	5280				Does Not Qualify
	30 Second	4224	20%			Does Not Qualify
	10 Minutes	4012	24%			Does Not Qualify
	20 Minutes	4224	20%			Does Not Qualify
150 rpm + Garam	Inlet	5280				Does Not Qualify
	10 Minutes	2956	44%			Does Not Qualify

In Table 5, it can be seen that the most optimum COD reduction condition is at a rotation speed of 150 rpm with a residence time of 10 minutes where the flow rate is 0.6 L/minute. The rotational speed at 150 rpm is said to be optimum for reducing COD levels because mixing Al^{3+} ions as a coagulant will work well, as well as a residence time that is too fast and too long can cause the failure of mixing of cations and anions produced by the electrode, a long residence time as well. It causes deposits on the electrodes to reduce the ability of the electrodes to be able to produce the required ions. The non-fulfilment of the quality standard criteria for tofu wastewater after the electrocoagulation process is caused by factors that affect the efficiency of electrocoagulation. Such as the electric current flowing in the wastewater and the electric voltage during the electrocoagulation process [5]. Addition of salt at a salt concentration of 1 g/L to increase the efficiency of removal, the salt content affects the amount of electrical conductivity in the water, with the addition of the salt to the most optimum variable, namely the stirring speed of 150 rpm with a detention time of 10 minutes. Set aside the COD content of 44%. There are results of the BOD allowance analysis with the same variation of variables, which are shown in Table 6.

Table 6. Result of BOD analysis.

Rotation Speed (Rpm)		BOD (Mg/L)	%Efficiency	Standard		Description
				PerGub DKI No.69/2013	PerMen LH No. 5/2014	
150 rpm	Inlet	4898				Does Not Qualify

IOP Conf. Series: Earth and Environmental Science **894** (2021) 012016	doi:10.1088/1755-1315/894/1/012016

Rotation Speed (Rpm)		BOD (Mg/L)	%Efficiency	Standard PerGub DKI No.69/2013	Standard PerMen LH No. 5/2014	Description
	30 Second	4080	17%			Does Not Qualify
	10 Minutes	3600	27%	100 mg/L	300 mg/L	Does Not Qualify
	20 Minutes	3740	24%			Does Not Qualify
100 rpm	Inlet	6120				Does Not Qualify
	30 Second	5680	7%			Does Not Qualify
	10 Minutes	5200	15%			Does Not Qualify
	20 Minutes	5460	11%			Does Not Qualify

Based on Table 6, the most extensive BOD removal was 27% at 150 Rpm stirring with a detention time of 10 minutes. It can be seen that there is an interaction between the variation of residence time and the speed of stirring rotation on the decrease in BOD levels in tofu wastewater treated with electrocoagulation. At speeds of 150 rpm and 100 rpm with variations of residence time, the most optimum is at 10 minutes with a stirring rotation speed of 150 rpm with the results of the removal efficiency of 27% and 15%. In reducing BOD levels, variations in residence time and variations in the speed of stirring rotation have a strong influence on decreasing BOD levels. In Table 7, the results of the analysis of Total Suspended Solid (TSS) are obtained.

Table 7. Results of TSS analysis.

Rotation Speed		TSS (Mg/L)	%Efficiency	Standard PerGub DKI No.69/2013	Standard PerMen LH No. 5/2014	Description
150 Rpm	Inlet	570				Does Not Qualify
	30 Second	277	51%			Does Not Qualify
	10 Minutes	236	59%			Does Not Qualify
	20 Minutes	257	55%			Does Not Qualify
100 Rpm	Inlet	378				Does Not Qualify
	30 Second	196	48%			Does Not Qualify
	10 Minutes	159	58%	100 mg/L	200 mg/L	Does Not Qualify
	20 Minutes	257	32%			Does Not Qualify
50 Rpm	Inlet	366				Does Not Qualify
	30 Second	268	27%			Does Not Qualify
	10 Minutes	167	54%			Does Not Qualify
	20 Minutes	236	36%			Does Not Qualify
150 Rpm + Garam	Inlet	**366**				Does Not Qualify
	10 Minutes	**135**	63%			Does Not Qualify

Electrocoagulation treatment based on Table 7 has proven successful in reducing TSS levels in tofu wastewater, with the best % removal at a stirring speed of 150 Rpm with a residence time of 10 minutes with a % removal value of 59%, reducing TSS with an initial value of 570 mg/L to 277 mg/ L, this result still exceeds the specified quality standard. However, it can be concluded that electrocoagulation can reduce the TSS content in tofu wastewater. However, in the variation of stirring of 100 Rpm and residence time of 10 minutes, although the % removal of TSS is more minor, it is 58%. This variation can reduce TSS levels not to exceed the quality standard of Minister of Environment Regulation No.

5/2014. The main form of nitrogen in natural waters is Nitrate. Besides that, Nitrate acts as the primary nutrient in the photosynthesis process of autotrophic organisms in the waters. It makes the nitrate content in the waters monitored not excessive and triggered algae blooms. To reduce the nitrate concentration in tofu wastewater, statistically, it can be seen that the nitrate concentration decreased after being given electrocoagulation treatment. As shown in Table 8, the results of the analysis of tofu wastewater, the nitrate concentration met the quality standards of DKI Jakarta Governor Regulation No. 69/2013 and Minister of Environment Regulation No.5/2014 both before after electrocoagulation.

Table 8. Result of nitrate analysis.

Rotation Speed		**Nitrate (Mg/L)**	**%Efficiency**	**Standard**		**Description**
				PerGub DKI No.69/2013	**PerMen LH No. 5/2014**	
150 rpm	**Inlet**	4				Qualified
	30 Second	2.29	43%			Qualified
	10 Minutes	1.784	55%			Qualified
	20 Minutes	2.253	44%			Qualified
100 rpm	**Inlet**	1.265				Qualified
	30 Second	0.798	37%			Qualified
	10 Minutes	0.789	38%	10 mg/L	20 mg/L	Qualified
	20 Minutes	0.726	43%			Qualified
50 rpm	**Inlet**	0.592				Qualified
	30 Second	0.527	11%			Qualified
	10 Minutes	0.328	45%			Qualified
	20 Minutes	0.472	20%			Qualified
150 rpm + Garam	**Inlet**	0.592				Qualified
	10 Minutes	0.246	58%			Qualified

Based on Table 8, the optimum condition for decreasing the nitrate concentration was at a stirring speed of 150 Rpm with a residence time of 10 minutes, with a % removal of 55%, in this case, the stirring speed had a more vital interaction in reducing the nitrate concentration compared to the residence time interaction. It depends on the amount of H_2 gas formed during the electrocoagulation process. The H_2 gas formed is strongly influenced by the formation of Al^{3+} from the aluminium anode. Organic matter can conduct electricity better, so the result of the % allowance is 58%, where this result is the most optimum result among other variables. Salt is an alternative that can increase the % removal of pollutant parameters because, at a low cost, it can get more optimum results [6].

4. Conclusion

Based on the research conducted, it can be concluded that electrocoagulation with Al-St as an electrode can reduce the pollutant content of COD, BOD, TSS and Nitrate in tofu wastewater even though its efficiency is low and does not meet quality standards, except Nitrate. The optimum variation to reduce COD, BOD, TSS, and Nitrate is at an emotional cycle of 150 Rpm and a residence time of 10 minutes. It is necessary to add salt because the 150 Rpm variable with a residence time of 10 minutes added salt at a concentration of 1gr/L can get a higher removal result for COD by 44%, in TSS by 63%, and in Nitrate by 58%.

References

[1] Effendi H 2003 Kanisius
[2] Nohong 2010 Jurnal Pembelajaran Sains 257-269
[3] Liu H Z 2010 Electrochemsitry for the environment
[4] Pusat Standarisasi Kementrian Lingkungan Hidup dan Kehutanan
[5] Hernaningsih T 2016 Jurnal Pusat Teknologi Lingkungan BPPT 31-46
[6] Iswanto B 2001 Tesis Magister Institut Teknologi Bandung

IOP Conf. Series: Earth and Environmental Science **894** (2021) 012017 doi:10.1088/1755-1315/894/1/012017

Study of tofu wastewater treatment using anaerobic baffled reactor: laboratory scale

H Trihidayanti, R Ratnaningsih* and B Iswanto

Department of Environmental Engineering, Universitas Trisakti, Jakarta, Indonesia

ratnaningsih@trisakti.ac.id

Abstract. One of the foods favorites by Indonesian people is tofu. In the process of making tofu, it produces wastewater. The tofu wastewater must be treated first to reduce environmental pollution. To treat the liquid waste, it is using biological treatment by an anaerobic baffled reactor with bio-ball. The purpose of this study was to determine the optimum time at 75% tofu wastewater concentration. The detention time used in this study was 12 hours, 24 hours, 36 hours and 48 hours. Seeding takes time about 45 days, with the obtained VSS was 5550 mg/L. Furthermore, acclimatization in the reactor takes about 30 days, with an efficiency of removing COD of 86.3%. The results obtained that at 12 hours of detention, there is a COD allowance of 79.8%, 24 hours of detention time there is a COD allowance of 85.4%, 36 hours of detention time 86.3%, and 48 hours of detention time 88.4 %. It shows that the optimum detention time in this anaerobic reactor is 48 hours. However, the COD level was 765.3 mg/L is still slightly above the quality standard according to the Minister of Environment Regulation No.5 2014, which is the maximum COD level of 300 mg/L.

1. Introduction

Tofu is one of the processed food products made from soybeans. The tofu industry in the processing process produces waste, both solid and liquid waste. Most of the wastewater treatment in the tofu manufacturing industry has not been appropriately treated. The tofu manufacturing industry only disposes of its liquid waste in the waters. The liquid waste generated from the tofu manufacturing process comes from the washing, boiling, pressing and printing processes. Therefore, the resulting liquid waste contains high organic content. Therefore, because the organic content in tofu wastewater is high and has high BOD and COD levels, if tofu wastewater is directly discharged into water bodies, it will reduce the environment's carrying capacity.

Tofu wastewater is organic waste and does not contain heavy metals. Thus, the processing can be done biologically. The biological treatment process is wastewater treatment by utilizing microorganisms in degrading pollutant content in the wastewater. The currently existing tofu wastewater treatment technology is generally waste treatment with an anaerobic system. It is because the operational costs are cheaper and allow it to be used on a large scale.

KOPTI or Indonesian Tofu-Tempeh Production Cooperative in Semanan was established in 1972. This cooperative was founded by tofu-tempe craftsmen, the majority of whom came from Pekalongan, Central Java. The Semanan Industrial Village was finally inaugurated in 1992 after receiving support from the DKI Jakarta Regional Government, the Minister of Industry, the Minister of Cooperatives and the Minister of Housing. In 1974 Kopti Semanan changed its name to PRIMKOPTI Swakerta, West Jakarta. Primkopti is a tofu and tempeh production cooperative located in Semanan, West Jakarta. This cooperative has a land area of 12 hectares used for members' homes and as a tofu and tempeh production house. The cooperative located in Semanan Primkopti has a shared kitchen which members use to produce tofu and tempeh. The location of the shared kitchen is right in front of the Semanan IPAL [1].

Organic materials contained in the waste of the Tofu Industry are generally very high. The organic compounds in the wastewater can be in proteins and fats, which are the largest in number [2]. The characteristics of tofu wastewater include total solids, suspended solids, temperature, color and odor. Chemical characteristics include organic materials, inorganic materials and gases. Wastewater temperature ranges from 37-45 °C, turbidity ranges from 535-585 FTU, color 2225-2250 Pt.Co, Ammonia 23.3 – 23.5 mg/l, BOD_5 6000-8000 mg/l and COD 7500- 14000 mg/l [3].

Anaerobic Baffled Reactor (ABR) is a suspended treatment system that occurs anaerobically in a bioreactor divided into several compartments. A series of bulkheads installed in the ABR makes the liquid waste flow up-flow and down-flow from the inlet to the outlet so that there is contact between the liquid waste and the active biomass [4]. The concentration of organic compounds varies throughout the ABR, resulting in different populations of microorganisms in each compartment. This condition depends on the specific environmental conditions produced by the decomposition product.

In waste treatment with ABR, it is necessary to pay attention to activated sludge. The higher the activated sludge added, the higher the decrease in COD levels produced by bacteria that decompose organic matter in wastewater. In addition, the type of activated sludge used in the ABR can affect the performance of the ABR unit itself. The content of microorganisms and the environment of microorganisms in their natural habitats affect the performance of activated sludge in treating influent waste in ABR [5].

The bacteria contained in the bioreactor float and settle horizontally towards the end of the reactor slowly, thereby increasing the cell retention time. Liquid waste is in contact with active biomass so that the effluent is free from biological solids. The configuration shows a high degree of removal [6].

The main advantages possessed by ABR, among others:

1. ABR can separate the processes of acidogenesis and methanogenesis longitudinally without any control problems and high costs [7].
2. Simple ABR design, relatively low constructive cost, and stability against hydraulic and organic shock loads [8].
3. The hydrodynamic pattern of ABR can maintain biomass without fixed media [4].

In this study, we want to know the optimal hydraulic detention time in the Anaerobic Baffled Reactor to know the efficiency of COD level and VSS levels in tofu wastewater.

2. Methodology

2.1. Location

The tofu industry, which is located in Semanan Village, Kalideres District, West Jakarta, as a place for sampling tofu wastewater and the location of the insulated Anaerobic reactor processing is carried out on a laboratory test scale located at the Environmental Laboratory of the Environmental Engineering Department, FALTL, Trisakti University.

2.2. Anaerobic Baffled Reactor

The reactor used is a continuous reactor that has four anaerobic zones. This study uses a laboratory-scale reactor. The reactor is made of acrylic glass, 5 mm thick with dimensions of 60cm × 20 cm × 35 cm. The media used in this reactor is bio-ball. The wastewater is put into a waste container with a capacity of 90 liters. Then, it enters the insulated anaerobic reactor from the discharge regulator. Wastewater from the discharge control tub entering the reactor is regulated using a pump with a discharge regulator channeled into the insulated anaerobic reactor according to the predetermined discharge. Discharge is obtained from the volume of treated water divided by the hydraulic residence time of each treatment. In the 12-hour treatment, it is known that the resulting discharge is 37.5 ml/minute; 24-hour treatment of 18.75 ml/minute; 36-hour treatment of 12.5 ml/minute; and 48 hours treatment of 9.4 ml/minute. Pictures of reactor equipment can be seen in Fig 1.

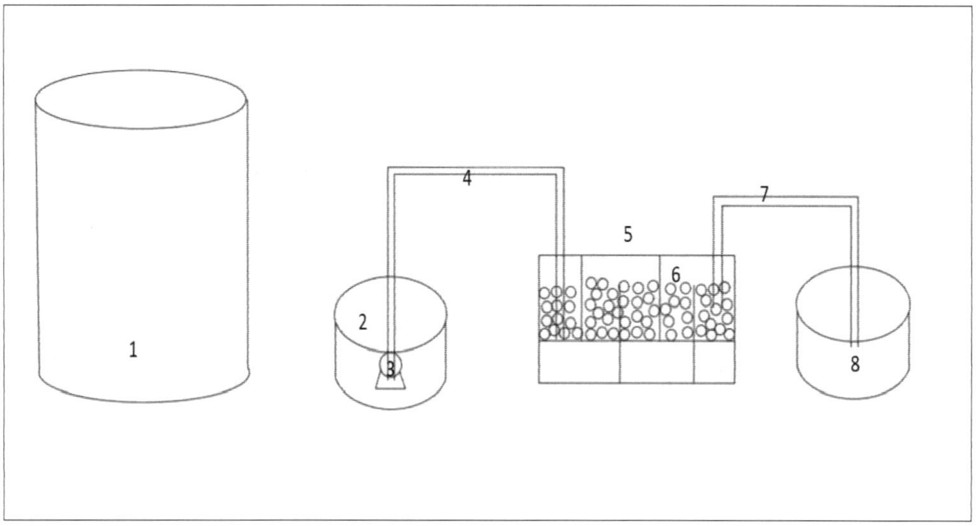

Figure 1. Reactor equipment.

Description:
1. Waste Container with a capacity of 90 liters
2. Discharge control basin
3. Discharge regulating pump
4. Influent hose
5. Insulated Anaerobic Reactor
6. Bioball Media
7. Effluent Hose
8. Effluent basin

2.3. Quality Standard

The wastewater used in this research is tofu industrial wastewater that does not comply with the quality standard requirements to discharge into the environment—the quality of tofu industrial wastewater before processing is shown in table 1.

Table 1. Quality of tofu industrial wastewater in Semanan.

Unit	Quality of wastewater (mg/L)	Quality Standards (mg/L)	
		PerGub DKI No 69 Year 2013	PerMen LH No 5 Year 2014
BOD_5	3648.5	75	150
COD	6912	100	300
TSS	767	100	200
pH	4.6	6-9	6-9
VSS	756		
N total	43.3		
P Total	1.9		

3. Result and discussion

3.1. Seeding
The seeding process is a process to breed bacteria to decompose or degrade wastewater which can form a biofilm layer contained in the buffer media, namely bioball. The seeding process is carried out for 45 days. On the 45th day, the VSS value obtained was 5550 mg/l. It shows that the VSS concentration and the amount of biomass are sufficient to proceed to the next stage, namely acclimatization.

3.2. Acclimatization
The acclimatization stage is a stage of adapting bacteria to the wastewater to be treated. In the acclimatization process, the tofu wastewater is given in stages, starting from a low concentration until it is entirely replaced by 100% pure waste. The addition of tofu wastewater continues to be carried out until it is entirely replaced by tofu wastewater by considering COD removal's efficiency until it is in a stable condition. Provision of tofu wastewater is carried out in stages to avoid a significant decrease in COD removal efficiency and prevent shock loading of bacteria.

In the acclimatization stage, a detention time (td) of 24 hours was used. It is because tofu wastewater can have a reasonably good contact time with the bioball media so that bacteria can grow attached by forming a biofilm layer. The acclimatization process was carried out with a recirculation system directly in an anaerobic insulated reactor attached to the bioball buffer media.

Acclimatization was carried out for 30 days. In the first stage, with a concentration of tofu wastewater at 20% for three days, the COD efficiency was 61.2%. Then, from day 4 to day 9, acclimatization was carried out with a concentration of tofu wastewater of 40%, obtaining a COD removal of 71.5%. From the 10th to the 17th day, tofu wastewater was used with a concentration of 60%, and the COD removal efficiency was 75%. From the 18th day to the 23rd day with 80% tofu wastewater concentration, the COD removal result was 77%, and the last stage of acclimatization, namely on the 24th to 30th day, the replacement of pure tofu waste was made to 100%, the COD removal was 86.3%.

3.3. Running
The insulated anaerobic reactor with bioball media is carried out continuously, which can be started after acclimatization ends with a stable condition, namely in a state of high COD removal efficiency. At this stage, COD and VSS research was carried out at the sampling point at the anaerobic inlet and anaerobic outlet. The hydraulic detention time is 12 hours, 24 hours, 36 hours and 48 hours. The start of the running process is from the longest detention time, which is 48 hours. It is due to avoid shock loading.

Table 2. Result of detention time against COD efficiency

Detention time	Inlet	Outlet	Efficiency Removal (%)
12	6268.25	1257	79.8
24	6735.25	981	85.4
36	6197.5	852	86.3
48	6567	765.25	88.4

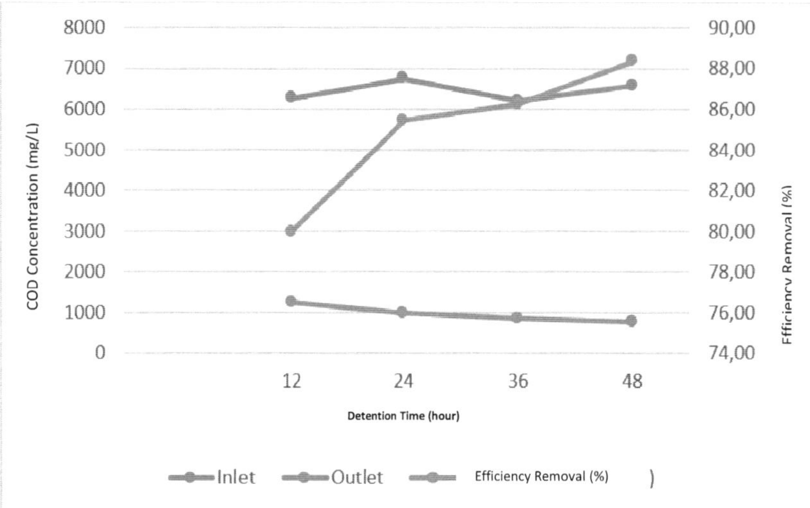

Figure 2. Detention time against COD Efficiency

Based on the analysis of COD removal for variation of detention time, it was found that the greatest efficiency was at detention time of 48 hours with an efficiency of 88.4%. While at the detention time of 36 hours, the results were obtained at an efficiency of 86.3%. Meanwhile, at the detention time of 24 hours, the removal efficiency was 85.4%, and at the detention time of 12 hours, the removal efficiency was 79.8%.

The study results prove that the longer or more fantastic the detention time of wastewater in the bioreactor, the better the removal efficiency. Because the longer the wastewater is in the reactor, the longer the contact time of wastewater with microorganisms. So that the time used to degrade the material is getting more and more, which will eventually be a lot in setting aside the levels of COD in the wastewater. The results of COD on detention time also show that there is an increase and decrease in removal efficiency. It is caused by the condition of tofu wastewater also fluctuating. However, the efficiency results can still be stable in terms of increase and decrease because the increase and decrease in efficiency are still within the difference of 5%.

Table 3. Result of detention time against VSS efficiency.

Detention Time	VSS Concentration (mg/L)		Efficiency Removal (%)
	Inlet	Outlet	
12	2612	444	83,0
24	2627	358	86,4
36	2587	269	89,6
48	2450	139	94,3

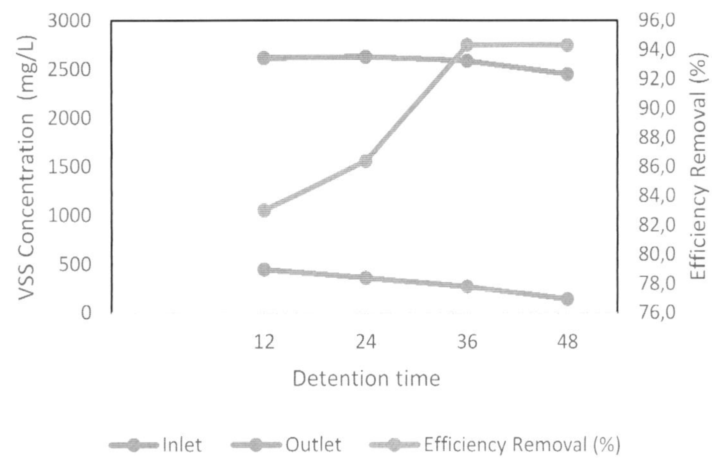

Figure 3. Detention time against VSS Efficiency.

From the VSS results above, it is known that the VSS level at the reactor inlet is greater than the VSS level at the reactor outlet; this is due to the large number of microorganisms attached to the bioball media and as a biofilter of the reactor, the bioball media can be able to eliminate the VSS concentration in wastewater quite well.

4. Conclusion

Based on the analysis of the results of the tofu wastewater treatment with the anaerobic baffled reactor, we concluded that the longer or the more significant the detention time of wastewater on the bioreactor, the better the removal efficiency. In this study, the optimum detention time was 48 hours to achieve the COD removal efficiency value of 88.9% with a COD value of 723 mg/L. So as with the VSS removal efficiency value of 94.3% with VSS value is 139 mg/L.

References

[1] Jayanti and Lia D 2011 *Evaluasi dan Peningkatan Kinerja Sistem Instalasi Pengolahan Air Limbah Semanan, Jakarta Barat*

[2] Sugiharto A, Setiawan Y, Saleh A A 2014 *HWTM Newsletter* **5**

[3] Amudyanto dan Nurhasan 1991 *Penanganan Air Limbah Pabrik Tahu* (Semarang: Yayasan Bina Lestari)

[4] Movahedyan H, A Assadi and A Parvaresh 2007 *Iran J. Environ. Health Sci. Eng.* **4 (2)** 77-84

[5] Amin F, Darda A, Sumantri I 2013 *Jurnal Teknologi Kimia dan Industri* **2(3)**

[6] Wang J, Huang Y, Zhao X 2004 *Bioresource Technol.* **93** 205-208

[7] Bell J 2002 *Treatment of Dye Wastewaters in The Anaerobic Baffled Reactor and Characterisation of The Associated Microbial Populations* (Ph.D. Thesis, School of Chem. Eng., Univ. of Natal, Durban)

[8] Eawag S 2008 *Sanitation Systems and Technologies* (Switzerland:Department of Water and Sanitation in Developing Countries Dubendorf)

IOP Conf. Series: Earth and Environmental Science **894** (2021) 012018 doi:10.1088/1755-1315/894/1/012018

Optimisation of degreemont water treatment package on Kedung Halang water treatment plant, Bogor, West Java

H Gabrielle[1], R A Kusumadewi[1*], Ratnaningsih[1]

[1]Department of Environmental Engineering, Faculty of Landscape Architecture and Environmental Engineering, Universitas Trisakti, Jakarta, Indonesia

rianaayu.kusumadewi@trisakti.ac.id

Abstract. To provide drinking water in Bogor Regency, PDAM Tirta Kahuripan has several Water Treatment Plants (WTP) Kedung Halang WTP. Kedung Halang WTP consists of two water treatment packages, Indisi and Degreemont, with an average production of 37.3 L/s and 83.9 L/s, respectively. The purposes of this research are to evaluate the performance of the operating unit and process on the Degreemont Package and to provide optimisation recommendations based on the evaluation result. Steps in this research include evaluating the existing condition, providing recommendations, and analysing the quality of water produced. The evaluation found that one of the parameters did not fulfil the design criteria: the G.Td value on coagulation. The evaluation result of maximum flow rate is 125 L/s on intake, 150 L/s on coagulation, 130 L/s on flocculation and sedimentation, and 200 L/s on filtration. Thus, the optimum flow rate that can be operated is 125 L/s. The turbidity of the water produced has met the drinking water standard according to Minister of Health Regulation No. 492 of 2010, with removal efficiency above 95%. According to the result, it can be said that Degreemont Package is qualified to be applied as a drinking water treatment.

1. Introduction

Water is the fundamental of human, and all other creatures, life. However, with the increasing number of populations, the water demand is also increasing. Population growth will limit the amount of water available per person [1]. The United Nations World Water Development Report 2021 stated that the global population experiencing water scarcity increased from 32 million people in 1900 to a projected 3.1 billion people by 2050 [2]. Besides the water quantity, but population growth also decreases water quality in natural water sources due to human activities and land use around the water sources [3]. Many of the world's major urban areas are withdrawing water from rivers recharged with wastewater from all the cities located upstream or even from the urban area itself [1].

The Water Treatment Plant (WTP) aims to ensure that the water distributed in society is safe for health. In general, the water treatment process consists of coagulation, flocculation, sedimentation, filtration, and disinfection [4]. In Indonesia, the water distributed should fulfil the drinking water requirements in quality, which is regulated in Minister of Health Regulation No. 492 of 2010 [5]. One way to fulfil the increasing water demand is optimising the drinking water supply system, including the WTP [6].

In recent studies, optimisation consists of measures to decrease the managing and monitoring expenses and improve the quality of produced water [4]. A study done by Sani et al. showed that optimisation of Cipageran WTP could increase the processing capacity to 210 L/s from the initial design capacity of 200 L/s [7]. In contrast, a study done by Ramadhan et al. showed that optimisation of Solar WTP could increase the processing capacity of 130 L/s from the initial design capacity of 100 L/s [6].

Like any other city or regency across Indonesia, the population in Bogor Regency is increasing, with a population growth rate of 2.45%. The drinkingPDAM Tirta Kahuripan manages the drinking water supply in Bogor Regency. To fulfil the drinking water demand, PDAM Tirta Kahuripan has several WTP. One of them is WTP Kedung Halang, with a design capacity of 130 L/s. WTP Kedung Halang consists of two water treatment packages, Indisi and Degreemont, with a design capacity of 40 L/s and 90 L/s. However, the average processing flow rate is only 37.3 L/s for the Indisi Package and 83.9 L/s for Degreemont Package. Both packages have processing units consist of intake, coagulation, flocculation, sedimentation, filtration, and disinfection.

Though many studies have discussed WTP optimisation, there have not been any studies mentioning Kedung Halang WTP yet. Thus, this research would propose an optimisation to maximise the processing capacity of Kedung Halang WTP, particularly the Degreemont Package, without compromising the quality of water produced.

The purposes of this research are (1) to evaluate the performance of operating unit and process on Degreemont Package, and to provide optimisation recommendations based on the evaluation result; and (2) to compare the quality of water produced with drinking water quality standards.

2. Methodology

2.1. Research stages

Stages on this research include literature review, data collection, and data analysis. The outcomes of this research consist of evaluation results of existing conditions, optimum processing flow rate, and existing performance based on turbidity and removal efficiency. The research flow chart can be seen in Figure 1.

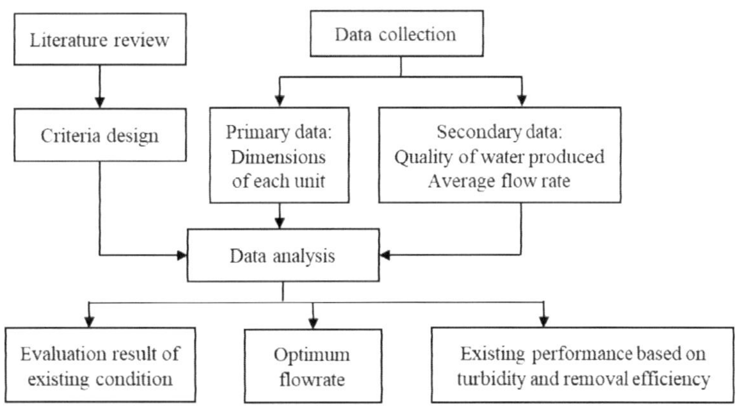

Figure 1. Research flow chart.

2.2. Location and time

This research was conducted on Kedung Halang WTP, which is located at Jl. Perumahan PPKN, Kedung Halang Ward, Bogor Regencies, throughout March 2021.

2.3. Data collection

There were two types of data used in this research, primary data and secondary data. Primary data includes the dimensions of each unit, which are measured directly on the installation. Secondary data consist of the average processing flow rate and the quality of raw water and water produced in turbidity. The secondary data was obtained from Kedung Halang WTP's archives for March 2021. Kedung Halang

WTP used a water meter to measure the processing flow rate, while turbidity was measured using a turbidimeter.

2.4. Data analysis

The dimensions of each unit and average processing flow rate data were used to calculate the value of design parameters important to the operational system. There are 8 design parameters calculated with the following equations [13]:

Pipe velocity:

$$v = \frac{Q}{A\ \text{pipe}} \tag{2.1}$$

Gradient velocity :

$$G = \sqrt{\frac{P}{\mu \times Vol}} \tag{2.2}$$

Detention time

$$T_d = \frac{Vol}{Q} \tag{2.3}$$

G.Td value

$$G.Td = G \times Td \tag{2.4}$$

Surface loading

$$S_o = \frac{Q}{A\ \text{sedimentation}} \tag{2.5}$$

Reynolds number

$$N_{Re} = \frac{v_o \times R}{v} \tag{2.6}$$

Froude number

$$N_{Fr} = \frac{{v_o}^2}{g \times R} \tag{2.7}$$

Filtration rate

$$v_f = \frac{Q}{A\ \text{filtration}} \tag{2.8}$$

Whereby,

Q = flow rate (m^3/s)
A = area (m^2)
P = power ($kg.m^2/s^3$)
μ = absolute viscosity (kg/m.s)
Vol = volume (m^3)
v_o = settling velocity on settler (m/s)
R = hydraulic radius (m)
v = kinematic viscosity (m/s^2)

Analysis of operational system was carried out by comparing the critical design parameters value between results of calculations by the actual condition and the design criteria obtained from the literature review. Analysis of the quality of water produced by comparing the turbidity parameter of water produced with drinking water quality standard contained in Minister of Health Regulation No. 492 of 2010. The removal efficiency was calculated with equations below:

$$\text{Removal efficiency: } \frac{\text{Raw water turbidity-treated water turbidity}}{\text{raw water turbidity}} \times 100\% \quad (2.9)$$

3. Results and discussion

Treatment processes of Degreemont Package on Kedung Halang WTP consist of intake, coagulation, flocculation, sedimentation, filtration, disinfection, and reservoir, with an average production of 83.9 L/s as the raw water source is the Ciliwung River, which is pumped to Kedung Halang WTP. The Degreemont Package processing flow chart is shown in Figure 2.

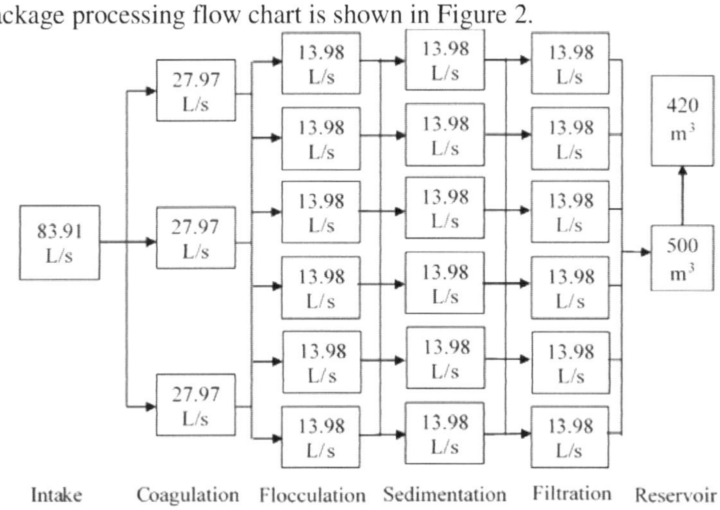

Figure 2. Processing flow of degreemont package.

3.1. Intake

Intake is a building made on the side of a water source, with the intention that some water from the source can be deflected used as desired [8]. On Kedung Halang WTP, the intake is equipped with a bar screen, sluice gate, intake well, and intake pump. There are four pumps with a maximum capacity of 50 L/s each. The water collected will be brought to WTP with a 300 mm diameter transmission pipe. The length of the transmission pipe is 120 m. The evaluation result of intake on Kedung Halang WTP can be seen in Table 1.

Table 1. Evaluation result of intake.

Parameter	Design Criteria	Source	Result	Remarks
Velocity in transmission pipe	0.9 – 1.8 m/s	Fair et al., 1968 [9]	1.51 m/s	appropriate

The evaluation result showed that the velocity in the transmission pipe had met the design criteria. This parameter was evaluated because the velocity that is too slow could cause deposition inside the transmission pipe, while velocity that is too fast could damage the pipe.

3.2. Coagulation

Coagulation, or rapid mixing, aims to disperse the added coagulant uniformly and trigger the collisions between coagulant and turbidity-causing particles [10]. The coagulant usually is a positively charged metal ion. Thus, it can neutralise the negative charges of colloidal particles, which are the cause of turbidity. Coagulants that are usually used are aluminium sulphate and Poly Aluminium Chloride (PAC) [11]. The type of coagulation used on the Degreemont Package is hydraulic mixing, where the water is dropped from a certain height so that it causes mixing power. The water is collected first in the receiving tank before being dropped to the coagulation tank. The coagulant used is aluminium sulphate, with an average dose of 29.10 mg/L. There are three coagulation tanks on Degreemont Package. The evaluation results of the coagulation process are shown in Table 2.

Table 2. Evaluation results of coagulation.

Parameter	**Design Criteria**	**Source**	**Result**	**Remarks**
Detention time	1 – 60 s	Schulz & Okun, 1984 [12]	10.69 s	appropriate
Gradient velocity	100 – 1000 /s	Qasim, 2000 [13]	594.06 /s	appropriate
G.Td value	$10^4 – 10^6$	Reynolds & Richards, 1996 [14]	6350.79	inappropriate

Gradient velocity and G.Td value should be calculated in coagulation and flocculation units because these numbers represent the mixing intensity. The evaluation results found that G.Td value on coagulation did not fulfil the design criteria. Theoretically, the G.Td value that is too low may cause the coagulant not to be wholly dispersed and eventually affect the floc formation process on flocculation. However, it did not cause any problem on the installation, as the flocs formed after flocculation are already large and dense.

3.3. Flocculation

The flocculation process is the aggregation of destabilising particles in the micro-flocs and larger particles called flocs. There are two types of flocculation, hydraulic and mechanical. The hydraulic flocculator is preferred in developing countries because of the ease of operation and maintenance [15]. There are six hydraulic flocculation units on Degreemont Package, with six rectangular-shaped compartments on each unit. The agitation occurs because of the opening between compartments. The evaluation results of the flocculation process are shown in Table 3.

Table 3. Evaluation results of flocculation.

Parameter	Design Criteria	Source	Result	Remarks
Detention time	10 - 60 min	Droste, 1977 [16]	15.85 min	appropriate
Gradient velocity	10 - 100 /s	Schulz & Okun, 1984 [12]	74.29 – 27.15/s	appropriate
G.Td value	10^4 - 10^6	Reynolds & Richards, 1996 [14]	44997.99	appropriate

The evaluation results showed that all the parameters on flocculation had fulfilled the design criteria. It indicates that the agitation on flocculation is slow enough to give the destabilised particle a chance to coalesce, but not too slow that it may cause sedimentation on the flocculation unit. The gradient velocity is gradually decreasing from one compartment to the next compartment, creating tapered flocculation. It intended to keep the flocs formed from breaking.

3.4. Sedimentation

Sedimentation is one of the most common particle separations. In water treatment, sedimentation aims to remove impurities, turbidity, and colour produced by the floc formed in the coagulation and flocculation process [17]. Plate or tube settlers could be installed in the sedimentation tank with a particular slope to increase the settling efficiency in the sedimentation tank [18].

There are six sedimentation units equipped with 60° slope tube settlers on Degreemont Package. Each unit has two sludge spaces at the bottom of the tank. The water is flowing out from the sedimentation tank using two gutters. The evaluation results of sedimentation are shown in Table 4.

Table 4. Evaluation results of sedimentation.

Parameter	Design Criteria	Source	Result	Remarks
Detention time	15 - 60 min	Darmasetiawan, 2004 [19]	24.08 min	appropriate
Surface loading	5.8 – 9.6 $m^3/m^2/h$	Schulz & Okun, 1984 [12]	6.18 $m^3/m^2/h$	appropriate
Reynolds Number	> 10^{-5}	Kawamura, 2000 [20]	2 x 10^{-5}	appropriate
Froude Number	$\leq$ 0.15	Kawamura, 2000 [20]	0,12	appropriate

The evaluation results showed that all the parameters on sedimentation had fulfilled the design criteria, creating an ideal situation for particles to settle. Surface loading is one of the essential parameters because when the value is too high, the particles would not settle ultimately and could burden the following unit. Detention time is calculated because its value is proportional to the number of particles that settle. Reynolds and Froude number that has fulfilled design criteria signify that the flow is calm and laminar so that the particles may settle.

3.5. Filtration
Filtration removes suspended and colloidal particles in an aqueous suspension that drains through a porous medium [21]. On Degreemont Package, six filtration units with dual media filters consist of anthracite and sand, with gravel below them as supporting media. The disinfection process is applied to the connecting pipe between the filtration unit and reservoir as a disinfectant is calcium hypochlorite, with an average dose of 1.48 mg/L. The evaluation results of filtration can be seen in Table 5.

Table 5. Evaluation results of filtration.

Parameter	Design Criteria	Source	Result	Remarks
Filtration rate	7.3 – 19.5 m/h	Reynolds & Richards, 1996 [14]	8.18 m/h	appropriate

The evaluation results showed that the filtration rate had fulfilled the design criteria. The filtration rate is the only parameter calculated infiltration because a filtration rate that is too high could reduce the quality of the water produced, while a filtration rate that is too low could decrease the effectiveness of the filtration process.

3.6. Optimisation of degreemont package
To optimise the Degreemont Package's production capacity, the maximum flow rate that each unit could process should be found. The optimisation is done by simulating various flow rates until the maximum value is found. The evaluation results on the maximum flow rate of each unit are shown in Table 6.

Table 6. Evaluation results on maximum flow rate on each unit.

Parameter	Maximum Flow Rate	Design Criteria	Existing Condition	Result	Remarks
		Intake			
Pipe velocity	125 L/s	0.9 – 1.8 m/s	1.51 m/s	1.77 m/s	appropriate
		Coagulation			
Detention time		1 - 60 s	10.69 s	9.2 s	appropriate
Gradient Velocity	150 L/s	100 - 1000 /s	594.06 /s	705.03 /s	appropriate
G.Td value		10^4 - 10^6	6350.79	6486.3	appropriate
		Flocculation			
Detention time	130 L/s	10 - 60 min	15.85 min	10.23 min	appropriate

Parameter	Maximum Flow Rate	Design Criteria	Existing Condition	Result	Remarks
Gradient Velocity		10 - 100 /s	74.29 – 27.15/s	92.47 – 33.85/s	appropriate
G.Td value		10^4 - 10^6	44997.99	36496.49	appropriate
		Sedimentation			
Detention time	130 L/s	15 - 60 min	24.08 min	15.54 min	appropriate
Surface loading		5.8 – 9.6 $m^3/m^2/h$	6.18 $m^3/m^2/h$	9.58 $m^3/m^2/h$	appropriate
Reynolds Number		> 10^{-5}	2 x 10^{-5}	71.64	appropriate
Froude Number		≤ 0.15	0.12	4.8 x 10^{-5}	appropriate
		Filtration			
Filtration rate	200 L/s	7.3 – 19.5 m/h	8.18 m/h	19.49 m/h	appropriate

From Table 6, it was shown that one of the parameters, which is the G.Td value of coagulation, did not fulfil the design criteria. It could happen because the G.Td value on the existing condition did not fulfil the design criteria from the beginning. However, because this does not cause any problem in the existing condition, the lowest G.Td value limiting the maximum flow rate is set on the G.Td value on the existing condition, 6350.79. Based on Table 6, the maximum flow rate on Degreemont Package is 125 L/s. The evaluation result with a maximum flow rate of 125 L/s can be seen in Table 7.

Table 7. Evaluation results on 125 l/s flow rate.

Parameter	Design Criteria	Existing Condition	Result	Remarks
		Intake		
Pipe velocity	0.9 – 1.8 m/s	1.51 m/s	1.77 m/s	appropriate
		Coagulation		
Detention time	1 – 60 s	10.69 s	11.04 s	appropriate
Gradient Velocity	100 – 1000 /s	594.06 /s	996.54 /s	appropriate
G.Td value	10^4 – 10^6	6350.79	7105.39	appropriate
		Flocculation		
Detention time	10 – 60 min	15.85 min	10.64 min	appropriate
Gradient Velocity	10 – 100 /s	74.29 – 27.15 /s	90.68 – 33.13 /s	appropriate
G.Td value	10^4 – 10^6	44997.99	36866.9	appropriate
		Sedimentation		
Detention time	15 – 60 min	24.08 min	16.16	appropriate
Surface loading	5.8 – 9.6 $m^3/m^2/h$	6.18 $m^3/m^2/h$	9.21 $m^3/m^2/h$	appropriate
Reynolds Number	> 10^{-5}	2 x 10^{-5}	68.88	appropriate
Froude Number	≤ 0.15	0.12	4.5 x 10^{-5}	appropriate
		Filtration		
Filtration rate	7.3 – 19.5 m/h	8.18 m/h	12.18 m/h	appropriate

From Table 7, it was obtained that the evaluation results of all parameters are appropriate. This show that the flow rate of 125 L/s is possible to be applied on the Degreemont Package.

3.7. Quality of water produced

The quality of the water produced and efficiency removal based on turbidity parameter on Degreemont Package can be seen in Table 8. Water turbidity measurements were carried out throughout March 2021. Table 8 show that the water produced has fulfilled drinking water standard according to Minister of Health Regulation No. 492 of 2010 [5] based on turbidity parameter, which is below 5 NTU.

Table 8. Turbidity and removal efficiency of degreemont package.

Description	Turbidity (NTU)		Removal Efficiency
	Raw Water	Treated Water	
Minimum	21.92	0.33	96.95%
Maximum	470.67	0.93	99.86%
Average	95.58	0.54	98.81%

4. Conclusion

Degreemont Treatment Package on Kedung Halang WTP has an average production of 83.9 L/s, with treatment processes consist of intake, coagulation, flocculation, sedimentation, filtration, disinfection, and reservoir. On existing conditions, the value of G.Td on coagulation did not fulfil the design criteria. However, this did not cause any problem on installation since the flocs formed were already large and dense enough. The maximum flow rate that can be operated on each unit is 125 L/s on intake, 150 L/s on coagulation, 130 L/s on flocculation and sedimentation, and 200 L/s on filtration. This show that the optimisation of the Degreemont Package could increase the processing capacity to 125 L/s from the initial design capacity of 90 L/s. The quality of water produced has fulfilled drinking water standards based on Minister of Health Regulation No. 492 of 2010, with turbidity on treated water ranged from 0.33 to 0.93 NTU and average removal efficiency of 98.81%. Therefore, it can be said that the Degreemont Package is qualified as a drinking water treatment.

References

[1] Thomas J -S 2003 Integrated Water Resource Management: looking at the whole picture. *Desalination* **156** 21-28

[2] UNESCO 2021 *The United Nations World Water Development Report 2021: Valuing Water* (The United Nations Educational, Scientific and Cultural Organization)

[3] Jannah F H S, Kusumadewi R A and Ratnaningsih R 2020 *International Journal of Scientific & Technology Research* **9**(01) 3500-05

[4] Farhaoui M and Derraz M 2016 *Journal of Water Resource and Protection* **8** 777-786

[5] Minister of Health Regulation of Republic Indonesia No. 492 of 2010

[6] Ramadhan F, Siami L and Winarni 2019 *Prosiding Seminar Nasional Pembangunan Wilayah dan Kota Berkelanjutan*

[7] Sani I K, Winarni and Kusumadewi R A 2020 *IOP Conference Series: Earth and Environmental Science* **426**

[8] Silotanga B and Hendry 2018 Perencanaan Hidrolis Pintu Pada Bangunan Pengambilan Air (Intake). *Jurnal Rekayasa Konstruksi Mekanika Sipil* **1(2)** ISSN 2614-5707

[9] Fair G M, Geyer J C and Okun D A 1968 *Water and Wastewater Engineering Vol. 2. Water Purification and Wastewater Treatment and Disposal* (John Wiley and Sons, Inc.)

[10] Winarni, Iswanto B and Karina C 2011 *Indonesian Journal of Urban and Environmental Technology* **5(6)**

[11] Aziz N, Effendy N and Basuki K T 2017 *Inovasi Teknik Kimia* **2(1)** 24-31

[12] Schulz C A and Okun D A 1984 *Surface Water Treatment for Communities in Developing Countries* (WASH Technical Report No. 29, Water and Sanitation for Health (WASH) Project of the United States Agency for International Development)

[13] Qasim S R and Motley G Z 2000. *Water Works Engineering: Planning, Design, and Operation.* (Prentice-Hall, Inc.)
[14] Reynolds T D and Richards P A 1996 *Unit Operations and Processes in Environmental Engineering (2nd ed.)* (PWS Publishing Company)
[15] Ghawi A H 2018 *Scientific Review – Engineering and Environmental Science* **27**(4) 438-451
[16] Droste R L 1997 *Theory and Practice of Water and Wastewater Treatment* (John Wiley and Sons, Inc.)
[17] Ghawi A H and Jasem Y I 2018 *Diyala Journal of Engineering Sciences* **11(1)** 33-37
[18] Crittenden J, Trussell R, Hand D, Howe, K and Tchobanoglous G 2012 *MWH's Water Treatment: Principles and Designs (3rd ed.)* (John Wiley & Sons, Inc.)
[19] Darmasetiawan M 2004 *Teori dan Perencanaan Instalasi Pengolahan Air* (Jakarta: Ekamitra Engineering)
[20] Kawamura S 2000 *Integrated Design and Operation of Water Treatment Facilities (2nd ed.)* (John Wiley & Sons, Inc.)
[21] Fernando G -A et al 2020 *Water-Energy Nexus* **3** 1-10

Comparison of pour point depressant (PPD) ethylene-vinyl acetate (EVA) and nano-montmorillonite

Z D Nurfajrin, B Satiyawira, R A Putra

Universitas Trisakti - Jl. Kyai Tapa No.1, Grogol Jakarta 11440, Indonesia

zakiahdarajat.zdn@trisakti.ac.id

Abstract. One of the problems with oil production is stuck of oil transportation flow in the pipeline caused by wax deposits. The high wax content in crude oil causes the oil viscosity value to increase so that crude oil has High Pour Point Oil (HPPO) properties. In this research, using crude oil sample with initial pour point of 31°C and viscosity of 556 cP. This samples are classified as heavy oil. Therefore, this sample is suitable for use as an experiment in this research. There are several methods to deal with HPPO. One chemical method used to overcome this problem is adding Pour Point Depressant (PPD) to crude oil. This PPD functions to decrease pour point value by binding wax crystals in crude oil. In this study, two types of PPD were compared, Ethylene-Vinyl Acetate (EVA) and Nano-Montmorillonite (Nano-MMT). Both can reduce the pour point value on crude oil, but PPD EVA shows a more excellent reduction than nano-MMT. The results show that EVA is better at binding crystals than nano-MMT. After the injection of 400 ppm EVA, the pour point value of crude oil could be decreased up to 24°C, and the viscosity also decreased by 185 cP.

1. Introduction

In general, the type of crude oil in Indonesia is a heavy oil classified as high pour point, some examples of crude oil with the HPPO category can be seen in Table 1. With the highest pour point value at a temperature of 39 °C. Operating conditions below the pour point temperature will impact the production shift that causes it to record the flow due to blockage from the wax deposits in the pipeline to the crude oil storage. If this happens, then the volatile components contained in crude oil will be evaporated so that the concentration of heavy fraction of crude oil increases. It is caused by a decrease in pressure which causes the driving efficiency and flow of heavy fractions to decrease. The flow of petroleum causes the possibility of forming wax deposits quickly because the crude oil contains wax or paraffin, asphaltene, and resin.

Table 1. Examples of petroleum fields in Indonesia that have HPPO characteristics.

No.	Name of Fields	Wax Content (%)	Pour point (°C)	References
1.	Pinag, Riau	n.a	39	[1]
2.	Minas, Riau	10.8	36 – 38	[2]
3.	Duri, Riau	16	22	[2]
4.	Sepanjang, East Java	24.9	49	[3]

No.	Name of Fields	Wax Content (%)	Pour point (°C)	References
5.	Udang, Natuna	n.a	38	[4]
6.	Kuda Laut, Natuna	n.a	38	[4]
7.	Handil, East Kalimantan	16.5	32	[4]
8.	Tanjung, Eat Kalimantan	31.9	38	[4]
9.	Cinta, West Java	30	41	[2]
10.	Tapian Timur, South Kalimantan	6.7 – 13.5	n.a	[5]

Paraffin or wax is a colorless, odorless, crystalline substance that can be solid or semi-solid. Paraffin is not easy to react with other chemical compounds (inert), but at high temperatures, a small part will be oxidized or broken (cracking), insoluble in water and alcohol but soluble in petroleum and benzene fractions. Paraffin is a highly saturated hydrocarbon compound (paraffin). In the distillation process, it is also distilled after gas oil. The resulting wax is classified into several types according to the melting point: 45-52°C, 55-57°C, and 63-66°C. Two main parameters affect the solubility of wax in oil under ambient conditions, namely temperature, and composition, while pressure has a minimal effect on wax formation in oil compared to the two parameters above. The flow of crude oil containing wax, with several indicators analyzed such as Wax Appearance Temperature (WAT), pour point temperature (PP) or cloud point temperature (CP), and gel strength.

This wax component can be dissolved in petroleum (crude oil) and condensate in a liquid phase. The solubility of paraffin wax is very sensitive to temperature changes. Temperature changes are factors that affect the process of formation of wax crystals. Paraffin wax remains dissolved in crude oil while in the reservoir and is in equilibrium with petroleum in thermodynamics. Similar to the asphaltene deposition event, when the thermodynamic equilibrium begins to be disturbed, such as a change in temperature or pressure, the paraffin will crystallize or begin to precipitate due to the loss of the volatile light end in petroleum, where the volatile fraction in petroleum seems to be -if acts as a solvent for paraffin wax. When this mixed fluid begins to cool, each wax component will separate (become insoluble) until finally, the wax component with a high molecular weight will solidify (solidify). The event where the wax crystals first form at a specific temperature is called the onset of wax crystallization or better known as the cloud point or Wax Appearance Temperature (WAT). When the reservoir fluid temperature drops to a temperature T, paraffin hydrocarbons with a solidification temperature greater than T will tend to precipitate and separate from the solution. The following is a schematic illustration of the thermodynamics of a wax deposit.

The P-T of crude oil describes about the process of wax formation, where the blue area with a low temperature is the area for the formation of wax. Provides an interesting review of the wax deposit measurement technique and states that there is a serious problem in the field with the presence of wax during production, so that plugging is often encountered in the tubing pipeline and some areas—surface production equipment [6]. Due to too many wax deposits in the piping system, pigging activities will be carried out more often. Wax crystals, when they appear, will change the flow behavior of an oil fluid from Newtonian to non-Newtonian conditions. Wax crystals will also cause the oil's viscosity to flow in the pipeline by increasing energy consumption and decreasing pumping capacity. In addition, wax deposits also increase the pipe's roughness and reduce the pipe's cross-sectional area, resulting in increased pressure drop in the pipeline system.

Three factors contribute to the formation of wax deposits in flowing systems including flow rate, temperature difference, cooling rate, and the surface condition [7]. The wax can form a precipitate or crystal as the flow temperature decreases below the cloud point or Wax Appearance Temperature (WAT). This phenomenon can impede flow in a piping system by triggering non-Newtonian fluid behavior. The viscosity of petroleum can increase as the operating temperature reaches the pour point (the minimum temperature at which the liquid phase becomes semi-solid and can flow). Therefore, the newly deposited paraffin wax layer begins to stick, which can further narrow the area of the pipe [8]. Of course, this can cause various technical problems, so it is necessary to prevent wax deposition in the piping system. One way that can be humiliated in dealing with wax deposition is to use Pour Point Depressant (PPD) as a chemical method that aims to reduce the value of the pour point and viscosity of crude oil with several advantages, such as saving costs, energy, and production time [9,10,11].

Wax inhibitors, commonly referred to as PPD (Pour point depressants), need to be added to crude oil before the wax crystallizes, which is wax inhibitors can only work effectively in a very narrow range, which also depends on the composition of the oil. Therefore, it is necessary to design different treatments for each field with different petroleum compositions. Fang, 2012 describes a strong relationship between the specific efficiency of PPD and the composition of crude oil. PPD can effectively change the structure and morphology of wax crystals to prevent wax deposits that can form a rigid three-dimensional network at low temperatures. Conventional PPDs are homo- or copolymers of a wide variety of monomers. Ethylene-vinyl acetate (EVA) copolymer is the most commonly used copolymer. In modern theory, the mechanism of EVA as PPD includes adsorption, co-crystallization, nucleation, and improving the solubility of wax in petroleum. The higher the vinyl acetate composition increased, the more polarity of the polymer and dissolved in the oil. However, increasing vinyl acetate can reduce co-crystallization in PPD with wax. EVA with 18-30% vinyl acetate is commonly used as PPD [13,14].

EVA (Ethyl Vinyl Acetate) is a copolymer made from two different types of monomers, namely ethylene and vinyl acetate. The first ethylene copolymers, including EVA, were synthesized and patented in 1930 IC-1 Great Britain in the same laboratory where polyethylene was discovered [14]. EVA copolymers are synthesized through the copolymerization process of ethylene and vinyl acetate at high pressure and temperature and by the bulk polymerization process [9]. Bulk polymerization is the most frequently used technique for producing EVA copolymers containing up to 50% vinyl acetate. Another copolymer used as PPD is nano-montmorillonite (MMT). The commonly used method is to use pure EVA or another PPD, namely nano-montmorillonite (MMT). A new generation of PPDs based on nanotechnology, including nanocomposites, has attracted much attention in recent years. Nano-silica and modified montmorillonite nano-clay have been very populated in recent studies [15]. Montmorillonite is a type of clay that is widely found in bentonite rocks with the chemical formula (Na, Ca)(Al, Mg)$_6$ (Si_4O_{10})$_3$ (OH)$_6$-nH_2O (Hydrated Sodium Calcium Aluminum-Magnesium Silicate Hydroxide) [16]. See the advantages of PPD EVA and MMT in reducing pour point and viscosity values of crude oil, with high pour point and viscosity values. so that we will compare the two performances of the PPD in order to increase the effectiveness of the production [17,18,19]

2. Methodology

Look at The advantages of PPD EVA and MMT in reducing pour point and viscosity values of crude oil, with high pour point and viscosity values. Therefore, in this study, we will compare this study. The following tools and materials were used, Viscometer Ostwald Cannon-Fenske for Transparent Liquid, measuring cup, oil bath, Cleveland Open Cup Apparatus, Magnetic Stirrer, Stirrer rod, sample crude oil, PPD Ethylene-Vinyl Acetate, PPD nano-Montmorillonite. Meanwhile, the initial step is to heat the crude oil sample to a temperature of 50oC, then pour it into 200 ml cylindrical bottles. Then add PPD into crude oil, with variables PPD EVA and MMT with the same dose of 0 ppm, 200 ppm, 300 ppm, 400 ppm, and 500 ppm. Next, pour point and viscosity analysis was performed on the oil sample—the two performances of the PPD in order to increase the effectiveness of the production.

For the preparation of PPD EVA material which has a melting point temperature of 80°C – 120°C, which means that for this initial step, the EVA that will be used in the study is up to a temperature above the melting point for 2 hours so that all EVA compounds can be ensured to melt completely. Then EVA is dissolved into petroleum with a predetermined variable dose. The initial treatment that needs to be done for PPD MMT is to put MMT into a beaker glass, pour toluene into a beaker glass that already contains MMT, then stir using a magnetic stirrer while heated at 80°C for 2 hours. After two hours, continued heating at a temperature of 120°C until the toluene in the beaker evaporated. After that, nano-MMT is obtained, which is ready to be mixed into the crude oil sample.

2.1. Pour point analysis

Pour point is defined as a temperature point when oil does not flow when cooled under certain conditions. The crude oil sample is placed in a closed tube then the tube is inserted into the cooling mixture (substance) where there is no direct contact between the tube containing the oil and the refrigerant used, namely ice and water [20]. Determination of pour point can be used equation,

$$\text{Pour Point} = \text{Temperature} + 3 \quad (1)$$

Cleveland open cup apparatus for pour point measurement on oil samples have a several part, there are:

1. Sample to be tested
2. Thermometer
3. Pour point detector
4. Test jar
5. Cooling medium
6. Inlet cooling medium
7. Outlet cooling medium
8. Jacket
9. Temperature detector of cooling medium
10. Air cavity

Firstly, the oil sample is filtered from dissolved substances in it can be removed, and then the sample is inserted into a test cylinder which is equipped with a cork and a thermometer to the mark. The capillary tip of the thermometer should be submerged approximately 3 mm from the surface of the oil sample. Furthermore, the test cylinder is put into a cooling bath with a temperature far below the pour point temperature of the oil sample. Every 3°C decrease in temperature, the biodiesel change was observed by flowing the oil sample horizontally. If there was no movement for 5 seconds, then this value was the measured pour point value.

2.2. Viscosity analysis

The fluid's viscosity is related to the resistance to the forces that move the fluid in one layer against another. The viscosity produced by force in the dyne striking two horizontal planes a distance of 1 unit between them flows a viscous fluid. The unit of viscosity is dyne second/cm^2 which is proportional to 1 poise (dynamic unit), while the kinematic viscosity has units of cm^2/second. Oil viscosity is highly dependent on temperature pressure. While the tool used to determine the viscosity is the Ostwald Viscometer type Viscometer Cannon-Fenske for Transparent Liquid. The sample to be tested for viscosity is inserted into the Ostwald viscometer, then blows the sample using a rubber bulb until the sample is at the top of the boundary mark, then the sample is allowed to drop, and records the time it takes to pass the two marks. The time it takes to pass that mark will be recorded as "t." This time is used to measure the viscosity of the sample solution at each concentration. Clean the viscometer tube with toluene, dry it and blow it with air to clean it from toluene.

3. Result and discussion

The crude oil sample used has properties with a property in high pour point value or can also be referred to as (High Pour Point Oil) of 31°C, and other physical characteristics can be seen in the Table 2.

Table 2. Properties of crude oil sample.

Properties	Methods	Analysis Result
Temperature of measurement		30 °C
SG		0.844
Density at 15°C	ASTM D-1298	0.8539
API Gravity at 60 °F	ASTM D-187	33.71
Water content (%vol)	ASTM D-95	-
BS&W (%vol)		0.2
Pour point	ASTM D-96	31 °C
Viscosity		556 cP

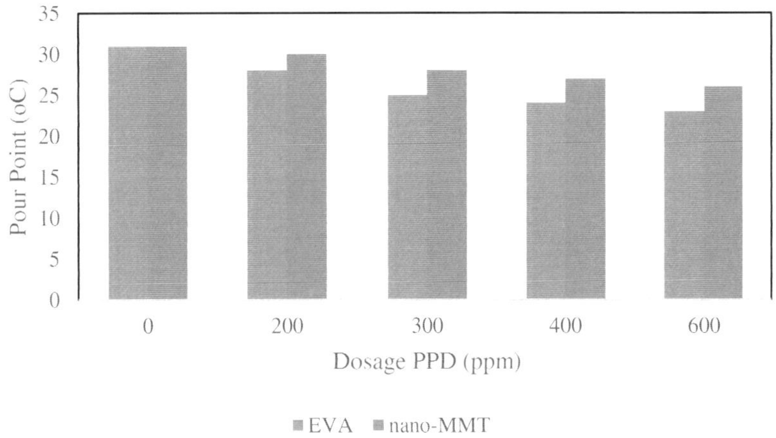

Figure 1. The relationship between EVA and nano-MMT PPD Dosage on Pour point in crude oil samples.

Figure 1 shows the effect of EVA and nano-Montmorillonite copolymer PPD doses, which reduced the pour point value significantly. So it can be concluded that the Pour point value decreases with the addition of pour point depressant EVA chemicals. The addition of a bit of chemical pour point depressant EVA can reduce the viscosity value of crude oil. The alkyl group causes it on the pour point depressant absorbs the wax crystal content in the oil and prevents the growth of wax crystal formation; in addition to the addition of chemical pour point depressant prevents the absorption of oil by the wax crystals. Wax can become crystals due to paraffin, which interacts and lock into each other at low temperatures.

Meanwhile, the PPD nano-MMT succeeded in reducing the pour point value in the crude oil sample. The pour point value of the crude oil samples decreased from 31°C to 30°C, 28°C, 25°C, and 24°C. These phenomena can happen because the nano-MMT molecules absorb the wax crystal content in crude oil and prevent hardened wax deposits from forming, hindering the transportation of crude oil.

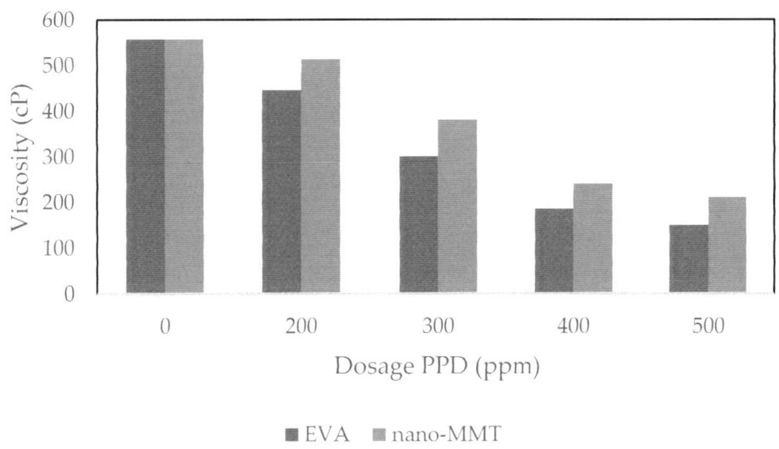

Figure 2. The relationship between EVA and nano-Montmorillonite PPD Dosage on the viscosity of crude oil samples

Viscosity is a measure of the viscous fluid which states the size of the impact in the liquid. The greater the viscosity value in a fluid, the more difficult it is for a fluid to flow and the harder it is for an object to move within it. In Figure 2 shows us that both of PPDs can reduce the viscosity of the crude oil sample. At a dose of 200 ppm, the viscosity is decreased to 445 cP for EVA PPD and 512 cP for nano-MMT PPD. Meanwhile, at 300 pm, there was a decrease in viscosity to 300 cP for PPD EVA and 380 cP for PPD nano-MMT. Then at a dose of 400 ppm, there was a decrease in viscosity to 185 cP for PPD EVA and 240 cP for PPD nano-MMT. Furthermore, at a dose of 500 ppm, there was a decrease in viscosity to 150 cP for PPD EVA and 210 cP for PPD nano-MMT. It can be concluded that there is a decrease in every PPD dose added. It is caused by the presence of PPD in the sample, thereby reducing the viscosity value.

The decrease in the viscosity value shown in nano-MMT was not as significant as the decrease in the value indicated by EVA because EVA molecules are better at binding wax crystals than nano-MMT. The wax crystals' content, abundant in crude oil, is the leading cause of hardened wax crystal deposits and clogging transportation pipelines. When PPD is mixed in oil samples, PPD will bind to wax crystals, preventing precipitation. In addition to PPD EVA to pour point and viscosity, the higher the dose of PPD given, the lower the viscosity and pour point of the sample. In general, the ambient temperature in the piping system is around 27-30oC, which means that pour point temperatures below 27oC can be recommended. Then the PPD EVA dose of 400 ppm can be chosen to be advised, which sees the pour point value, which decreases by 26%, and viscosity by 185 cP with a decrease of 67%, where crude oil with a viscosity can flow well. While the addition of PPD nano-MMT to the pour point and also the viscosity of the sample. The more doses added to the oil mixture, the lower the pour point and viscosity. Similar to EVA, injection at a dose of 400 ppm is recommended because it is seen from the 19% decrease in pour point and 57% decrease in viscosity which is the best result among the other doses. In addition to PPD EVA to pour point and also viscosity.

4. Conclusion

The viscosity value of the oil will decrease according to the number of doses of PPD mixed into the oil. more PPD molecules because it binds to wax crystals in the oil content. PPD ethylene-vinyl acetate (EVA) showed better results in lowering the pour point of oil than nano-Montmorillonite (nano-MMT). Crude oil added the optimum dose of PPD EVA at 400 ppm with a decrease in pour point of 26% and a decrease in viscosity of 67%.

References

[1] Tang Y, Shuler P J, Cheug S K, Goodgame J A, Hsu J J, and Padilla A V 2003 *in SPE International Symposium on Oilfield Chemistry* Texas, 2003.
[2] Environment Canada Database 2001 Available from: http://www.etc-cte.ec.gc.ca/databases/Oilproperties.
[3] Kitamura T, Zaki H, Akbar K, Palao F B, Sosa JCL, Toralde JSS, and Nas S 2010 *in SPE/IADC Managed Pressure Drilling and Underbalanced Operations Conference and Exhibition*, Kuala Lumpur.
[4] Irani CA and J. Zajac 1981 *in 58th Annual Fall Technical Conference and Exhibition of the Society of Petroleum Engineers of AIME,* Texas.
[5] Hardikin M I, Indriyono E S, and Hariyono 2009 *in Simposium Nasional IATMI,* (Bandung, 2009).
[6] Kosta L, Efsratios G 2011 *Offshore Technology Conference* (Texas)
[7] Bott TR, and Gudmundsson JS 1977 *Deposition of Parafin Wax from Flowing System* (Chemical Engineering Department, University of Birmingham. Birmingham)
[8] Yaang F, Zhao Y, Sjonlom J, Li C, Paso K G 2014 *Journal of Dispersion Science and Technology* **36**(2), 213-215
[9] Chala G T, Sulaiman S A, Japper-Jaafar A 2018 *Journal of Non-Newtonian Fluid Mechanism* **251** 69-87

IOP Conf. Series: Earth and Environmental Science **894** (2021) 012019 doi:10.1088/1755-1315/894/1/012019

[10] Kazemi M A, Nazar A R S, Karimi Y 2020 *Journal of Petroleum Science and Technology* **10,** 67-74

[11] Rehan M, Nizami A S, Taylan O, AL-sasi B O, Demirbas A 2016 *Petroleum Science and Technology* **34(9),** 799-804

[12] Fang L, Zhang X, Ma J, Zhang B 2012 *Ind Eng Chem Res* **51**: 11605–11612.

[13] Marenov K S, Nadirov M K, Zhantasov R K, Nadirov 2020 *Hindawi International Journal of Chemical Engineering* (Kazakhstan)

[14] White M, Pierce K, Acharya T 2018 *SPE Production e-operation* **33(3)** 476-485

[15] Kazemi M A, Nazar A R S, Karimi Y 2020 *Journal of Petroleum Science and Technology* **10** 67-74

[16] Fisli A and Yusuf S 2007 *Sintesis Nanokomposit Magnetik Berbasis Bahan Alam Untuk Adsorben Thorium* LIPI No 536/D/2007

[17] Sousa A L, Matos H A, Guerreiro L P 2019 *Journal of petroleum exploration and production technology* **9,** 2091-2107

[18] Singh P, Venkatesan R, Fogler H S and Nagarajan N 2000 *AIChE Journal* **46(5)**, 1059–1074

[19] Garcıa M d C 2000 *Energy & Fuels* **14**(5), 1043–1048

[20] Yully M, Dian A, Shamayanthie I 2016 *The 2nd Conference on Innovation and Industrial Aplication.*

Saline water desalination using solar energy: performance analysis and implementation barriers

R A Kusumadewi[1], Y Wijayanti[2*], M Fittkow[3]

[1]Department of Environmental Engineering, Faculty of Engineering, University of Trisakti, Jakarta, Indonesia
[2]Department of Civil Engineering, Faculty of Engineering, University of Bina Nusantara, Jakarta, Indonesia
[3]Department of Economics, Hochschule Ruhr West, Mulheim, Germany

yureana.wijayanti@binus.ac.id

Abstract. Solar water desalination system is one of the alternatives for supplying fresh water. On the other hand, some people are reluctant to use and rely on water produced by the system. Also, there are other barriers to the adoption of this system in Indonesia. This research aimed to conduct a preliminary study on the performance of solar desalination systems and identify factors that hindered its implementation. The methodology applied in this study consist of two approaches. First, the desalination system performance is determined by the quantity of freshwater produced and the quality of feed water, distilled water, and brine compared to Indonesia's water quality standard. Second, the identification of implementation barriers was evaluated in the structural analysis method using MICMAC (the Impact Matrix Cross-Reference Multiplication Applied to a Classification) software. The performance analysis result shows low efficiency based on desalinated water produced, which ranges from 2.59% up to 9.67%. The water quality parameters of pH, turbidity, TDS, chloride, Fe, and hardness met the Indonesian water quality standard. The factor of "uncertainty of subsidies," "niche market," "poor solar energy data," and "lack of policy/regulatory framework" were determined as critical barriers which have the most significant influence and highest independence.

1. Introduction

The process of processing seawater into water is ready to be one of the solutions for providing clean water. The process of processing seawater into freshwater is known as the desalination process [1]. The whole desalination process requires energy to remove salt from seawater. If desalination is carried out using conventional technology, it will require burning large amounts of fossil fuels (the production of 1000 m^3/day of clean water requires 10,000 tons of oil per year). As the availability of these fuels is decreasing, then other energy sources are needed. One energy source for desalination is using solar power [2]. It is based on solar energy being renewable energy, safe, accessible, and pollution-free. Solar power is a promising solution to save costs. In addition, Indonesia is a country that has abundant solar intensity, which is an average of 4.8 kWh/m^2.day [3].

Solar desalination can be divided into direct and indirect technologies. In the direct method, the solar energy collector and desalination component are integral units, for example, a solar still. A general rule of thumb for simple solar stills is that about 1 m^2 of sunlight collection area must produce 3 - 5 L of water per day. Thus, a large sunlight collection area is required at a high capital cost. Multi-effect solar stills are used to improve the performance of solar stills [4]. Solar water desalination system is one of the alternatives for supplying fresh water.

On the other hand, some people are reluctant to use and rely on water produced by the system. Also, there are other barriers to the adoption of this system in Indonesia. A transition to renewable energy is a complex process that involves technical [5,6], social, political, and economic factors [7], and environmental [8] aspects. Therefore, this research aimed to conduct a preliminary study on solar desalination systems' performance and identify factors that hindered its implementation.

Several studies that used single solar still have been conducted. Safitri [9] reported that distilled water produced was 0.2 – 0.9 L/m^2 with an evaporation area of 0.75 m^2, measurement duration of 4 hours, and collector inclination angle of 30°. Other researchers used the evaporator and condenser under vacuum conditions to treat saline water. Kusumadewi *et al.* [10, 11] reported that distilled water produced was 4 – 11.6 L/m^2 with an evaporation area of 0.0576 m^2, measurement of 8 hours, and collector inclination angle of 30°. This paper reports on the performance of multiple trays tilted still, which is the development of a desalination unit made by Safitri [9]. In this research, a styrofoam layer was used, which functions as a heat insulator to remain in the distillatory box. The distillatory box was covered with glass with a thickness of 5 mm and dimensions of 85 cm × 100 cm with silicon to form a line pattern as the distillate water effluent system. The distillatory box was tilted at an angle of 25°. There are 5 levels with an estuary on the gutter made of a 1.5 inch PVC pipe to drain brine water.

2. Methodology

It was a saline water desalination using solar energy study conducted using experimental work at Environmental Engineering Laboratory, University of Trisakti, Jakarta. A modified MICMAC has been used to analyse enablers and barriers factors in renewable energy studies [8,12]. This study will use free software LIPSOR-EPITA-MICMAC version [13].

2.1. Experimental work

2.1.1. Water distillation system. This distillatory design results from the development of the multiple trays tilted still type distillatory in a previous study [10,11]. In this study, there is an addition of absorbent material in the form of black cloth. The purpose of using a fabric absorber in this design is to increase the absorption surface area, accelerating the evaporation rate. The capillarity effect given by the black cloth as an absorber allows the air in the layer to spread evenly across the surface of the distillation. It is composed of a water input reservoir, where the distillation process occurs, then the distillate water and the resulting brine are accommodated using a container (Figure 1).

The distillatory body is composed of plywood measuring 85cmx100cmx15cm for width, length and height, respectively. It covers with aluminium and black cloth as the heat absorber. The cover of the distillatory is clear glass with a thickness of 5 mm. After the body is completed, it is covered with an insulator in the form of styrofoam measuring 4 cm. Then the distillatory body is placed on an iron frame with a distillatory tilt position of 25°. This distillatory has a distributor pipe. Unevaporated water (brine) contains high levels of salt. This brine also has a high heat that can heat the input water or feed water. The water outcome of the distillatory is then measured and recorded every 1 hour. So, the efficiency of the distillatory can be analysed and calculated. The distillatory test was carried out from 9.00 WIB to 15.00 WIB in the Pademangan Village area, East Pademangan District, North Jakarta. Water quality testing from the desalination process was conducted at the Trisakti University Environmental Laboratory, West Jakarta, on November 10 - 15, 2020. The ambient air temperature, wind speed and humidity range were 30.6 – 34.2°C, 0.2 – 2.7 m/s, and 51 – 71%Rh, respectively.

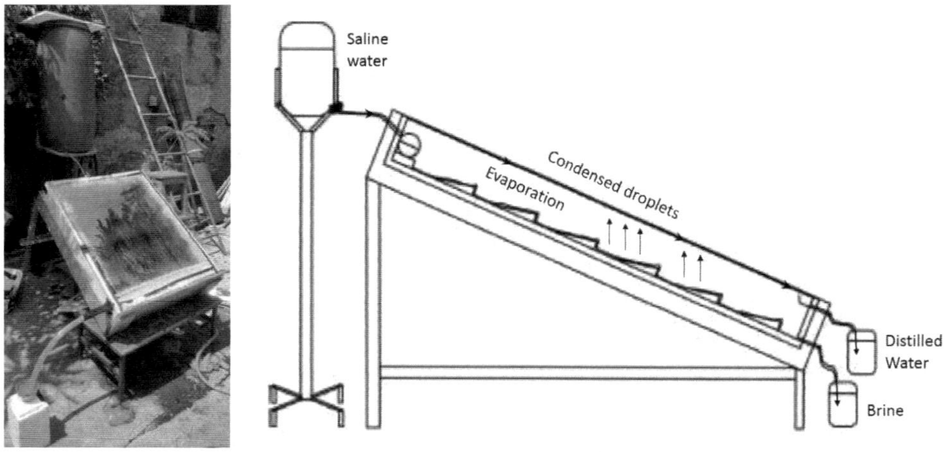

(a) (b)

Figure 1. Experimental set-up (a) and schematic diagram (b) of the multiple trays tilted still distillatory.

2.1.2. The efficiency of the distillation system. This study compares the theoretical efficiency and the actual efficiency of the distillatory [14]. The actual efficiency is calculated using equation (1), the ratio of the volume of water leaving and entering the distillatory. The theoretical efficiency was calculated by using equations (2), (3) and (4), as follows [15]:
Efficiency:

$$\frac{\text{Volume out}}{\text{Volume in}} \text{x} 100\% \tag{2.1}$$

$$\eta_c = \frac{Q_u}{Q_{in}} . 100\% \tag{2.2}$$

Where Q_u is the energy supplied by the solar collector (Watt), and Q_{in} is the radiation energy in the absorbing plate (Watt).

$$Q_u = m.C_p.\Delta T + m.h_{fg} \tag{2.3}$$

Where m is the mass of distilled water (litre), C_p equal to 4180 J/kgK [16], ΔT is the different temperature between feed water and brine, h_{fg} is the enthalpy of evaporation of water which equals 22.6 105 J/kg.°C [16]

$$Q_{in} = \alpha.IT_{tot}.A_c \tag{2.4}$$

Where IT is solar intensity (W/m^2), Ac is an area of the absorbing plate or solar collector (m^2), α is absorptivity of the solar collector (0.50)

2.2. Implementation barriers identification

2.2.1. Data collection. Identifying barriers that could hinder the realisation performed through literature review. A literature review is an excellent approach to form a solid foundation and integral part of research [17]. Due to insufficient literature in the context of Indonesia rural sector, this study utilised references from other developing countries.

Table 1. Barrier Factors [8,18,19,20,21,22,23].

Dimension	Barrier	Description
Economic	High Initial Capital Cost (E1)	Cost of installation for the distillation system
	Uncertainty of subsidies (E2)	The subsidies that available from the government to cover the initial installation of the distillation system
	Absence of financing institution (E3)	The companies that are interested to invest
Technology	poor solar energy data (T1)	Reliable information to design distillation technology may influence Analysis of cost and efficiency.
	Unskilled professionals (T2)	The distillation system needs a skilled professional in repairing and maintaining the system.
Social	the reluctance of people (S1)	People perceive that wave energy has a significant dependence on weather conditions. So, it raised doubt on its reliability.
	Niche market (S2)	There is uncertainty on potential consumers and the availability of information and knowledge about markets.
Political	Lack of Policy/ regulatory framework (P1)	Assurance of solid enforcement of policies, precise regulatory mechanism and legal support

2.2.2. Impact matrix multiplication applied to classification (MICMAC). MICMAC technique is conducted by mapping variables into a relationship of the influence variables and dependent variables in the matrix [24,25]. The procedures consist of three main steps: classification of the variables, evaluation of the correlation among the variables, and determination of the critical variables. In this study, the classification of the variables is performed to categorise the model under economic, technology, social and political dimensions. Then the Analysis enables the researcher to understand the distribution of the impacts by using the loops and paths [26] using the iteration process. This process simulates the behaviour of the higher-order system. The iteration process is usually completed within 7-8 iterations [27]. There are 8 (eight) variables distributed among sustainability dimensions of the economy, technology, social and policy (Table 2). Table 2 shows the matrix of direct influence. It allows the development of an influence map. Each of these 8 variables was evaluated to develop an 8 x 8 cross-impact matrix called the Matrix of direct Influence (Table 2).

Table 2. Matrix of direct influence.

		Influence							
		E1	E2	E3	T1	T2	S1	S2	P1
Dependence	E1	0	0	3	0	0	2	0	0
	E2	3	0	3	2	2	3	0	0
	E3	1	0	0	0	1	1	0	0
	T1	3	2	1	0	0	1	0	2
	T2	0	0	0	3	0	0	0	0
	S1	0	0	0	0	0	0	3	3
	S2	1	1	1	0	0	3	0	2
	P1	3	3	3	1	1	1	1	0

3. Results and discussion

3.1. Saline water desalination performance analysis

3.1.1. Effect of solar radiation on the distilled water volume. The sun's intensity dramatically affects the temperature of the distillate water and the temperature of the glass in the distillatory, which appears to increase with the increase in the intensity of the sun (Table 3).

Table 3. Solar radiation intensity, feed water, distilled water, ambient air and glass temperature.

Parameter	Observation day-					
	1	2	3	4	5	6
Solar radiation intensity (kfc)	35.9	30.41	50.02	50.19	48.5	53.84
Feedwater (°C)	45.54	49.39	49.19	37.03	47.84	40.76
$T_{distilled\ water}$ (°C)	34.78	36.53	37.17	37.03	35.62	35.88
$T_{ambien\ air}$ (°C)	32.07	32.73	32.84	32.67	33.01	32.9
Glass (°C)	41.79	44.99	43.47	41.41	39.8	39.5

The value of the temperature's distillate water and the glass in the distillatory increases with the sun's intensity. It is the plate that absorbs the sun's energy as it receives it at that time and uses it to evaporate the feed water contained in the distillatory. Black's Principle states that the heat absorbed is equal to the heat received by the substance and will be proportional to the increase (change) in the temperature of the substance [15].

3.1.2. The efficiency of the distillatory. The actual efficiency of the distillatory for 6 hours of observation can be seen in Table 3. The actual efficiency of the distillatory in this study tends to be low—the lowest efficiency is 2.59%, and the highest is 4.94%. Only a tiny part of the feed water becomes distillate water, and most of it becomes brine (saltwater). It can be caused by the use of solar energy and the evaporation process that is not optimal, the duration of observation is not long, the surface area of the glass in the distillatory is not large, and the feed water discharge is too fast.

Table 4. Actual distillatory efficiency.

Feedwater volume (ml)	Distilled water volume (ml)	Efficiency (%)
11,388.53	373	3.28
13,008.57	509	3.91
14,046.21	464.5	3.31
11,537.98	570	4.94
13,100.33	398	3.04
14,988.41	388	2.59

Table 5. Energy balance of distillation process (theoretical distillatory efficiency).

Day	Radiation energy in the absorbing plate (Watt)	The energy supplied by the solar collector (Watt)	Efficiency (%)
1	14,034.94	858.63	6.12
2	12,162.62	1175.85	9.67
3	20,006.53	1071.54	5.36
4	20,075.23	1288.2	6.42
5	19,401.16	918.46	4.73
6	21,534.88	884.26	4.11

Table 4 shows the value of energy balance in the distillation process. It can be seen that for 6 days, only a tiny part of the incoming energy is used for the distillation process. The energy lost during the distillation process is estimated to occur. Due to the heat's conduction and convection processes, either on the evaporator's walls or the absorbent plate. In addition, the energy that enters the absorbing plate is not all used, but some of the energy is reflected, so it is assumed as energy lost [10].

Table 5 shows that the efficiency of the distillation apparatus has a linear relationship with the amount of energy that enters the evaporator. The calculated efficiency value is the daily efficiency value of the distillatory's ability to produce fresh water. Based on the experiment and observation of the distillatory for six days, the highest efficiency value was obtained on the second day with 9.67%, while the lowest efficiency was on the sixth day with a value of 4.11%. Several factors that affect the efficiency value include climatic and weather conditions (sun intensity, irradiation time, ambient temperature, wind resistance, heat transfer rate, and so on) that affect heat or temperature, both in the distillation system and in the environment [10]. Other factors that are also considered influential in the distillation process are the dimensions of the distillation apparatus and the volume of distilled water.

3.1.3. Distilled water quality analysis. Brackish water and seawater have different characteristics, so it is necessary to analyse the quality of the distilled water produced from the two types of feed water. The quality of the distilled water produced from artificial brackish water and artificial seawater is shown in Table 6. It shows that the distillate water produced from artificial brackish water and artificial seawater has met the drinking water quality standards [28] for the parameters of pH, turbidity, electrical conductivity (DHL), total dissolved solids (TDS), salinity, chloride (ion Cl^-), iron (Fe), and total hardness. The distillation process is a physical process where the processing is based on the difference in the boiling point of the liquid so that it can set aside these parameters. The product of the desalination process is generally water with a dissolved salt content of less than 500 mg/L, which can be used for domestic, industrial, and agricultural purposes [28]. This study obtained distillate water with a TDS content of 9 – 83 mg/l for artificial brackish water and 17 – 376 mg/L for artificial seawater. It indicates that the proposed desalination system has met the criteria for a desalination system in general.

Table 6. Distilled water quality of feed water from artificial brackish water and seawater.

Parameter	unit	Brackish water		Seawater		Water quality standard [28]
		Feedwater	Distilled water	Feedwater	Distillated water	
pH	-	7.19 - 7.57	6.12 - 6.94	7.3 - 7.63	5.63 - 6.69	6.5 - 8.5
Turbidity	NTU	1.17 - 2.42	1.21 - 1.41	22.70 - 66.6	0.8 - 2.38	5
DHL	μS	22,710 – 24,100	20 - 172	55,400 – 59,300	35 - 601	-
TDS	mg/l	11,200 – 12,100	9 - 83	27,600 -29,700	17 - 376	500
Salinity	‰	15	0	33.9 - 38	0	-
ion Cl-	mg/l	760.84 - 1408.36	0.81 - 1.62	1991.12 - 1748.3	2.27 - 4.69	250
Fe	mg/l	0 - 0.032	0 - 0.586	0 - 0.16	0 - 0.055	0.3
Total hardness	mg/l	72 - 744	0	104 - 291.2	0	500
Ca hardness	mg/l	25.6 - 44.8	0	30.4 - 46.4	0	-
Mg hardness	mg/l	0 - 480	0	168 - 3672	0	-

The distillation process will produce freshwater that is very high in purity [9]. It can be seen from the shallow salt content in the distillate water resulting from this desalination system. It is evident from the parameters of pH, turbidity, electrical conductivity (DHL), total dissolved solids (TDS), salinity, chloride (Cl^- ions), iron (Fe), and total hardness which is far below the drinking water quality standard. In addition, [12] states that the distillation process produces high-quality water with a TDS (Total Dissolved Solids) value in the range of 1.0 – 50 ppm. From the results of this study, the TDS content in the distillate water was 9 – 83 mg/L for artificial brackish water and 17 – 376 mg/L for artificial seawater. IN SOME OBSERVATIONS, the TDS content in this distillate water reached more than 50 ppm but still meets the drinking water quality standards. It can be caused by the condition of the reactor that is not clean so that there is a high TDS content of the feed water in the distillatory and affects the distillate water formed.

3.2. Implementation barriers

The variable of “uncertainty of subsidies”, “niche market”, “poor solar energy data," and "lack of policy/regulatory framework” are classified as dominant variables in quadrant I (Figure 2). Variables in this quadrant have the most significant influence and the most vital influence among other variables. These variables are categorised within economic (uncertainty subsidies), social (niche market), technology (insufficient solar energy data) and political (lack of policy/regulatory framework) aspects.

These dominant variables have a low dependency and strong influence on other variables. Therefore, priority should be applied to overcome these barriers to implement a solar energy desalination system. The certainty of subsidy for both the investment and maintenance budget is one of the implementation successes. The variable under quadrant II of relay or critical variable is “reluctance of people”. This barrier factor has domino effects throughout all factors in the model.

These factors are considered unstable factors because they have a feedback effect. It means that any action taken to these factors will impact other factors. It might show in the lack of people’s participation in disseminating the system due to a lack of information about the desalination system and its benefit. The people perception and/or acceptance of wave energy cannot be achieved in a short time. The government authorities must conduct frequent campaigns to enhance understanding and knowledge of solar energy desalination systems. There are two (2) dependence variables in quadrant III, which are “high initial cost” and “absence of financing institution”. It indicated that high initial cost and absence of financing institution relatively unimportance factors hinder the adoption of desalination technology. The reason is that the cost of a small scale distillation system is not very expensive. Also, the operational and maintenance of this system is quite simple. Hence the role of financing institutions is not so urgent.

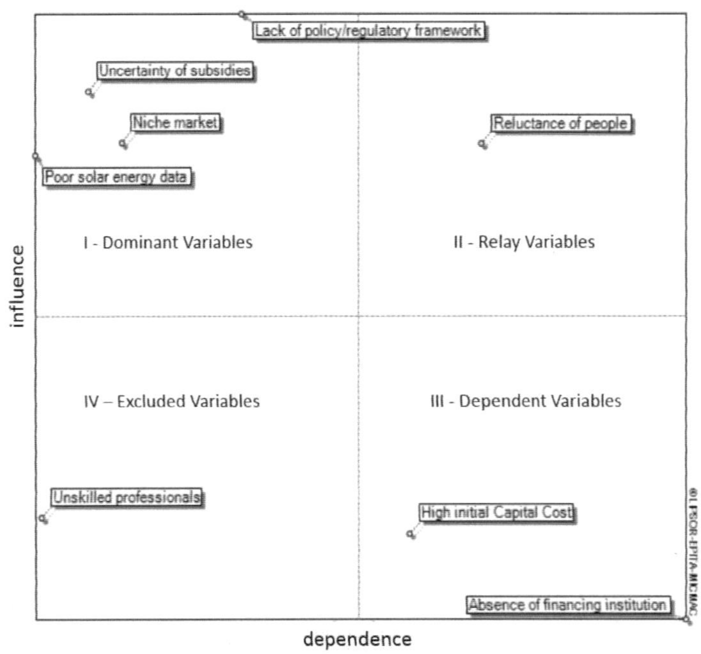

Figure 2. Barriers variable map of direct influence-dependence (MICMAC).

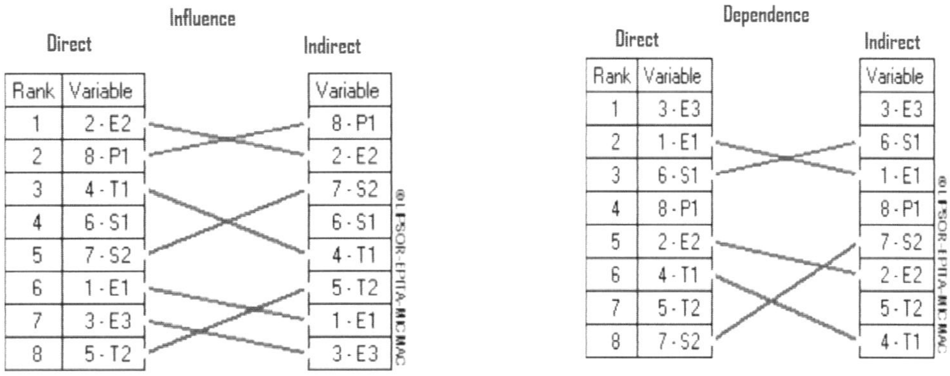

(a) (b)

Figure 3. Barrier variables rank based on influence (a) and dependence (b).

Finally, we can determine the level of dependence and influence among variables in the long term. Figure 3a presents the rank of influence comparison between the existing condition (Direct) and future condition (indirect) among the variables. "Uncertainty of subsidies" (E2) was in rank 1 under the Direct matrix, but in future, it is in rank 2 under the Indirect matrix. It might be due to the "uncertainty of subsidies" (E2) in the future has less influence on other barriers compare to "lack of policy" (P1).

Figure 3b shows factors of high initial cost (E1), absence of financing institution (E3), and the reluctance of people (S1) are consistent in the highest level of dependence variables, both for the direct and indirect time frame. This condition aligns with the fact that these barriers depend on the "uncertainty of subsidy" and "lack of policy" for adopting a solar energy desalination system.

4. Conclusion

The solar water desalination system was designed and tested under actual environmental conditions in Jakarta, Indonesia. The actual efficiency of the distillatory in this study tends to be low, with the lowest efficiency being 2.59% and the highest being 4.94%. Only a tiny part of the feed water becomes distillate water, and most of it becomes brine. Based on the experiment and observation of the distillation apparatus for six days, the most significant theoretical efficiency value was obtained on the second day with 9.67%, while the lowest efficiency was on the sixth day with a value of 4.11%. The efficiency of the distillation system has a linear relationship with the amount of energy that enters the evaporator. The calculated efficiency value is the daily efficiency value of the distillatory's performance to produce fresh water. Further research should include more experimental set-up variations, such as the fabric colour of the absorber, inclination angle, and feed water discharge.

Eight barrier factors of the implementation of a solar energy water desalination system have been analysed. It shows that "uncertainty of subsidies", "niche market", "poor solar energy data," and "lack of policy/regulatory framework" are the key barriers. Future studies should be proposed to analyse more variables within the four dimensions to determine an effective strategy for implementing the solar energy desalination system in Indonesia.

References

[1] Deng R, Xie L, Lin H, Liu J, Han W 2010 *Energy* **36** 4368-4374
[2] Yilmaz I and Soylemez M S 2012 *Desalination* **291** 23-40
[3] Astawa K 2008 *Jurnal Ilmiah Teknik Mesin Cakram* **2(1)** 34 - 41
[4] Fath H E S, Elsherbiny S M, Hassan A A, Rommel M, Wieghaus M, Koschikowski, J, Vatansever M 2008 *Desalination* **225** 58-69

[5] Razmjoo A A, Sumper A, Davarpanah A 2019 *Energy sustainability analysis based on SDGs for developing country Energy Sources, Part A: Recovery, UtilisationUtilization, and Environmental Effects*
[6] Wijayanti Y, Anda M, Safitri L, Tarmadja S, Juliastuti, Setyandito O 2020 *IOP Conf. Series: Earth and Environmental Science* **426** 012058
[7] Shatat M, Worrall M, Riffat S 2013 *Sustainable Cities and Society* **9** 67-80
[8] Sindhu S, Nehra V, Luthra S 2016 *Renew. Sustain.Energy Rev.* **62** 70-88
[9] Safitri M A 2011 *Penelitian Kualitas dan Kuantitas Air Destilat dengan Pengembangan Multiple Trays Tilted Still (Studi: Penambahan Kolektor Kain Hitam)* (Thesis Program Sarjana Teknik Lingkungan ITB)
[10] Kusumadewi R A, Notodarmodjo S, Helmy Q 2018 *MATEC Web of Conferences* **197** 13015
[11] Kusumadewi R A, Notodarmodjo S, Helmy Q 2018 *Indonesian Journal of Urban and Environmental Technology* **2**(1) 66-87
[12] Ansari M F, Kharb R K, Luthra S, Shimmi S L, Chatterji S 2013 *Renew. Sustain. Energy Rev.* **27** 163–174
[13] Godet M 2003 *MICMAC Methodology*
[14] Dewantara, Yogi I G, Suyitno, Muliawan B and Lesmana I G E 2018 *Jurnal Teknik Mesin (JTM)* **07(1)**
[15] Iqbal S, Sukmawaty P, Guyup M D, Setiawati D A 2019 *Jurnal AGROTEK* **6(1)**
[16] Keenan J H 1969 *Steam tables: thermodynamic properties of water, including vapour, liquid, and solid phases* (English units)
[17] Kaushik A, Kumar A, Luthra S, Haleem A *Int. J. Technology, Policy and Management* **14** 2
[18] Saxena A and Deval N 2016 *Journal of Engineering*, Article ID 7937696
[19] International Renewable Energy Agency (IRENA) 2012 Water Desalination using Renewable Energy *Technology Brief*
[20] International Renewable Energy Agency (IRENA) 2013 Water Desalination using Renewable energy *Insights for Policymakers*
[21] Taha M and AlSaed R 2017 *International Journal of Environmental Studies*
[22] Widiasa I N and Yoshi L A 2016 *Internat.J.Scie.Eng* **10**(2) 51-57
[23] Azevedo F D A S M 2014 *Renewable energy powered desalination systems: Technologies and market analysis Dissertation* (University of Lisboa)
[24] Godet M Arcade J Meunier F and Roubelat F 1999 *Structural Analysis with the MICMAC method and Actors'Strategy with MACTOR method*
[25] Fauzi A 2016 *Teknik Analisis Keberlanjutan* (Gramedia)
[26] Godet M 1994 *From Anticipation to Action: A Handbook of Strategic Prospective* (Dunod, Paris: UNESCO Publishing)
[27] Delgado-Serrano, M del M, Ambrosio-Albalá M and Amador F 2015 *Rural* **12** 35-56
[28] Indonesia Ministry of Health 2010 Indonesia Drinking Water Standard No 492

IOP Conf. Series: Earth and Environmental Science **894** (2021) 012021 doi:10.1088/1755-1315/894/1/012021

Investigating the effect of PET plastic bottle strips on the strength and compressibility properties of clayey soil

J B Niyomukiza[1,2*], A Bitekateko[1], J Nsemerirwe[1], B Kawiso[1] and M Kiwanuka[1,2]

[1]Department of Civil Engineering, Faculty of Engineering, Ndejje University, Kampala, Uganda

[2]Department of Civil Engineering, Faculty of Engineering, International University of East Africa, Kampala, Uganda

niyojayb1992@gmail.com

Abstract. The production of plastic bottles by the manufacturing industry has increased drastically over the last six decades across the globe. This rapid production has led to the generation of many waste plastic bottles, thus causing environmental pollution. About 180 tonnes of plastics are generated daily in Kampala, the capital city of Uganda, and around 50% is dumped into the Kiteezi landfill. Instead of putting pressure on the landfill, these plastic bottle wastes could be reused in stabilizing soils with poor engineering properties. The current study investigates the engineering properties of clayey soil reinforced with Polyethylene-terephthalate waste plastic bottle strips. In order to achieve the objectives of the study, the geotechnical and engineering properties of the soil reinforced with waste plastic bottle strips at 0.1, 0.2, 0.3 and 0.4% of the dry unit weight of the soil and non-stabilized soil were determined by conducting laboratory tests, such as particle size distribution, Atterberg limits, compaction test and California Bearing Ratio. The results revealed that the California Bearing Ratio of the soil reinforced with Polyethylene-terephthalate waste plastic bottle strips increased with the increase in the percentage of Polyethylene-terephthalate waste plastic bottle strips up to 0.3%. Beyond 0.3%, a drop in California Bearing Ratio was observed. It indicates that 0.3% Polyethylene-terephthalate waste plastic bottle strips is the optimum percentage for stabilizing low plasticity clayey soils.

1. Introduction

Plastic waste management is becoming a significant challenge in developing countries like Uganda. Data from various sources reveal that out of 180 tons of plastics that are generated [1,2], 40-50% are transported and dumped to the Kiteezi landfill that is located in Kampala, Uganda [3]. It means that the rest of the waste plastic bottles are disposed in a manner that is not safe. Even those dumped on landfills increase pressure or are poorly managed, as shown in Figure 1 [1]. It is noted that since the population is increasing, the number of wastes generated will also keep on increasing. It calls for looking for alternative ways of utilizing plastic wastes in other sustainable ways. Among the ways of reusing waste plastics could be stabilizing soil. Different solid waste materials, like waste tyres, industrial wastes, incinerator ash, and others, have been used to stabilize soils with poor engineering properties, showing positive results [4–6]. The basic idea about soil stabilization is about improving or altering the index properties, improving the gradation, and improving the soil's strength properties to be suitable for constructing the engineering infrastructure. These improvements occur due to a series of chemical reactions when a stabilizer is introduced in the soil [7–9].

Several studies have been conducted on Polyethylene-terephthalate (PET) waste plastic bottles as additives in concrete [10–14], and it has been shown that the strength properties increase upon the inclusion of PET waste plastics. For example, the study by Borg et al. [13] in determining the early age performance and mechanical characteristics of concrete reinforced with recycled PET bottles found out that 1% by volume of PET plastic deformed fibres of length 50 mm gave the best results, as cracks were not observed. Thus the study concluded that PET plastic fibres play a significant role in structures delicate to cracking. The study by Ongpeng et al. [14] showed that PET plastic waste bottles showed an increase in the ultimate compressive strength of concrete. Besides, some studies were conducted on stabilizing soil with PET waste plastic bottles also showed significant results. Mishra and Gupta [15] used a combination of fly ash and PET waste plastic bottles, having ratios ranging from 0% to 20% (fly ash) and 0% to 1.6% (PET waste plastic bottles) of the weight of the soil. The results obtained from various tests showed improvement in the index and engineering properties—for example, plasticity index decreases, CBR and shear strength increase. The optimum percentage of the materials from their study was 1.2%, and 15% recycled PET waste plastic bottles, and fly ash, respectively. Bozygit et al. [16] determined the effect of PET plastic bottle strips on the mechanical properties of cement stabilized kaolin clayey. Their study found out that the mechanical properties of cement stabilized kaolin soil increased after the inclusion of PET waste plastic bottle strips. Based on the above findings, it can be concluded that PET waste plastic bottle strips could successfully be used in the field. However, studies on using PET plastic studies in Uganda are still limited. The current study tries to address the gap in knowledge by checking the possibility of using PET plastic bottle strips to stabilize clayey soils. Their usage is two way, that is to say, improving the geotechnical properties of the soil, and reducing environmental pollution that could happen if they are disposed in an inappropriate way. When plastics are dumped in landfills, they do not break up or decay fast, and as a result, they interact with water and form hazardous chemicals [17]. Therefore, it becomes essential to modify the properties of expansive soils with a cheaper alternative method like plastic waste bottles which reduces plastic waste pollution and its effects.

The desirable properties that enable PET plastic bottles to be effective stabilizers could be the high modulus of elasticity (around 2950 MPa) and high resistance to acid and alkaline [16]. Since PET plastic bottles have a high modulus of elasticity, once reinforced in soil enables it to resist deformation when subjected to loading.

Figure 1. Poor disposal of wastes at Kiteezi landfill [1].

2. Methodology

2.1. Materials
The materials used in the current study include; expansive soil and PET (Polyethylene-terephthalate) waste plastic bottle strips. Both the materials were obtained locally.

2.1.1 Clayey soil. The clayey soil sample used in the current study was collected at Kyitetika village in Kasangati Town Council, Wakiso District, Uganda, at a coordinate of 0°24'40"32°35'21"E. The samples were picked at a depth of 0.5 m. The sample was then taken to the laboratory for testing. Physical and index properties were determined following the British Standards.

2.1.2 PET (Polyethylene-terephthalate) plastic waste bottles. PET raw plastic bottle strips were used with a length and width of 5 mm and 0.5 mm, respectively. Accordingly, four different percentages of 0.1, 0.2, 0.3, and 0.4% by weight of the soil were used. These sizes and percentages of plastic strips were adapted from previous studies conducted by Akbulut et al. [18].

2.2. Methods
The methodology used was conducting laboratory tests on both PET stabilized soil and non-stabilized soils. All the tests were conducted following British Standards, as shown in Table 1[19]. Sieve analysis was considered to determine the particle size distribution of the soil. Particle size distribution significantly impacts other properties like strength, the solubility of the mixture, and surface area properties. The liquid limit (LL) was obtained using the cone penetration test. After doing the Particle size distribution and Atterberg limits test, the soil was classified using the American Association of State Highway and Transportation Officials (AASHTO) and Unified Soil Classification system (USCS) methods. According to Das [20], under cone penetration test, the LL is obtained as a moisture content where a standard cone of apex angle 30^0 and a weight of 0.78 N penetrates a distance of 20 mm in 5 seconds. Based on how the test is conducted, cone penetration could yield better results than Casagrande's method. Heavy compaction was done to determine the maximum dry density (MDD) and optimum moisture content (OMC), and California Bearing Ratio (CBR) test was conducted under wet conditions.

Table 1. Test and the standard [19].

No	Test	Purpose	Standard
1	Particle size distribution	Grading	BS1377: Part 2
2	Atterberg Tests	Plasticity index	BS1377: Part 2
3	Compaction Test	To determine OMC content and MDD	BS1377: Part 3
4	California Bearing Ratio	To determine the bearing capacity of soil	BS1377: Part 9

3. Results and discussion

3.1. Geotechnical Properties of the soil understudy
A brief description of the geotechnical properties of the soil used in the current study is summarized in Table 2. The soil that was used was light greenish-grey clays with black organic matter. It was classified as A-6 using the AASHTO soil classification system and low plastic clay under USCS [20]. According to AASHTO, the soil under class A-6 possesses poor engineering properties, hence a need for stabilization.

Table 2. Test and the standard.

No	Property	Value
1	Percentage passing BS No 20 Sieve	66.7%
2	Natural moisture content	20.2%
3	Liquid limit	25%
4	Plastic limit	13.3%
5	Plasticity index	11.7%
6	AASHTO classification	A-6
7	Unified Soil Classification System	CL

3.2. Particle size distribution

The results show that the soil sample under the current study was uniformly graded, as seen in Figure 2. The percentages of different types of soil were determined as Gravel= 3.3%, Sand = 29.6%, Silts and Clays = 66.7%. It showed that the soil is dominated by silts and clays since they have a higher percentage content.

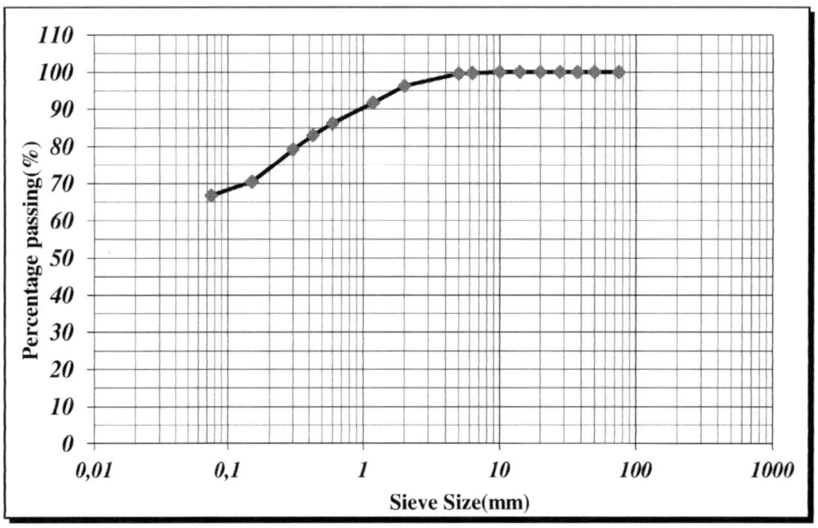

Figure 2. Particle size distribution

3.3. Atterberg limits

The liquid limit (LL) was determined using the cone penetrometer apparatus, and the results are seen in Figure 3. The moisture content obtained at a cone penetration of 20 mm was taken as the liquid limit, which is 25%. According to Whitlow [21], the soil under study exhibits low plasticity. The plastic index was calculated as the difference between the liquid limit and the plastic limit, as seen in Table 3. Based on the results, the soil is low plastic clayey because the plasticity index is 11.7%, which is less than 35%.

Table 3. The results from the Atterberg test

No.	Parameter	Results (%)
1.	Liquid limit (LL)	25.0
2.	Plastic limit (PL)	13.3
3.	Plasticity Index (PI)	11.7

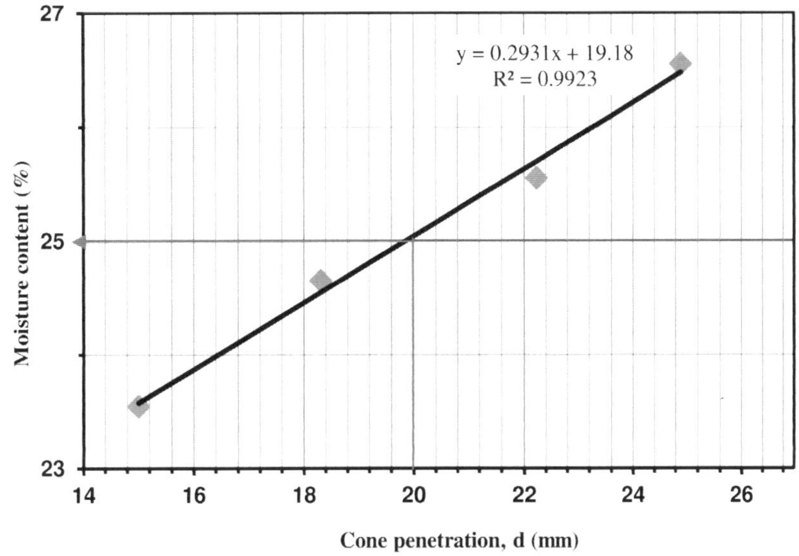

Figure 3. Determination of liquid limit

3.4. Compaction Test

The values for the dry density of the soil samples starting from 0% of PET plastic bottle fibre strips to 0.4% PET plastic bottle fibre strips reinforced samples are 1946, 1960, 1970, 1981 and 1972 kg/m^3, respectively. This increasing pattern shows that the density of the soil kept increasing with increasing percentages of plastic inclusion. While for 0.4% inclusions of plastic strips, the value for dry density decreased because the plastic content is found to be in large amounts in the plastic soil matrix. The maximum dry density and the optimum moisture content were 1946 kg/m^3 and 9.1%, respectively, as seen in Figure 4. The vital role of compacting is to increase on shear strength of the soil, decrease permeability, and increase density [15].

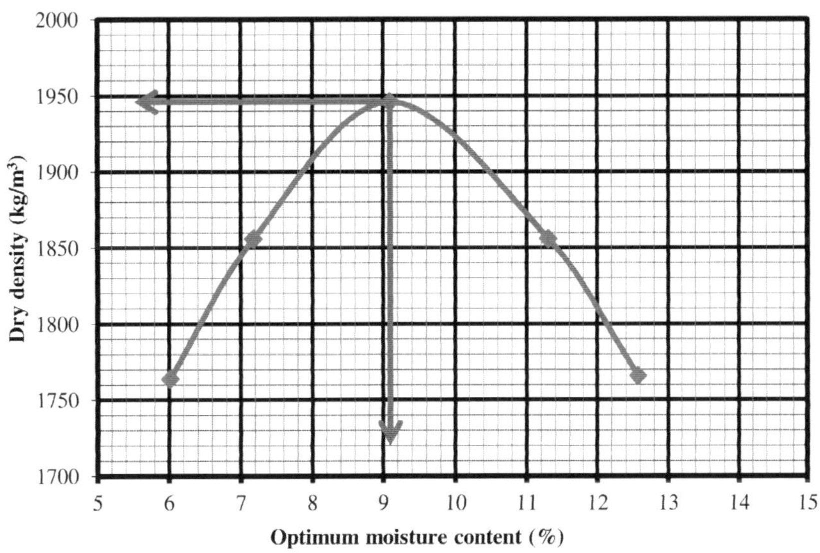

Figure 4. Compaction test results

3.5. California Bearing Ratio (CBR)

The CBR test was conducted on samples soaked for four days. The CBR value of unreinforced expansive clayey soil obtained was 12.2%, while the highest CBR achieved for the PET plastic bottle strips reinforced soil is 16.2%, as seen in Figure 5. After reinforcing the soil with different percentages of PET plastic bottles, the CBR values increase up to 16.2% at 0.3%. An increase in CBR is a significant indication of improved strength of the soil [9,15]. However, a drop was observed at a percentage after reinforcing the soil with 0.4% PET plastic bottle strips. It could be attributed to the large quantities of waste plastics that could have weakened the soil. Based on the highest CBR achieved after reinforcing the soil with PET waste plastics, it can clearly be stated that the optimum percentage of stabilizing the soil under study is 0.3%.

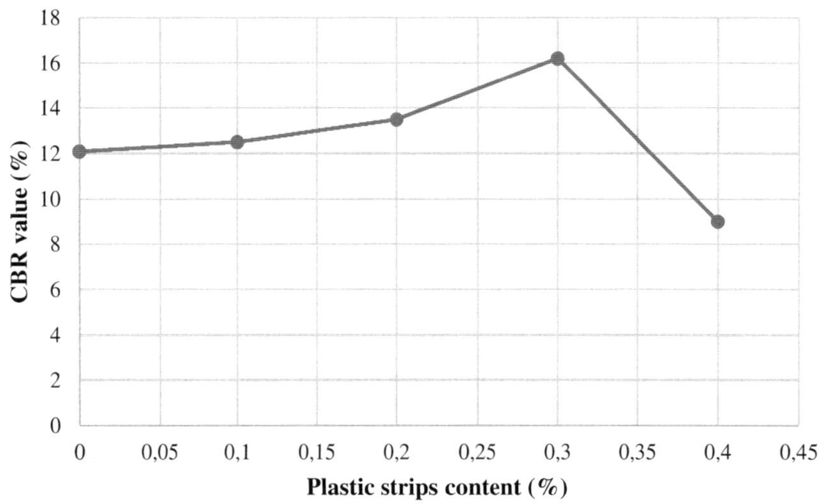

Figure 5. The effect of plastic strips on CBR values

4. Conclusion

The current study investigated the effect of waste PET plastic bottle strips on soil strength and compressibility parameters. Based on the findings, the soil under study was classified as A-6 under AASHTO classification, hence possessing poor engineering properties. Considering CBR values, soils reinforced with 0.3% plastic strip bottles yielded higher CBR values than other percentages. It shows that 0.3% PET plastic bottle strips inclusion could stabilize clayey soils under study. However, the study recommends conducting durability tests to determine how waste PET bottle strips reinforced soil will behave in the long run since the design life for engineering structures is relatively long.

References

[1] Kinobe J R, Gebresenbet G, Niwagaba C B, and Vinnerås B 2015 *Waste Manag.* **42** 82-92
[2] Komakech A J, Banadda N E, Kinobe J R, Kasisira L, Sunberg C, Gebresenbet G, and Vinnerås B 2014 *J. Air Waste Manage. Assoc.* **2247** 340-348
[3] WaterAid 2011 *Solid Waste Management Arrangements and its Challenges in Kampala: A Case Study of Bwaise II Parish*, Kawempe Division
[4] Niyomukiza J B, Wardani S P R and Setiadji B H 2021 *IOP Conf. Ser. Earth Environ.* Sci. **643** 1-9.
[5] Tiboti P, Niyomukiza J B, Kiwanuka M, Mbujje J W, Baiga J, and Hadudu A F 2021 *IOP Conf. Ser. Earth Environ. Sci.* **802** 1-8
[6] Niyomukiza J B, Wardani S P R. and Setiadji B H 2020 *IOP Conf. Ser. Earth Environ. Sci.* **448**

1-10.
[7] Amadi A A and Okeiyi A 2017 *Int. J. Geo-Engineering* **8**(1) 1-13
[8] Gowthaman S, Nakashima K and Kawasaki S 2018 **11**(4) 553-576
[9] Niyomukiza J B, Wardani S P R. and Setiadji B H 2020 *In 2nd International Symposium on Transportation Studies in Developing Countries (ISTSDC 2019)* **193** 157–161
[10] Al-hadithi A I, Tareq A, and Khairi W 2019 *Compos. Struct.* **224** 111021
[11] Foti D 2013 *Compos. Struct.* **96** 396–404
[12] Adnan H M and Dawood A O 2020 *Case Stud. Constr. Mater*, **13** 1–19
[13] Borg P R., Baldacchin O, and Ferrara L 2016 *Constr. Build. Mater. J.* **108** 29–47
[14] Ongpeng J M, Barra J, Carampatana K, Sebastian C, Yu J J, Aviso K B, and Tan R R 2021 *Eng. Sci. Technol. an Int. J.* **24** 405–413
[15] Mishra B and Gupta K M 2018 *Constr. Build. Mater.* **190** 95–107
[16] Bozyigit I, Bulbul F, Alp C, and Altun S, 2021 *Eng. Sci. Technol. , an Int. J.* **24** 1090–1101
[17] Adane L and Muleta D 2011 *J. Toxicol. Environ. Heal. Sci.* **3** 234–248
[18] Akbulut S, Arasan S, and Kalkan E 2007 *Appl. Clay Sci.* **38** 23–32
[19] British Standard 1990 *Methods of test for Soils for civil engineering purposes*
[20] Das B M 2006 *Principles of Geotechnical Engineering*, Fifth Edit (Stamford, CT: Thomson Learning College)
[21] Whitlow R 1995 *Basic Soil Mechanics*, Third Edit (London: Longman Group Limited)

Hydrocyclone optimisation to separate oil and water in the separator

P T Wastoyo[1], D A Mardiana[1], R Setiati[1]*, T Fathaddin[1]

Master of Petroleum Engineering, Faculty of Earth Technology and Energy, Universitas Trisakti, Jakarta, Indonesia

rinisetiati@trisakti.ac.id

Abstract. The conventional 3-phase separator installed cannot anticipate the increase in fluid flow so that the oil content carried into the produced water from the separator consistently exceeds the upper operating threshold. This study aimed to test the application of a hydrocyclone device to a 3-phase horizontal separator. Hydrocyclones are widely used as auxiliary devices to optimise oil separation by minimising oil carried into the produced water stream. This study made a comparison between installing a hydro cyclone at the inlet with a 3-phase horizontal separator. Applying a hydrocyclone at the inlet of a 3-phase horizontal separator increases the efficiency of the separation process in production. Proper design improvements with Hysys and flow characteristics with CFD can reduce the oil content carried in the produced water stream below 20 mg/l. The results of this study can support de-bottlenecking to increase production to a production target of above 375 kbps. A robust application of engineered hydrocyclones with correct production and operating shrouds has been experienced to optimise the separation process by up to 92%.

1. Introduction

Located in the North Caspian Sea, the XXX field production rate has not yet reached the production plateau by the design of the production equipment. One reason is that the oil content carried out into the produced water flow from the separation process in three phases horizontal separators still exceeds the upper operating threshold of 20 mg/l or off-spec.

High carried oil content becomes a burden for the downstream water treatment unit. Due to the design limitations of the process equipment, the required specifications of the incoming water are to be below 20 mg/l [1]. Otherwise, the treated water cannot meet the local government environmental specifications stipulates that water discharged into Evaporation Ponds must have an oil content of less than 5 mg/l [2].

The XXX field has made several planned efforts to increase production: secondary recovery with raw gas injection, followed by scheduled desander and descaling pipeline. Further, installing a demulsifier in 2020 has increased the efficiency of separation by successfully reducing the oil content to a level of 27 mg/l from above 40 mg/l to the same water cut-rate and production rate. Nevertheless, it still cannot able to reduce oil content below 20 mg/l. Therefore, looking at the designing and optimising the separator condition self would get the most optimum operations [3].

This research focuses on defining correct operating and production envelopes and refining the design by applying the hydrocyclone separation technique at the inlet of the installed separator, whereas this technique has been proven successful in other oil fields.

2. Methodology

Benchmarking data from successful separation optimisation techniques applied, and production, operations, and design data are collected from a secondary source and the XXX field as a primary source. The methodology used is divided into the main steps below.

The literature reviews the most applicable techniques for optimisation of three-phase gravity separation processes firstly. Hydrocyclone application with proper engineering design and correct production and operating parameters have been experienced in optimising the separation process up to 92% [4, 5].

And then review the inlet type of hydrocyclone in terms of most functionality applied shown in table 1 as follows.

Table 1. Performance comparison of inlet types.

Inlet Type Tool Inlet Function	Inlet Type Diverter Plate	 Vane Inlet	 Inlet Hydrocyclone
Reduces feed flow momentum and ensure good distribution of gases and liquids	Good/Poor	Good	Good
Separating high pouring liquids	Poor	Good	Good
De-foam	Poor	Average	Good
Prevents re-entrainment / scaling of separated fluids	Average/Poor	Good	Good

Collect data from such oil fields successfully on hydrocyclone inlet type applied with typical water cut, production, and oil content. As in the Angsi field from Peninsular Malaysia Operation (PMO) PETRONAS Carigali Sdn, Bhd. (PCSB) has proven that a modified separator is installed by installing a hydrocyclone inlet separator itself [6]. After that, collect historical data from the XXX field on water cut, production, and oil content as shown in graphic 1 as follow;

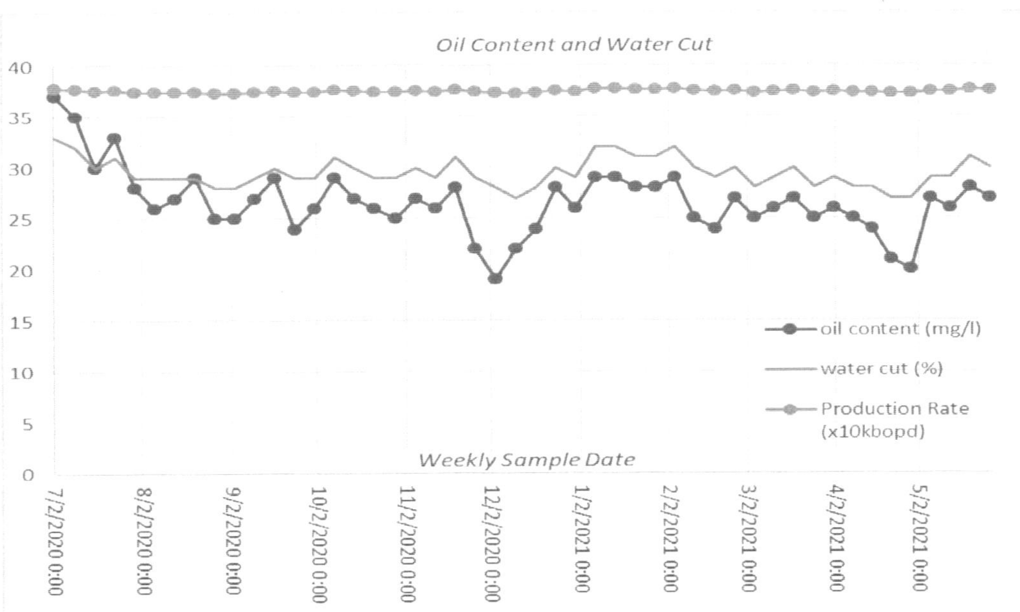

Figure 1. Oil content to water cut and production.

Re-run simulation with the defined production and operating envelopes resulted from earlier simulation by applying hydrocyclone at the inlet of three-phase separators. Perform fine adjustment aiming at optimum efficiency of separation. And then simulate flow characteristics by CFD on the applied hydrocyclone to fine-tune the design. Finally, evaluate the efficiency of separation for all scenarios.

3. Results and Discussion

Stokes Law is used as a separation theory to design a separator design with a gravity deposition method where the velocity of rising oil droplets is based on the size and density of the oil droplets. The larger the oil droplet, the faster the oil rise rate or the separation rate will be. Therefore, the separator is designed based on the difference in density between oil and free water. Based on these design criteria, free water will be deposited at the bottom of the separator, while oil with a lighter density will rise to the top of the separator due to differences in density [7].

$$Vt = K * G * D^2 * (\rho 1 - \rho 2)\boldsymbol{\mu}^{-1} \tag{3.1}$$

Where;
Vt = Increase in oil droplet velocity
G = Gravity force for conventional separator
D = Droplet diameter
$\rho 1$ = Density of continuous phase (water)
$\rho 2$ = Density of dispersed phase (oil)
$\boldsymbol{\mu}$ = Viscosity of continuous phase (water)
K = Constant

On the current data, water cut is directly proportional to the production rate and inversely proportional to the retention time of water in the separator, making the determinants of the separation process in the separator unable to work optimally, so that it becomes inefficient. It makes it one of the bottlenecks for increasing production from wells [8]. Due to the high water-oil ratio, having water as the continuous phase causes the water layer to be thicker than the oil layer. Hence, the larger oil droplets require to travel more than usual from the continuous water phase, resulting in a longer retention time to coalescence. Thus, at a high water cut, the acceleration on centrifugal force provided by the hydrocyclone might assist in breaking the emulsion; hence the water layer is lesser, thus improve the retention time when operated with a higher production rate [9].

The Stokes Law tells the separation efficiency is directly proportional to the size of the droplets that can be manipulated and maintained at all times. The larger oil droplets have less dense need less time to more significant coalescence with oil by emerging rapidly to the interface of the oil emulsion. Acceleration of liquid flow rate contributes to separating the oil from the water phase in less time because allowing the emulsion to break up into an oil and water larger droplets. As the liquid flow rate increases, the high-speed rotating stream creates the centrifugal acceleration of liquid flow in hydrocyclone, allowing gross separation between liquids with different densities [10]. Herewith this justification, simulation with changes of elevated production rate might optimise the oil-water separation efficiency. Thus, oil content might be lower.

Thus, by following the Stokes Law, the separation efficiency is inversely proportional to the continuous phase (water) viscosity. By supplying heat to the fluid causes, the viscosity of the continuous phase (water) is reduced, and the velocity of oil droplets will be faster, resulting in easier separation [3, 11]. The intermolecular forces between the oil-water molecules in the emulsion are reduced by means that the attraction between the two molecules becomes weaker by increasing the temperature of the feed stream and improved overall performance, and decreased the concentration of oil in the water discharge stream [12]. Because water is less viscous from oil, the attraction between water and oil droplets is quickly released with the vortex aids from the hydrocyclone and caused more water to separate from the emulsion. Herewith this justification, simulation with changes of elevated temperature might optimise the oil-water separation efficiency. Thus oil content might be able to lower.

As the separator pressure increases, many light and intermediate components will stay in the gas phase in the separator, and the liquid flow rate out of the separator increases. Pressure at the inlet of the hydrocyclone is an important indicator of where the separation point will take place [13,14]. The separation point is the size at which liquid chance of either to the underflow or overflow. Because water is less viscous from oil, a higher pressure will send more water as fine droplets into the underflow, inversely oil as larger droplets quickly released with the vortex aids from the hydrocyclone and causes more water to separate from the emulsion [15]. Herewith this justification, simulation with changes of elevated pressure might optimise the oil-water separation efficiency. Thus oil content might be able to lower [16].

4. Conclusion

In conclusion, by defining correct operating and production envelopes and refining the design, application of the hydrocyclone separation technique at the inlet of the installed separator resulted from the simulation based on the hypothesis above would be able to optimise the oil-water separation efficiency and would be able to lower oil content in produced water within 20 mg/l.

Acknowledgement

The author would like to thank Universitas Trisakti, International Conference on Environmental and Sustainability Context (ICOESCO) 2021, and the publishers who have published this paper.

References

[1] Neff J, Lee K, De Blois E M 2011 *Produced Water* 3-54
[2] BenDor T K, Metcalf S S and Mark P 2011 *Sustainability* **3** 914-936
[3] Adewumi M 2016. *I*, 141. https://www.e-education.psu.edu/png520/m20_p3.html
[4] Patterson P A, Munz R J 1989 *The Canadian Journal of Chemical Engineering* **67(2)** 321–328
[5] Chin R W, Stanbridge D I, Schook R 2002 *Proceedings - SPE Annual Technical Conference and Exhibition* 1437–1442
[6] Naim M 2008 Certification of approval. *January*
[7] Ahmed T, Makwashi N, Hameed M 2017 *Journal of Emerging Trends in Engineering and Applied Sciences 8*(**3**) 143–153
[8] Walsh J M 2015 *Oil and Gas Facilities* **4(02)** 073–086
[10] Liu H F, Xu J Y, Zhang J, Sun H Q, Zhang J, Wu, YX 2012 *Journal of Hydrodynamics* **24**(**1**) 116123
[11] Guo, Boyun, Lyons, William, Ghalambor A 2007 *ISBN: 0750682701 Publisher: Elsevier Science & Technology Books*
[12] Klasson K T, Taylor P A, Walker J F, Jones S A, Cummins R, Richardson SA 2004 *Petroleum Science and Technology* ***22*(9–10)** 1143–1159
[13] Minghu J, Lixin Z 2007 *SPE Production and Operations Symposium, Proceedings*, 195–199
[14] Arnold K E, Paragon S P E, Services E, Paragon L F 1999 *SPE 56644*
[15] Belaidi A, Thew M T, Munaweera SJ 2003 *Canadian Journal of Chemical Engineering* **81(6)** 1159–1170
[16] Schook R, Van Asperen V 2005 *SPE Middle East Oil and Gas Show and Conference, MEOS, Proceedings* 331–337

IOP Conf. Series: Earth and Environmental Science **894** (2021) 012023 doi:10.1088/1755-1315/894/1/012023

The Covid-19 pandemic impact on indigenous people livelihoods in the peat swamp forest ecosystem in Central Kalimantan Indonesia

D Suwito[1*], Suratman[2], E Poedjirahajoe[3]

[1]Doctoral Program on Environmental Science, Universitas Gadjah Mada, Indonesia
[2]Faculty of Geography, Universitas Gadjah Mada, Indonesia
[3]Faculty of Forestry, Universitas Gadjah Mada, Indonesia

david.suwito@mail.ugm.ac.id

Abstract. Global tropical peat swamp forest ecosystems mostly are found in Indonesia. However, these unique yet fragile ecosystems are subject to deforestation and degradation due to forest fires and land conversion. Peat swamp forests are well known for their ecosystem services related to global climate mitigation because they store carbon inside and are very vulnerable to forest fires. Another direct function is providing various needs for the indigenous people livelihoods who live close to and in the areas. This study was conducted qualitatively to assess the forest utilization and the impact of the COVID-19 pandemic on the livelihoods of indigenous people. The study results show that the Dayak Ngaju Communities utilize the forest for timber, non-timbers, medicinal plants, foods, and religious culture. There is no significant impact of COVID-19 on forest product collection activities, but the pandemic impacts the marketing and sales of forest products. The marketing of fish and other non-timber forest products has been negatively affected due to decreased demand and limited transportation. On the contrary, the use, need, and demand for medicinal plants for the traditional medicine of COVID-19 is increasing significantly and improving local livelihoods.

1. Introduction

Indonesia has the most extensive tropical peatland globally, around 21 million hectares or 47% of the total global peatland area [1]. However, based on Ritung et al. measurements, mapped peatlands in Indonesia are 15.4 million hectares, covering Sumatra, Kalimantan, and Papua [2,3]. The existence of this massive peatland makes Indonesia play a crucial role in peatland management on a global scale for various purposes, such as reducing carbon emissions and mitigating global climate change [4].

In Indonesia, the largest peatland is in Riau province, with 3,867,413 or 60% of the total peatland on Sumatra Island. At the same time, on Kalimantan Island, Central Kalimantan has the largest peatland with an area of 2,659,234 or 56% of the total land area. Peat on the island of Borneo. However, what is distinctive in that the peatlands in Central Kalimantan are mostly **peat swamp forests**. The part of peatland that is still a stretch of land dominated by trees is called Peat Swamp Forests (PSFs). Peat swamp forest in Southeast Asia, including Indonesia, is a lowland ecosystem with a critical role in biodiversity, carbon storage, and a resource used for human life [5,6].

Peat swamp forest part of peatland has a vital role in human life with various ecological and economic benefits [7,8,9,10]. Specifically, peat swamp forests are of global importance because of their high carbon stores [11], which means they are associated with global carbon emissions when burned [12,13, 14]. In addition, these ecosystems also have high biodiversity, including various endemic, rare and endangered species [15,12,16,6] and a source of livelihood for local communities [17,18,19].

In Central Kalimantan, the reduction of peat swamp forest started with the project of clearing 1 million hectares of peatland to pursue food self-sufficiency in the 1990s by clearing forests into agricultural land through the transmigration program. This land-use conversion then causes several aftershocks such as drought and land conversion to various utilization and fires. The leading causes consist of extensive land transformation to plantations and sizeable industrial-scale agriculture [20,21,22], massive canal/drainage building activities [23,11,24], and land and forest fires [25,26,11].

Land and forest fires are among the leading causes of peat swamp forest deforestation and peatland degradation in Central Kalimantan. Most fires are used to clear land for conversion to other uses, especially for large-scale agriculture and pulp and paper plantations or HTI [27,28,29]. However, a forest fire is not a single factor but is related to other factors such as drought, drainage opening, land conversion, and other human activities [30,31].

Peatland and Peat Swamp Forests have many functions on local, regional, and global scales. The first function of the peat swamp forest ecosystem is to store carbon and is linked to climate change. Although tropical peatlands, especially peat swamp forests dominated by trees, are essential for biodiversity, carbon, and water protection [23,32]), peat swamp forest differs from other peatlands in the form of organic matter. Although tropical Peat Swamp Forest mainly in the state of decomposition from tree roots, stems, and branches, peat swamp forest is of concern to scientists because of its high carbon stock and its strong influence on global climate change [31,7]. The second function is a habitat of conservation and habitat for biodiversity. Peatlands, especially peat swamp forests, have high conservation value and biodiversity [6]. The existence of peat swamp forests supports many high conservation value organisms, including endemic animals such as orangutans and tigers [34]. Peat swamp forest is the essential habitat for protected primates in Indonesia, such as forest orangutans (*Pongo Pigmeus*) and probosci's monkeys (*Nasalis lavartus)* [35]. Peat swamp forest is also a habitat for several types of small and rare vertebrates [7,14].

In addition to supporting animal diversity, peat swamp forest also has endemic flora species such as *Meranti Rawa* or *Shorea Balangeran*. Furthermore, several protected woody species such as jelutung and ramin[36] and various types of giant fungi, mammals, birds, fish, and invertebrates are also found in peat swamp forests [37,38,39]. The ecosystem also supports the biodiversity of grass and ferns that grow as pioneers in forest fires, including rare *kantong semar (Nephentes Sp.)* species [13].

The third function of peat swamp forest is hydrological benefits. Peat swamp forest is a unique ecosystem with black water so that it becomes the primary water source (aquifer) for river water flow during the dry season [40,41,2]. In addition, peatland is a freshwater source, wherein 1 million hectares of peatland with an average depth of 2 meters stores 1.2 million m3 of freshwater [11,42,44,29].

Peat swamp forest is a water regulator because, in the rainy season, water is absorbed into the peat soil, which functions like a sponge that absorbs moisture and holds water to keep it wet in the dry season if the hydrological balance is good [41, 42,] In addition, peat swamp forest also functions as a barrier to seawater intrusion, especially in coastal peat swamp forests [43,44]

In terms of economic function, peat swamp forest has high economic value in terms of goods and services. Peat swamp forest is an ecosystem where people depend on their livelihoods by collecting forest products in the form of wood, rattan, latex, game animals, fish, and mushrooms as their source of livelihood [17, 32,8].

Peat swamp forests provide extraordinary natural resources, including various types of woody plants that have high economic value, such as ramin (*Gonystylus bancanus*), jelutung (*Dyera costulata*), and meranti (*Shorea* spp). Several socio-economic studies show that the dependence of local communities on peat swamp forests can reach up to 80% (eighty per cent), which is higher than their dependence on agriculture [45,46,9,47,48].

The most important and often under-appreciated part globally is the economic benefits of peat swamp forests for local communities. Local communities live, utilize, and depend on the peat swamp forest

IOP Conf. Series: Earth and Environmental Science **894** (2021) 012023 doi:10.1088/1755-1315/894/1/012023

ecosystem to meet their daily need [9]. With high biodiversity and various ecosystem services [34], peat swamp forests are closely linked to the livelihoods of local communities.

This complexity is also related to various social and cultural aspects in the community that depend on forest products and ecosystem services from forest areas, thus requiring more equitable and acceptable policies [10].

The Covid-19 pandemic condition from the beginning of 2020 had a terrible impact on local communities due to restrictions on many social activities and limited distribution of goods and services. We study qualitatively the uses of forest areas and ecosystem services and how this covid-19 pandemic affects their livelihoods, especially the local Dayak Ngaju people who have traditionally depended on tropical peat swamp forests.

2. Methodology

The research was conducted in 3 villages around the peat forest, namely *Katunjung*, *Tumbang Mangkutup* and *Tuanan* Village, Kapuas District, Kapuas Regency, Central Kalimantan. The study was carried out from June 2020-January 2021 in three villages near the Kapuas Kahayan Protected Forest Area, where local communities primarily depend on the tropical peat swamp forest for their livelihoods.

This research was the qualitative-exploratory approach with multi-stakeholders as the sources of information. Our source of information was all communities who use forests and forest areas, village officials, traditional leaders, buyers of forest products at the village level, and buyers of forest products from outside the region. The research was conducted using the method of extracting information through in-depth interviews. The study would be terminated if no new or different information was found.

With limitations due to the Pandemic period, few interviews used Whatsapp media, while direct interviews were conducted in person by observing the health protocol. We also checked information by participating with the community in the forest to take photos and see firsthand.

The information collected is then interpreted because there is much information in the local area (Dayak Ngaju), then triangulated with different data by different sources of information, then grouped according to research categories. The categories of data include Forms of use, types of forest products used, types of ecosystem services utilized, types and forms of utilization of medicinal plants, and the impacts of the COVID-19 pandemic on the livelihoods of those who use forests and forest products. The tools used are writing instruments, laptops, calculators, cameras, and recorders during our study.

3. Result and discussion

3.1. The utilization of forest product from peat swamp forest ecosystem

We listed some forest product that utilized by local people in the Table 1.

Table 1. Forest products used by indigenous people of Dayak Ngaju.

Forest Products Utilized	Part collected	Usage
Timber Products (*Shorea balangeran, Shorea Sp, Litsea Sp.)*	log	Meubel, family use, building material for wooden house
Rattan	Rattan stick	for trade
Bamboo	Bambo bar	Meubel, for husbandry purposes
Bambang (*Donax canniformis)* and Purun (*Lepironia articulata)*	Stem, Stick, bar	For handicrafts, personal tools
Latek (*Dyera Lowii* dan *Hevea brasiliensis)*	Latex (liquid and solid)	for trade

Local communities around peat swamp forests, *Dayak Ngaju*, have been living and depending on forests and forest products for generations. Hunting and gathering are among the local tradition of the Ngaju Dayak people around the peat swamp forest. In addition, various forest products are used and utilized by the community, both timber and non-timber forest products.

Previously, the collection of timber forest products was the most valuable product of high economic value and was mainly carried out by local communities in the rural areas of Central Kalimantan. However, timber harvesting activities are currently rare because wood/timber harvesting must be carried out with complex permits, management, and technologies. Local people only collect and use timber products for their own needs or villages, such as building a wooden house as their residency. Almost all rural homes of local communities living in areas bordering peat swamp forests use wood as the primary material for their houses, with simple processing into boards and blocks.

Rattan (*Calamus trachycoleus)* is the most important non-timber forest product for local communities. The best-selling type of rattan is economical rattan (which can be sold at the village level for 200,000 IDR/1 quintal or 20 USD/100 kgs. One family can collect up to 200 kilograms of rattan per day with a workforce of 2 people, and the production costs are about of the total income. However, rattan collection cannot be carried out throughout the year because it takes time to regenerate. Usually, they move to the location of the rattan harvest every two years of harvest. Therefore, one family needs a large area of forest area to collect rattan. Rattan is the most critical non-timber forest product for local people in the area [17] and should be the part of forest area management, as it has a strong relationship with local communities [9].

In addition to rattan, non-timber forest products collected are bamboo. Bamboo in Kalimantan is not a commercial product but is collected for various local community needs, namely to make chicken drums, ducks, pigs, and fish cages. Therefore, bamboo is never sold. The bamboo primarily consists of small and thin species, which is not suitable for constructing large buildings like on the island of Java. In addition, there are not many bamboos in the peat swamp forest area.

Figure 1. Bamban (*Donax canniformis) is* used both for handicraft material and medicinal plants.

Figure 2. Purun (*Lepironia articulata*) harvesting from Peat Swamp Forest.

Purun and *bamban* (See Figure 1 and 2) are natural materials of understorey plants collected to make various goods, both daily goods and handicrafts. Purun (*Lepironia articulata*) is a species of peat swamp grass that is most commonly used to make kitchen utensils like bags, baskets, and containers for food and souvenirs. Most of the time, it is also used as handicrafts with high aesthetic value and sold at reasonable prices. On the other hand, bamban makes mats, bags, baskets, and various household utensils. Bamban is usually used for daily needs, but sometimes some buyers buy it at a reasonably high price.

Another non-timber forest product is latex. The latex collected consists of two types, namely rubber latex and jelutong latex. Rubber latex is very common and is sold in solid form for 8000-10000 IDR/kg or 80 cent-1 USD/kg and is an essential forest product for community livelihoods. On the other hand,

IOP Conf. Series: Earth and Environmental Science **894** (2021) 012023 doi:10.1088/1755-1315/894/1/012023

Jelutung (Dyera lowii) latex is much more expensive but difficult to market because it is usually only for handicrafts and industry.

3.2. The utilization of ecosystem services

Besides direct forest products, we also listed any ecosystem service and products for ecosystem service utilized by local people of Dayak Ngaju as listed in the Table 2.

Table 2. Forest product from ecosystem service and ecosystem service used by local people.

Forest Products Utilized	Part collected	Usage
Water	water	Household usage
Habitat for fish	fish	More than 30 species, including nine main species with the highest production and economic value
Habitat for medicinal plants	Leaves, roots, fruit, bark, or bar	More than ten species, including six main species using for Covid-19 treatment and medicine
Religious Places for local believe	River creeks	The socio-cultural and religious functions
Habitat for Plant-based foods	Fruit, stem, seedling	There are more than ten species, including three main species using for daily consumption.

Tropical peat swamp forests are known for their outstanding ecosystem services and are much higher in value than timber forest products. The most well-known ecosystem service and has always been the concern of various parties at the global level is the ability of ecosystems to store carbon and the function of these ecosystems in mitigating global climate change. Still, on the other hand, these indirect benefits are not felt economically at the household scale for local communities.

Peat swamp forest is a water storage aquifer. Peat water is usually red and black but is clear and with high acidity. However, many local communities use clean water because peat water is relatively clean compared to the ample river water that flows around community villages. Large river water can be contaminated with mercury from traditional gold mining, *e-coli* bacteria mixed with mud. Still, water from rivers in peat colour is usually free from pollution.

Peat swamp forest is a unique ecosystem because it is a swamp that is inundated during the rainy season and partially inundated throughout the year. Swamp with water plants rasau (Pandanus helicopus) is a habitat for more than 30 endemic species of peat swamp forest. 6 main species are used by the community, namely *papuyu, kakapar, patung, mehau, haruan, tauman, tempahas, sapat, and lais*. These nine species have high economic value, which is traditionally processed and marketed to local markets.

Figure 3. Swamp with water plants rasau (*Pandanus helicopus*) is a habitat for more than 30 endemic fish species of peat swamp forest.

Fish is a by-product of peat swamp forest, which is very important for the livelihood of local communities. However, fish harvest is usually in the dry season, where fish are collected in lakes where the water is receding or in artificial ponds called "*beje*."

In addition, the peat swamp forest and peatland ecosystems also provide various medicinal plants (See Figure 3). The use of these medicinal plants is done traditionally with local knowledge passed down across generations. There are more than 17 species they use from peat swamp forests, and nine species of which are medicinal plants used during the COVID-19 pandemic and have high economic value.

The peat swamp forest ecosystem is also connected with local beliefs. The Ngaju Dayak people believe that particular places at the river creeks are sacred places associated with the local religion, namely the *Kaharingan. The Kaharingan* religion is closely connected with the use of the surrounding forest, and they always protect the forest.

Peat swamp forest is a unique ecosystem and can provide a variety of vegetables for local communities. More than 13 peat swamp forest plants are commonly used as vegetables, such as *kalakai*, ferns, rattan grass, lilies, and various types of mushrooms. Mushrooms play a crucial role for livelihood as they are sold with high economic value, such as bantilung mushrooms, which sell for 50,000/kg (4 USD/KG).

3.3. High economic value of medicinal plant for covid-19 treatment

We also identified and listed every medicinal plant used by indigenous people of Dayak Ngaju for the treatment and medicine for Covid-19, as presented in the Table 3.

Table 3. Medicinal plants for Covid-19 by indigenous knowledge of Dayak Ngaju.

Species	Part collected	Usage
Bajakah (*Spatholobus littoralis),* and various unidentified species)	Stem, log	For various cancer treatmentFor traditional medicinal covid-19 with cancer
Asam Limpasu (*Baccaurea lanceolata)*	fruit	For hypertension (high blood pressure)For covid-19 with blood pressure problems
Asam Ramania (*Bouea Marophylla*)		

Species	Part collected	Usage
Kayu Daha (*Corymbia gummifera*)	Root	For Covid-19 with breathing Problems
Akar Kuning (*Arcangelisia flava*)	Stem, log	For Covid-19 with lungs and liver problems
Bamban (*Donax canniformis*)	Bark and root	For mother post-natal or a mother who had just delivered a baby To maintain the health and immunity during covid-19
Daun Sungkai (*Peronema canescens Jack)*	Leaves	For covid treatment, it is believed as an anti-virus, the most famous traditional plant during pandemic covid-19.
Tatunjuk Langit (*Helminthostachys zeylanica)*	Leaves and Stem	For mother post-natal or just delivery baby To maintain the health and immunity during covid-19
Pohon Taya (*Nauclea orientalis*)	Flower and young leaves	For mother past-natal To maintain health and immunity during covid-19.

Figure 4. Sungkai, *Peronema canescens Jack),* the most famous and widely used medicinal plant for covid-19 by indigenous people of Dayak Ngaju.

Figure 5. Pohon Taya, *Nauclea orientalis,* is used both as a medicinal plant and consumed as a vegetable.

The medicinal plants from peat swamp forests are products with a high economic value during the Covid-19 Pandemic. The selection of medicinal plants is made with local wisdom from village elders. This study explored what medicinal plants are traditionally used by local communities, especially during the COVID-19 pandemic, and are used to treat COVID-19.

The first Medicinal is the "bajakah," a climbing and twining plant consisting of many unnamed/unidentified species. Bajakah was viral some time ago because it is believed to treat cancer. The local government of Central Kalimantan had banned the collection of pirates because of over-exploitation and forest destruction. Local people usually collect a small amount according to their needs.

IOP Conf. Series: Earth and Environmental Science **894** (2021) 012023 doi:10.1088/1755-1315/894/1/012023

Bajakah decoction is an anti-virus for patients with covid and is still used as a cancer drug. Currently, the demand for bajakah from outside the region is also high.

Asam Limpasu is a fruit plant that grows in peat areas that are not flooded or parts that are not swampy. The fruit of this plant has recently been used in patients with COVID-19 because it can cure ammonia or loss of sense of smell. On the other hand, it is also used to prevent covid 19 because it contains high vitamin C, increasing endurance. *Limpasu* fruit is also used as a traditional medicine for high blood pressure.

Daha wood is a type of wood whose roots are used by the community to treat respiratory disorders in asthma and COVID-19. This wood has a red sap, just like fresh blood in humans. The part used is the root, and it is boiled with hot water low heat method until turn on a red liquid similar to tea, to be consumed by COVID-19 sufferers.

Akar Kuning (*Arcangelisia flava*) has been used for a long time in patients with impaired liver function and bile disorders. However, during a pandemic, this wood is also used to treat COVID-19 symptoms by boiling, and the water is used for bathing or boiled for traditional saunas, and it can also be drunk if you have liver problems. In addition, this plant is believed to be anti-virus.

Bamban root (*Donax canniformis*) is a root taken from the roots of a shrub that grows on the edge of a river or the edge of a swamp. The stems and bark of this plant are used for making various household items and handicrafts, while the roots are used for postpartum mothers after giving birth. In addition, the roots of this plant can be used to prevent viruses and increase the body's resistance to viruses. Sometimes the root of bamban is used together with the root of Tunjuk Langit (*Helminthostachys zeylanica*) by boiling and drinking the boiled water.

Sungkai (*Peronema canescens Jack*) is increasingly being used to treat covid-19 by various indigenous tribes of Kalimantan (See Figure 4). If other plants are used as prevention, then *sungkai* leaves are often used as an ingredient for saunas to expel sweat and kill the virus. In addition, some drink the stew of sungkai leaves. This species is the most famous medicinal plant for covid treatment and medicine. However, the effectiveness of this medicine is still in indigenous knowledge, not yet verified by pharmacology research. The sungkai leaves are also sold and trade-in a reasonable price and blessing in disguise for local people.

The next tree is the Taya tree (*Nauclea orientalis*). This riverside tree has corona-shaped flowers and looks like the coronavirus when enlarged (See Figure 5). The Ngaju Dayak people use young leaves as vegetables for a mixture of fish and meat. The flower and old leaves are used for medicine, while the young leaves are used for vegetables served with local fish in local food.

3.4. Pandemic Covid-19 impacts on local livelihoods

We also assessed any impacts of pandemic covid-10 on local people's livelihood, as they depend on peat swamp forest products and ecosystem service. The summary result can be seen in the Table 4.

Table 4. Pandemic Covid-19 impacts on local livelihood.

Activities	Impacts of Covid-19 Pandemic	Additional information
Timber Forest harvesting	Less Impacted	Rare activities due to forest status and legal requirement
Non-Timber Forest Harvesting	Less Impacted	Harvesting based on demand and reasonable price
Medicinal plants harvesting and collecting	Less Impacted	Harvesting based on demand and reasonable price

Activities	Impacts of Covid-19 Pandemic	Additional information
Traditional Fishery	Less Impacted	Harvesting based on demand and reasonable price
Marketing and Distributing of products	Highly impacted	Almost nothing, it is about regulation

We also interviewed local people to get firsthand information on how the COVID-19 pandemic is affecting their livelihoods. In general, the pandemic does not directly impact their activities in and out of the peat swamp forest. Still, it has a significant impact when they market and distribute products outside the village due to many restrictions on community activities.

The collection of forest products, both timber, and non-timber forest products, does not significantly impact the COVID-19 pandemic. Local people usually enter the forest area independently and independently by using a *kelotok* or water boat. There are two types: large water boats to transport forest products and small water boats for hunting and fishing. Some communities also have cottages categorized as temporary houses for staying and working for a few days. This halfway house is usually a 4x5 meter hut and can be used for cooking, eating, and resting. Activities that take days, such as harvesting rattan and collecting fish, are carried out by staying at a halfway house; the goal is to save fuel oil for their *kelotok*.

Capture fisheries and fish processing activities are also not affected by Covid-19, where they have no difficulty getting into the forest because existing regulations require restrictions on social activities. At the same time, they usually go to the forest areas in a small group of people in one household.

The most significant impact is marketing difficulties due to restrictions on activities; usually, they market forest products or processed fish and fish to sub-districts or districts 2-3 times a week, now usually only once a week. However, for villages that are far apart, the number of confirmed COVID-19 is deficient.

Another impact is the decline in people's purchasing power, for example, products that are not primary needs, such as handicrafts, fashionable rattan bags, and various other village products.

While fish and processed fish products and people's purchasing power are not affected by the pandemic, there is a slight decline in prices in June-August due to abundant fish production, but returns to normal in another month. Sales of rattan and rubber also experienced a decrease in prices at the beginning of the pandemic, but moving back to normal now, it is just that buyers come with a lower frequency due to social restrictions.

Some communities enjoy increasing demand and prices for traditional medicinal plants, where they sell medicinal plants at reasonable prices and increase their income.

In general, the Covid-19 pandemic affected people's livelihoods due to limited sales and distribution difficulties. However, most of them agreed that the pandemic had no significant impact as long as rubber, rattan, and fish buyers could come to buy their products at standard prices.

4. Conclusion

Our study concludes that The Ngaju Dayak people depend on forests and forest products for their livelihoods from tropical peat swamp forests. The community utilizes timber forest products, non-timber forest products, and ecosystem services. Rattan and rubber are the two non-timber forest products that provide the most economic benefits to local communities. Peat swamp forest also provides ecosystem services for various biodiversity habitats. However, the essential products for local people's livelihoods are fish habitat, forest food habitat, and habitat for rattan.

Although community activities utilizing peat swamp forests during the pandemic were not affected due to the COVID-19 pandemic. Still, the real impact was felt on the marketing of forest products, fish, craft materials, which experienced a decline in demand and fluctuating prices. On the other hand, the COVID-19 pandemic has also triggered local people to utilize medicinal plants with high demand and price. The two most frequently used species are *sungkai* and *limpasu* fruit. In addition, the pandemic condition has also created much demand for traditional medicines that are purchased at reasonable prices.

Acknowledgement
Indonesia Endowment Fund for Education supports this work), the Ministry of Finance, Kapuas Kahayan Forest Management staff, BOSF Foundations staff, villagers in Katunjung, Tumbang Mangkutup and Tuanan, and the local government of Kapuas and Central Kalimantan.

References
[1] Page S E, Rieley J O and Banks C J *Global Change Biology* **17(2)** 798–818
[2] Ritung S and Sukarman 2014 *Kesesuaian Jalan Gambut Untuk Pertanian (Peatlands Suitabilty for Agriculture)* **August 2016** 61–81
[3] Wahyunto N K, Ritung S and Sulaeman Y 2014 *Proceeding Lokakarya Kajian Dan Sebaran Gambut Di Indonesia* **2014** 81–96
[4] Murdiyarso D, Lilleskov E and Kolka R 2019 *Mitigation and Adaptation Strategies for Global Change* ***24*(4)** 493–505
[5] Rudiyanto, Setiawan B I, Arief C, Saptomo S K, Gunawan A, Kuswarman, Sungkono and Indriyanto H 2015 *Procedia Environmental Sciences* **24** 152–157
[6] Yule C M 2010 *Biodiversity and Conservation* **19(2)** 393–409
[7] Harrison M E, Ottay J B, D'Arcy L J, Cheyne S M, Anggodo, Belcher C, Cole L, Dohong A, Ermiasi Y, Feldpausch T, Gallego-Sala A, Gunawan A, Höing A, Husson S J, Kulu I P, Soebagio S M, Mang S, Mercado L, Morrogh-Bernard H C and van Veen F J F 2020 *People and Nature* ***2*(1)** 4–28
[8] Silvius M and Diemont H 2007 *Institute for Environmental Studies* **January 2007**
[9] Suwarno A, Hein L, Weikard H P, van Noordwijk M and Nugroho B 2018 *Land Use Policy* **75(March)** 340–351
[10] Uda S K, Schouten G and Hein L 2020 *Land Use Policy* **99(September 2017)** 103300
[11] Kӧnönen M, Jauhiainen J, Laiho R, Spetz P, Kusin K, Limin S and Vasander H 2016 *Wetlands Ecology and Management* **24(6)** 717–731
[12] Giesen W, Wijedasa L S and Page S E 2018 Unique Southeast Asian peat swamp forest habitats have relatively few distinctive plant species *22* 1–13
[13] Posa M R C, Wijedasa L S and Corlett R T 2011 *BioScience* **61(1)** 49–57
[14] Wijedasa L S, Vernimmen R, Page S E, Mulyadi D, Bahri S, Randi A, Evans T A, Lasmito, Priatna D, Jensen R M and Hooijer A 2020 *Forest Ecology and Management* **461(January)** 117868
[15] Fujita M S, Samejima H, Haryadi D S, Muhammad A, Irham M and Shiodera S 2016 *Ecological Research* **31(2)** 275–285
[16] Indriani F, Siregar U J, Matra D D and Siregar I Z 2019 *Biodiversitas* **20(2)** 482–488
[17] Afentina, McShane P and Wright W 2020 *Agroforestry Systems* **94(2)** 639–650
[18] Ruysschaert D and Hufty M 2020 *Land Use Policy* **99(April 2017)** 103359
[19] Water Management Institute (IWMI) 2014 *Wetlands and people* **2014** 202
[20] Hooijer A, Page S, Jauhiainen J, Lee W A, Lu X X, Idris A and Anshari G 2012 *Biogeosciences* **9(3)** 1053–1071
[21] Miettinen J, Hooijer A and Wang J 2012 *Regional Environmental Change* 729–737
[22] Wijedasa L S, Lupascu M, Sloan S, Page S E, Clements G R and Evans T A 2018 *Global Change Biology* **February** 4598–4613
[23] Jaenicke J, Wösten H, Budiman A and Siegert F 2010 *Mitigation and Adaptation Strategies for Global Change* 223–239
[24] Lindsay R 2016 *The Wetland Book* Lagg Fen 1–4
[25] Boehm H-D V, Siegert F, Rieley J O, Page S E, Jauhiainen J, Vasander H and Jaya A 2001 F

22nd Asian Conference on Remote Sensing **November** 5–9
[26] Giesen W and Meer P v d 2009 Guidelines for the Rehabilitation of degraded peat swamp forests in Central Kalimantan - Master Plan for the Conservation and Development of the Ex-Mega Rice Project Area in Central Kalimantan **Issued October** Alter
[27] Saharjo B H 2006 *Biodiversitas, Journal of Biological Diversity* **7(2)** 154–158
[28] Tapia J F D, Doliente S S and Samsatli S 2021 *Land Use Policy* **102(November 2020)** 105187
[29] Tonks A J, Aplin P, Beriro D J, Cooper H, Evers S, Vane C H and Sjögersten S 2017 *Geoderma* **289** 36–45
[30] Dohong A, Aziz A A and Dargusch P 2017 *Land Use Policy* **69(October)** 349–360
[31] Hapsari K A, Biagioni S, Jennerjahn T C, Reimer P M, Saad A, Achnopha Y, Sabiham S and Behling H 2017 *Quaternary Science Reviews* **169** 173–187
[32] Osaki M and Tsuji N 2015 *Tropical Peatland Ecosystems* **1**–651
[33] Posa M R C, Wijedasa L S and Corlett R T 2018 *BioScience* **61(1)** 49–57
[34] Morrogh-Bernard H C 2009 *Orang-utan behavioural ecology in the Sabangau peat-swamp forest, Borneo* ***February*** 338
[35] Mirmanto E 2010 *Journal of Biological Diversity* **11(2)** 82–88
[36] Dohong A, Abdul Aziz A and Dargusch P 2018 *Wetlands* 1–18
[37] Miettinen J and Liew S C 2010 *Status of Peatland Degradation and Development in Sumatra and Kalimantan* 394–401
[38] Strack M 2008 *International Peat Society*, Vapaudenkatu 12, 40100 Jyväskylä, Finland.
[39] Sinclair A L, Graham L L B, Putra E I, Saharjo B H, Applegate G, Grover S P and Cochrane M A 2020 *Science of the Total Environment* **699** 134199
[40] Page S E, Rieley J O, Shotyk Ò W and Weiss D 1999 *Philosophical Transactions of the Royal Society B: Biological Sciences*.
[41] Wösten H, R Jack and Page S 2008 *Restoration of Tropical Peatlands* **ALTERRA** Wageningen
[42] Hooijer A, Page S, Canadell J G, Silvius M, Kwadijk J, Wösten H and Jauhiainen J 2010 *Biogeosciences* **7(5)** 1505–1514
[43] Ritzema H, Limin S, Kusin K, Jauhiainen J and Wösten H 2014 *Catena* **114** 11–20
[44] Budiman I, Bastoni, Sari E N, Hadi E E, Asmaliyah, Siahaan H, Januar R and Hapsari R D 2020 *Global Ecology and Conservation* **23** e01084
[45] Page S, Hosciło A, Wösten H, Jauhiainen J, Silvius M, Rieley J, Ritzema H, Tansey K, Graham L, Vasander H and Limin S 2008 *Ecosystems* **12(6)** 888–905
[46] Tata H L 2019 *IOP Conference Series: Earth and Environmental Science* **394(1)**
[47] Verchot L V, Petkova E, Obidzinski K, Atmadja S, Yuliani E L, Dermawan A, Murdiyarso D and Amira S 2010 *Center for International Forestry Research* **1–12**
[48] Taufik M, Veldhuizen A A, Wösten J H M and van Lanen H A J 2019 *Geoderma* **347(February)** 160–169

IOP Conf. Series: Earth and Environmental Science **894** (2021) 012024 doi:10.1088/1755-1315/894/1/012024

New species *Pseudomonas capeferrum* TDA1 as a plastic monomer utilizer and a PHA native producer

R N C Utomo[1], H J Heipieper[2], C Eberlein[2], and L M Blank[1*]

[1]Institute of Applied Microbiology - iAMB, Aachen Biology and Biotechnology – ABBt, RWTH Aachen University Worringerweg 1, 52074, Aachen, Germany
[2]Department of Environmental Biotechnology, Helmholtz Centre for Environmental Research - UFZ, Permoserstr. 15, 04318 Leipzig, Germany

lars.blank@rwth-aachen.de

Abstract. Over the decades, global plastic production has been exponentially increasing with a significant increase of plastic waste as well. Consequently, our environment has suffered a lot because synthetic plastic is less biodegradable or even not completely biodegradable. On the other hand, the conventional recycling rate and plastic management in the top ten plastic contributors are still low to reduce the contamination and pollution from plastic waste. Particularly, Indonesia, one of the world's most outstanding emerging market economies and has the most contribution on plastic waste in ASEAN, should consider breakthrough and novel technology to fight global plastic waste. Polyhydroxyalkanoates (PHA) might have the closest relation to plastic waste upcycling because this compound can be used as the primary material to synthesize bioplastic, so-called plastic, to the bioplastic process. Many *Pseudomonads* can natively produce PHA as their extracellular product. This study qualitatively shows that the new strain *Pseudomonas capeferrum* TDA 1 natively produces PHA from various sole carbon sources, including plastic monomers. This finding gives significant insight for many improvements to the "plastic to bio-plastic" process on an industrial scale.

1. Introduction

Resin, which is made from crude oil, is the core ingredient for almost all plastic products. High-density polyethylene (HDPE), low-density polyethylene (LDPE), polypropylene (PP), polystyrene (PS), polyvinylchloride (PVC), polyethylene terephthalate (PET), and polyurethanes (PUR) are categorized as polymer resin-based plastics [1,2]. Since plastics are used in many sectors due to versatility, plastic waste has been exponentially accumulated in the environment. The significant increase of plastic waste in the environment is due to their low biodegradability, low recycling rate, and mismanagement, which brings many environmental and health issues, such as water, soil contaminations, and diseases [3–5].

Particularly, polyurethanes (PU), mainly synthesized from polyols, isocyanates, and chain extenders, share a global production of around 27 million metric tons, making them the 6th most abundant polymer worldwide. However, many PU is highly stable, which is apparent that only half of the PU amount product has been generated as waste [2]. Their varied mixture and composition also cause a lack of recycling of PU. Not limited to those factors, 2,4-toluenediamine, a common precursor and putative degradation intermediate of PU, is considered a toxic compound [6,7]. While enzymatic depolymerization of ester-based PU is possible by several enzymes, such as esterase, urease, amidase, and protease, ether-based PU is quite recalcitrant towards enzymatic depolymerization [8,9]. As alternatives, solvolysis, pyrolysis, and hydrolysis are reported for the chemical recycling of PU [10–12]. To follow up the recent studies on PU depolymerizations, the advanced technology on upcycling PU monomers to high-value products needs further investigation. Therefore, we tried to enable a new strain *Pseudomonas capeferrum* TDA1 to utilize PU monomers in this study [6,7]. Since not only plastic industry contributes annually million tons of waste, in this study, we also investigated the ability of this strain to grow on medium-chain fatty acids, which are possibly derived from oil waste of various industries [13].

2. Methodology

2.1. *Chemical and materials*

All monomers, namely adipic acid, 1,4-butanediol, ethylene glycol, 2,4-toluenediamine, terephthalic acid, sodium decanoate, and sodium octanoate, were purchased from Sigma-Aldrich Chemie GmbH. Sulfuric acid (98%) for the HPLC buffer was purchased from Carl Roth GmbH & Co. KG (Karlsruhe, Germany). Dichloromethane for GC-sample preparation was purchased from Carl Roth GmbH & Co. KG (Karlsruhe, Germany).

2.2. *Microorganism and culture condition*

Pseudomonas capeferrum TDA1 was isolated from soil rich in brittle plastic waste in a former landfill (Paunsdorf, Leipzig, Germany).[7] The strain was grown in a mineral salt medium (MSM) [14] using the different plastic monomers or monomer mixture as carbon source(s) as indicated.

2.3. *Adaptive laboratory evolution*

Initially, adaptive laboratory evolution (ALE) was performed by cultivating P. *capeferrum* TDA1 in MSM with 20 mM adipic acid as a pre-culture. The pre-culture then was used to inoculate 250 mL clear glass Boston bottles with Mininert valves (Thermo Fisher Scientific, Waltham, MA, USA) containing various compositions of monomers for the adaptation on ethylene glycol. Serial transfers were reinoculated several times after the cultures reached an OD600 of at least 0.5, with a starting OD600 of 0.1. After growth was detected (usually overnight), single colonies were isolated from ALE cultures by streaking samples on LB agar plates. After ALE on ethylene glycol, one strain (ALE TDA1) out of 10 strains was selected according to the highest growth rate and biomass in MSM with 30 mM ethylene glycol using the Growth Profiler (R) 960 (Enzyscreen, Heemstede, The Netherlands).

2.4. *Analytical methods*

2.4.1. Monomer analysis. Three monomers, adipic acid, 1,4-butanediol, and ethylene glycol, were analysed with High-Performance Liquid Chromatography (HLPC) using a Beckman System Gold 126 Solvent Module equipped with autosampler, column oven (Beckman Coulter GmbH, Krefeld, Germany), and a Smartline 2300 refractive index detector (Knauer, Berlin, Germany). Analytes were separated using a 300 x 7.8 mm, Metab -AAC, BF-Series column (Isera, Düren, Germany) and a 40 x 8 mm organic acid resin precolumn (CS Chromatographie, Langerwehe, Germany). The eluent was 5 mM H_2SO_4 at a flow rate of 0.7 ml/min. 2,4-toluenediamine was quantified by Gas Chromatography (GC) equipped with Flame Ionization Detector (FID) (Thermo Scientific-Trace GC Ultra). A ZB-wax plus column was used with helium as a carrier gas. For quantification, 2,4-toluenediamine was extracted using dichloromethane with a sample to dichloromethane ratio of 2:1 (0.4 ml sample + 0.2 ml dichloromethane). Afterward, the solution was vortexed for 30 mins at 400 rpm, centrifuged for 5 mins at 13,000 rpm, and GC-FID analysed the lower phase. The split ratio was set to 20 according to the manual of Dikma Technologies Inc [15].

2.4.2. Polyhydroxyalkanoates staining and microscopy observation. For PHA qualitative analysis by staining, 0.2% of a stock solution of 0.25 mg Nile red (Sigma-Aldrich Chemie GmbH) per ml dimethylsulfoxide (DMSO) was added into the sterilized medium to give a final concentration of 0.5 μg/ml. The agar plates were exposed to ultraviolet light (312 nm) after appropriate cultivation periods to detect the accumulation of PHAs.[16] For qualitative analysis by microscopy observation, the culture from appropriate cultivation periods was observed under Leica DM1000 Led microscope using dark field (DF) contrast with 1,000 folds of magnification

3. Result and discussion

3.1. Plastic monomer and preferable substrate utilization

Previously, *P. capeferrum* TDA1, which can utilize adipic acid, 1,4-butanediol, and 2,4-toluenediamine, was evolved to utilize ethylene glycol [6,7]. At the end of 19-day adaptive laboratory evolution, this strain can utilize 30 mM of ethylene glycol with OD_{600} of 1.8 (unpublished data) and now so-called ALE TDA1 strain. It has been proved in many studies that PHA is produced by bacteria when the environmental condition is not optimal due to the limitation of nitrogen or other nutrients and an excess of carbon source. Furthermore, nitrogen limitation plays a vital role in enhanced PHA granule formation in most Pseudomonads. Therefore, we first observed the growth of ALE TDA1 strain on several PU monomers (adipic acid, 1,4-butanediol, and ethylene glycol) and medium-chain fatty acids (Decanoate and octanoate) as a sole carbon source, while one-tenth of the usual amount of ammonium sulfate solution was used as a limited nitrogen source [17,18]. Decanoate and octanoate were used in this study as two of the best alternative substrates in PHA production [17,19]. Each substrate used in this experiment, gave equal C-molar concentration of 0.333 mol/L with 55 mM adipic acid, 83 mM 1,4-butanediol, 166 mM ethylene glycol, 33 mM decanoate, and 42 mM octanoate.

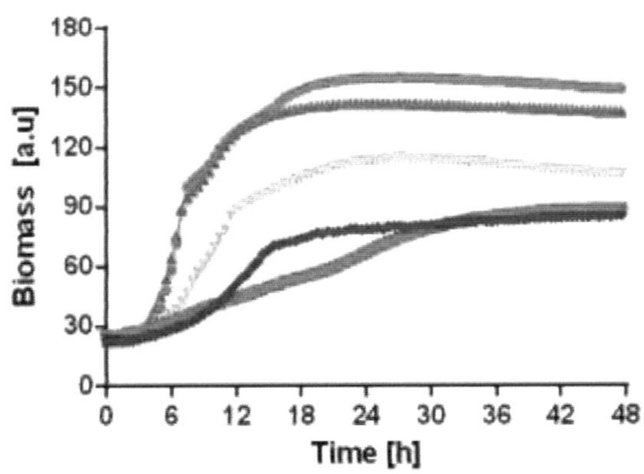

	μ max (h^{-1})
ALE TDA1 on AA + limited N	0.12 ± 0.01
ALE TDA1 on BDO + limited N	0.06 ± 0.01
ALE TDA1 on EG + limited N	0.20 ± 0.00
ALE TDA1 on decanoate + limited N	0.34 ± 0.01
ALE TDA1 on octanoate + limited N	0.38 ± 0.01

Figure 1. The growth of ALE TDA1 in MSM with different single monomers as a sole carbon source. A growth profiler accessed growth in 2 ml medium in 24-square-well plates. Errors depict the deviation of the mean (n = 2).

Based on the growth on AA, BDO, EG, decanoate, and octanoate with nitrogen limitation, we could see that growth on EG gave the highest maximum growth rate and the highest biomass among plastic monomers. On the other hand, growth on Decanoate and octanoate outperformed growth on plastic monomers with maximum growth rates of 0.34 and 0.38 h-1, respectively. As preferable and unique alternative substrates, decanoate, octanoate, and a mixture of fatty acids can be utilized by Pseudomonads, for instance, *P. putida* KT2440 and *P. putida* mt-2 for mcl-PHAs production [17,20].

IOP Conf. Series: Earth and Environmental Science **894** (2021) 012024 doi:10.1088/1755-1315/894/1/012024

3.2. PHA granules formation from plastic monomer and preferable substrate as sole carbon sources

Even though plastic monomers were less preferred as sole carbon sources than the reference substrates (Decanoate and octanoate), we further investigated the growth of *P. capeferrum* TDA1 on the mixture of PU monomers with 10 mM adipic acid, 15 mM 1,4-butanediol, and 30 mM ethylene glycol (60 C-mM of each). Since more 2,4-toluenediamine (TDA) gave a toxicity effect, only 1 mM of TDA was used in this study to see the minimum effect on both 4M and 4M + nitrogen limitation medium. The mixture of PU monomers with- and without nitrogen limitation was used as substrates to see the growth behaviour of *P. capeferrum* TDA1 and consumption rate on those two different conditions.

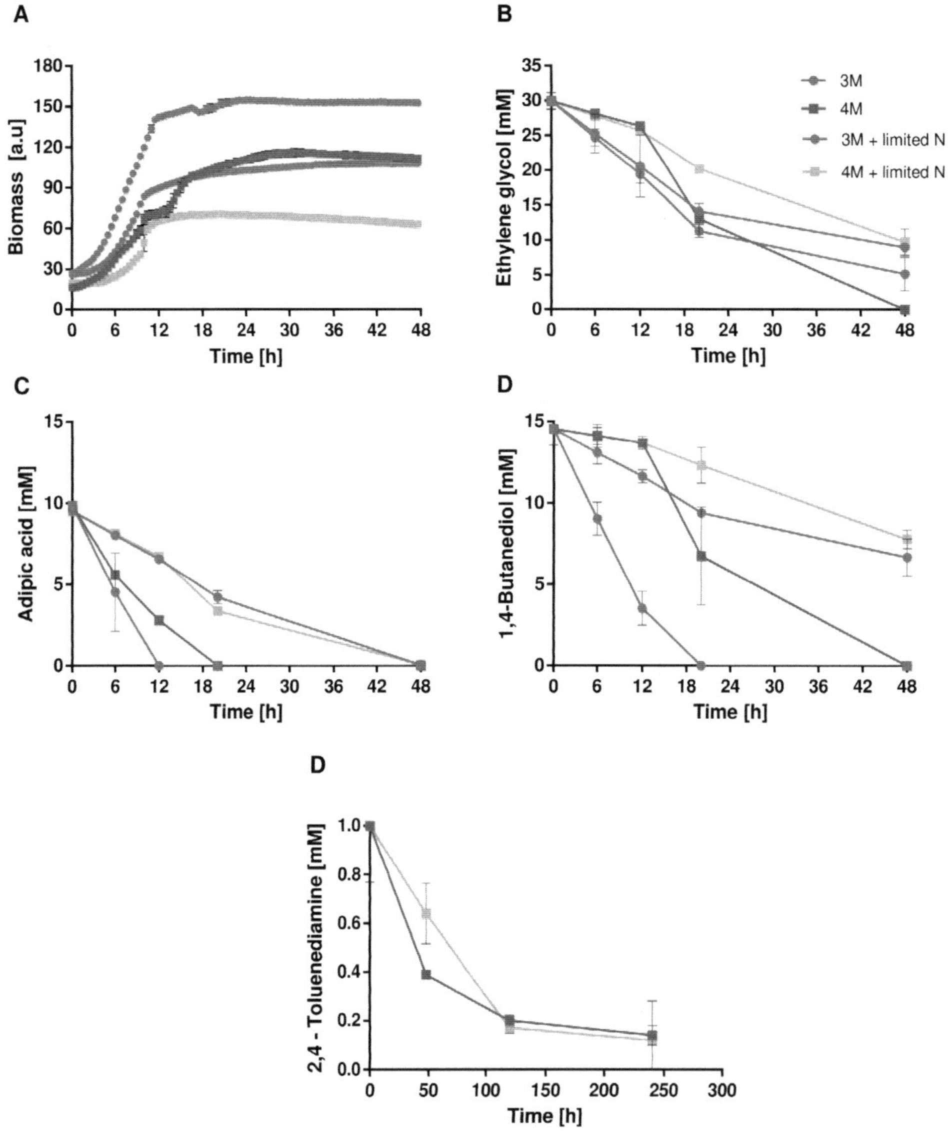

Figure 2. The growth of ALE TDA1 in MSM with 3M (adipic acid, 1,4-butanediol, and ethylene glycol), 4M (adipic acid, 1,4-butanediol, ethylene glycol, 2,4-toluenediamine), 3M + nitrogen limitation, and 4M + nitrogen limitation as sole carbon sources. A growth profiler accessed growth in 2 ml medium in 24-square-well plates. Errors depict the deviation of the mean (n = 2).

It is depicted in figure 2 that nitrogen limitation in 3M and 4M lowered the biomass than the control even though the maximum growth rate between with- and without nitrogen limitation was similar (table 1). The lower biomass showed that nitrogen source plays a vital role in the growth of the ALE TDA1 strain. Nitrogen source is essential for the growth and biochemical synthesis of nucleic acid, lipids, and protein in bacteria. Nitrogen also regulates product formation, in which the biosynthesis of secondary metabolites is limited by the nitrogen sources favouring cell growth, such as ammonium sulfates and certain types of amino acids. Biodegradation of TDA was only observed until 120 h incubation by ALE TDA1, showing better utilization in 4M than in 4M with nitrogen limitation.

Table 1. Effect of nitrogen limitation on MSM medium with 3M and 4M as sole carbon sources towards the growth of *P. capeferrum* TDA1 and consumption rates of each monomer.

Substrate	μ Max (h^{-1})	q_s (mmol g_{cdw}^{-1} h^{-1})
		AA : 1.48 ± 0.24
3M	0.20 ± 0.00	BDO : 2.13 ± 0.04
		EG : 1.81 ± 0.02
		AA : 1.29 ± 0.06
4M	0.23 ± 0.00	BDO : 1.02 ± 0.24
		EG : 2.22 ± 0.08
		AA : 1.02 ± 0.01
3M with nitrogen limitation	0.21 ± 0.00	BDO : 0.99 ± 0.15
		EG : 1.68 ± 0.11
		AA : 1.51 ± 0.09
4M with nitrogen limitation	0.21 ± 0.01	BDO : 0.55 ± 0.04
		EG : 1.19 ± 0.06

It is also interesting to see the consumption rates of adipic acid, 1,4-butanediol, and ethylene glycol in those various mixtures. While the consumption of adipic acid and ethylene glycol was relatively similar in 3M and 4M, the consumption rate of 1,4-butanediol was lower in 4M and further lower in 3M with nitrogen limitation and 4M with nitrogen limitation, respectively. Even though few studies of bioconversion of plastic into valuable products, the prominent studies are from upcycling of PET to PHA and PU to rhamnolipids. From Kenny et al., initially, PET was pyrolyzed at 450°C resulting solid fraction of terephthalic acid, which *P. umsongensis* GO16 used as a substrate for PHA production [21]. In a further study, it was genetically identified that responsible genes involved in ethylene glycol metabolism also exist in *P. umsongensis* GO16. After the strain was validated to use terephthalic acid and ethylene glycol as sole carbon sources, the PHA production from both monomers was also proved [22]. From Utomo et al., the mock PU hydrolysate consisting of adipic acid, 1,4-butanediol, ethylene glycol, and 2,4-toluenediamine combined with- and without reactive extraction 2,4-toluenediamine was used by defined mixed culture of engineered *P. putida* KT2440 strains as sole carbon sources for rhamnolipid production [6].

As proof of concept, we continued to investigate the PHA granule formation in nitrogen-limited medium with plastic monomers and medium-chain fatty acids as sole carbon source(s) via qualitative analysis, namely plate fluorescence and microscopy analysis.

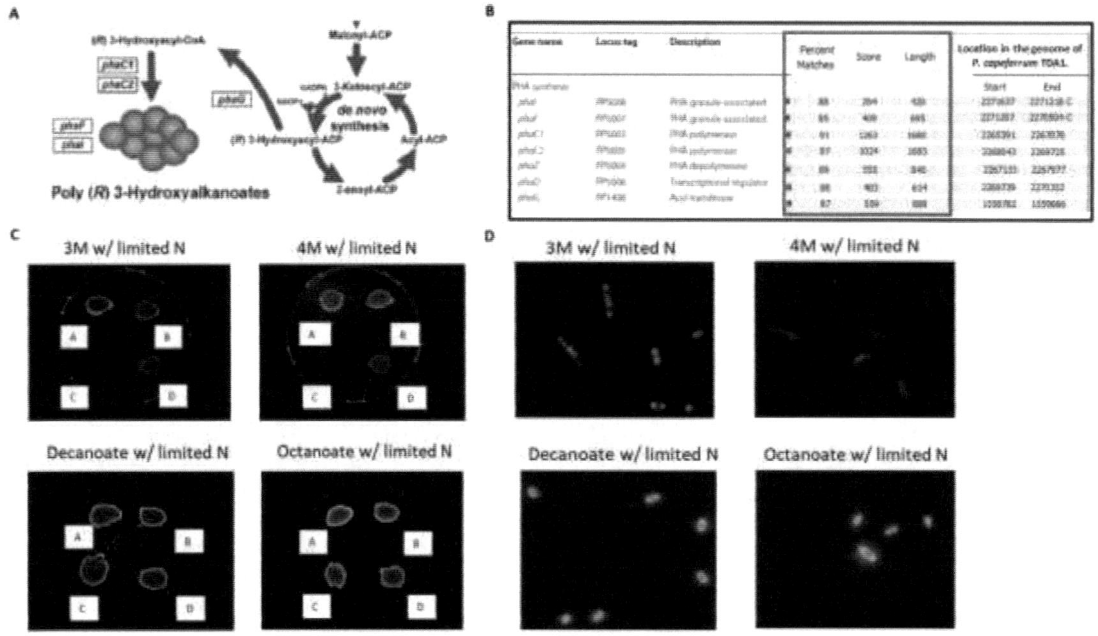

Figure 3. Proof of concept of *P. capeferrum* TDA1 as a native PHA producer. A: Biosynthetic pathway of PHA production in *P. putida* KT2440 as a reference [23]. B: BLAST results of PHA encoding genes between *P. putida* KT2440 and *P. capeferrum* TDA1. C: Plate fluorescence staining for PHA production from 3M (adipic acid, 1,4-butanediol, and ethylene glycol), 3M+2,4-TDA (4M), alternative substrate namely Decanoate and octanoate as sole carbon source with nitrogen limitation with different pre-cultures, namely A: glucose, B: glucose + little N, C: Decanoate + limited N, D: 3M + limited N. D: Microscopy observation of PHA granule formation from the different substrate.

Qualitative measurement for PHA production was conducted with a viable - colony staining method using Nile red for direct screening of PHA accumulation. From figure 3 C, pre-culture in MSM with different substrates were used, namely glucose without nitrogen limitation (A), glucose with nitrogen limitation (B), decanoate with nitrogen limitation (C), and 3M with nitrogen limitation (D). After that, the growing cultures were streaked out on Nile red agar plate with 3M (10 mM adipic acid, 15 mM 1,4-butanediol, and 30 mM ethylene glycol), 4M (3M + 1 mM 2,4 TDA), 55 mM glucose, 33 mM decanoate, and 42 mM octanoate as sole carbon sources. The growth and fluorescent emission were then observed after 1-day incubation.

The PHA production on 3M and 4M mediums was observed when glucose was used as a sole carbon source in the pre-culture, while PHA production on Decanoate and octanoate was shown from all pre-culture conditions. It can be concluded that the PHA production from PU monomers needs more favorable conditions in pre-culture. The PHA granule-forming cells were found from all conditions, even after adding 1 mM 2,4-TDA, but the granules inside the cell were relatively less than the other conditions. 2,4-TDA might inhibit not only the growth but also relatively too little carbon to be taken up [16].

According to the study from Beckers et al., several genes are responsible for PHA granule formation in *P. putida* KT2440, namely *phaC1* and *C2*, for encoding pha polymerases, *phaZ* for encoding pha depolymerase, *phaD* for encoding transcriptional regulator, *phaG* for encoding acyltransferase, and *phaI* and *F* for encoding PHA-forming enzymes. By aligning the local similarity between two molecules (*P. putida* KT2440 and *P. capeferrum* genomes) using clone manager (Sci-Ed software), it is confirmed that the responsible genes for PHA synthesis are the PHA synthesis are also present in *P. capeferrum* TDA1 [23].

4. Conclusion

From this study, ALE TDA1 strains can be used to utilize PU monomers and the other alternative substrates, such as decanoate and octanoate, making this strain a great candidate for cell factory in the production of the high-value product, namely polyhydroxyalkanoates (PHA). Under nitrogen limitation, the biomass from cultivation on a mixture of plastic monomers was lower than that without nitrogen limitation. However, the maximum growth rates among both conditions were relatively similar. As proof-of-principle, the PHA granule formation was qualitatively observed from a mixture of PU monomers with nitrogen limitation, demonstrating its applicability as a carbon source for a biotechnological conversion. This study brings to further promising investigation, such as strain, fermentation, and process engineering, and to get closer to the so-called plastic to bioplastic process. It is also shown in this study that strain *Pseudomonas* TDA 1 can also natively produce PHA. For sure, many improvements are still needed to bring the "plastic to bio-plastic" process on an industrial scale. For diversification of products, no doubt that this strain also can be used for producing varied high-value products, such as rhamnolipids and phenazine, which have many applications.

Acknowledgement

All authors contributed equally to this work. All authors have read and approved the final version of this manuscript, analyzed and interpreted the data. RNC Utomo gratefully acknowledges the financial support of the Indonesia Endowment Fund for Education (LPDP). Under grant agreement no, LM Blank and HJ Heipieper received funding from the European Union's Horizon 2020 research and innovation program. 633962 for the project P4SB. LM Blank acknowledge funding from the European Union's Horizon 2020 research and innovation program under grant agreement no. 870294 for the project MIX-UP.

References

[1] Plastics Europe 2020 *Plastics – the Facts 2020*
[2] Geyer R, Jambeck J R and Law K L 2017 *Sci. Adv.* **3** 25–9
[3] Modak P, Pariatamby A and Seadon J 2017 *Asia Waste Management Outlook*
[4] Barnes D K A, Galgani F, Thompson R C, and Barlaz M 2009 *Philos. Trans. R. Soc. B Biol. Sci.* **364** 1985–98
[5] Wierckx N, Narancic T, Eberlein C, Wei R, Drzyzga O, Magnin A, Ballerstedt H, Kenny S T, Pollet E, Avérous L, O'Connor K E, Zimmermann W, Heipieper H J, Prieto A, Jiménez J, and Blank L M 2018 *Plastic Biodegradation: Challenges and Opportunities Consequences of Microbial Interactions with Hydrocarbons, Oils, and Lipids: Biodegradation and Bioremediation* (Cham: Springer International Publishing) pp 1–29
[6] Utomo R N C, Li W-J, Tiso T, Eberlein C, Doeker M, Heipieper H J, Jupke A, Wierckx N and Blank L M 2020 *ACS Sustain. Chem. Eng.* **8** 17466–74
[7] Cárdenas Espinosa M J, Blanco A C, Schmidgall T, Atanasoff-Kardjalieff A K, Kappelmeyer U, Tischler D, Pieper D H, Heipieper H J and Eberlein C 2020 *Front. Microbiol.* **11** 1–10
[8] Magnin A, Pollet E, Perrin R, Ullmann C, Persillon C, Phalip V, and Avérous L 2019 *Waste Manag* **85** 141–50
[9] Magnin A, Pollet E, Phalip V and Avérous L 2020 *Biotechnol Adv* **39** 107457
[10] Xanthos M and Patel S H 1998 Solvolysis *Frontiers in the Science and Technology of Polymer Recycling* (Dordrecht: Springer Netherlands) pp 425–36
[11] Shakorfow A M 2016 *Int J Sci Res* **5** 13–25
[12] Garrido M A and Font R 2015 *J Anal Appl Pyrolysis* **113** 202–15
[13] Guneser O, Demirkol A, Yuceer Y K, Togay S O, Hosoglu M I and Elibol M 2017 *Brazilian J Microbiol* **48** 275–85
[14] Hartmans S, Smits J P, Van der Werf M J, Volkering F and De Bont J A M 1989 *Appl Environ Microbiol* **55** 2850–5
[15] Dikma Technologies Inc 2018 ProElut™ LLE+ (Liquid-Liquid Extraction)

[16] Spiekermann P, Rehm B H A, Kalscheuer R, Baumeister D, and Steinbüchel A 1999 *Arch Microbiol* **171** 73–80
[17] Poblete-Castro I, Escapa I F, Jäger C, Puchalka J, Chi Lam C, Schomburg D, Prieto M, and Martins dos Santos V A 2012 *Microb Cell Fact* **11** 34
[18] Ciesielski S, Możejko J and Przybyłek G 2010 *J Ind Microbiol Biotechnol* 511–20
[19] Borrero-de Acuña J M, Aravena-Carrasco C, Gutierrez-Urrutia I, Duchens D and Poblete-Castro I 2019 *Process Biochem* **77** 23–30
[20] Fontaine P, Mosrati R and Corroler D 2017 *Int J Biol Macromol* **98** 430–5
[21] Kenny S T, Runic J N, Kaminsky W, Woods T, Babu R P, Keely C M, Blau W and O'Connor K E 2008 *Environ Sci Technol* **42** 7696–701
[22] Narancic T, Salvador M, Hughes G M, Beagan N, Abdulmutalib U, Kenny S T, Wu H, Saccomanno M, Um J, O'Connor K E and Jiménez J I 2021 *Microb Biotechnol*
[23] Beckers V, Poblete-Castro I, Tomasch J and Wittmann C 2016 *Microb Cell Fact* **15** 73

IOP Conf. Series: Earth and Environmental Science **894** (2021) 012025 doi:10.1088/1755-1315/894/1/012025

Pesticide residues in food and potential risk of health problems : a systematic literature review

Yura W F[1*], Muhammad F R[1], Mirza F F[1], Maurend Y L[1], Widyantoro W[1], Farida S S,[1] Aziz Y P[2], Desti A,[3] Edy W[4], Septy M[5], Sutra D[6], Alifia I B[7], Nanda I V M[7], E Fikri[8]

[1] Master Program of Environmental Health, Faculty of Public Health, Diponegoro University, Semarang, Central Java, Indonesia.
[2] Master Program of Electrical Engineering, Faculty of Engineering, Diponegoro University, Semarang, Central Java, Indonesia.
[3] Master Program of Health Promotion, Faculty of Public Health, Diponegoro University, Semarang, Central Java, Indonesia.
[4] Laboratory of Microbiology, Faculty of Medicine, University of Northern Sumatra, Medan, North Sumatra, Indonesia
[5] Master Program of Environmental Science, Gajah Mada University, Yogyakarta, Indonesia
[6] Undergraduate Program of Health Entomology, Faculty of Public Health, Diponegoro University, Semarang, Central Java, Indonesia.
[7] Undergraduate Program of Environmental Health, Faculty of Public Health, Diponegoro University, Semarang, Central Java, Indonesia.
[8] Majoring Environmental Health, Health Polytechnic Ministry of Health, Bandung, West Java, Indonesia.

firmansyahyura@gmail.com

Abstract. Pesticide residue in food has been investigated since the growing demand of food safe. The determination of pesticides residues in food is becomes an essential requirement for consumers, producers, and authorities responsible for food quality control. Pesticides can poison humans through the mouth, skin, and breathing. Often unwittingly these toxic chemicals enter a person's body without causing sudden pain and causing chronic poisoning. This study aimed to investigate the impact of pesticides residue to health problems from meta-synthesize, sourced from the Scopus and Sinta indexed articles and obtained 12 indexed articles that were used as references. Meta-synthesize result showed that there are some type of pesticide who used by farmers such as chlorphenapir, emamctin benzoate, abamectin, chlorpyrifos, mankozeb, chlorotalonil, and propineb. Continous use of pesticides can cause such as fatigue, excessive saliva, hard breathing, frequent urination, blurred vision, dizzinesss, and fingerpain. At the end, pesticides residue is adverse effect on human health problems.

1. Introduction

Pesticides are nonetheless an essential a part of agricultural sports across the world. Pesticides are broadly used to enhance the productiveness and first-rate of agricultural products. Pesticides are compounds used to shield flora through killing, repelling, or controlling pests earlier than and after harvest. Insecticides will harm the goal organism. However, the mode of motion of pesticides isn't always species-specific. The maximum normally used pesticides are organophosphate, carbamate and pyrethroid pesticides [1]. The outcomes of pesticide publicity have emerged as an international environmental problem, including fatigue, immoderate salivation, trouble breathing, common urination, blurred vision, dizziness and finger pain. The expanded vulnerability of farmers to pesticide poisoning is because of a lack of understanding on a way to use insecticides properly and appropriately [1],[2].

Pesticides in the food chain is one of major human health concern, several studies already analysed pesticide residues content in rice grains [3]. Consuming food contaminated with pesticides poses a potential health risk to the human body. Recently, a cumulative risk assessment study on the possibility of exposure to multiple pesticides may cause unforeseen adverse effects on human health, which is a key issue in many fields [4]. In addition, trying out merchandise infected through insecticides is likewise critical for shielding the ecosystem. One of the bad outcomes of artificial insecticides, particularly organochlorine and organophosphorus (OP) insecticides, is meals contamination. In the Seventies and early 1980s, the excessive toxicity and residue tiers of insecticides utilized in flora had been the principle problems, with organochlorine (OC) and organophosphate dominating (OP) [5],[6]. Organochlorines is a persistent toxic chemical, which is a persistent organic pollutant. Exposure of humans and other life forms to these chemicals is reported to have a variety of carcinogenic, reproductive, neurological, immunological, and other adverse effects [7],[8].

Various analytical strategies had been used to evaluate the ability dangers of insecticides to human fitness and the surroundings. Reducing using insecticides for useful functions and minimizing the dangers to human fitness and the surroundings is a project and a intention really well worth pursuing [9]. It is important to see the correlation of reducing pesticide with reduction of the possible adverse effects of pesticides on human health and the environment [10].

Due to the full-size use of insecticides in agricultural practices, this has turn out to be a meals protection issue. Research on pesticide residues in rice could be very critical to shield the fitness of consumers. This have a look at evaluated the evaluation of pesticide publicity in meals. The file goals to show records approximately insecticides used, pesticide residues, the effect of number one processing on pesticide residues, human fitness threat assessment, and focus of pesticide infection in meals. This records allows keep away from terrible results on humans.

2. Methodology

Metasynthesis is a technique of reading facts accrued from preceding studies articles as a way to draw new conclusions or bridge the distance among preceding studies. Meta-synthesis includes collecting, summarizing, reading, and reviewing facts from preceding studies literature. Meta-synthesis is a part of the systematic literature studies method. This article makes use of 15 studies articles to check the capability dangers of pesticide residues and fitness troubles in food. The article choice procedure is primarily based totally on the subsequent inclusion criteria:

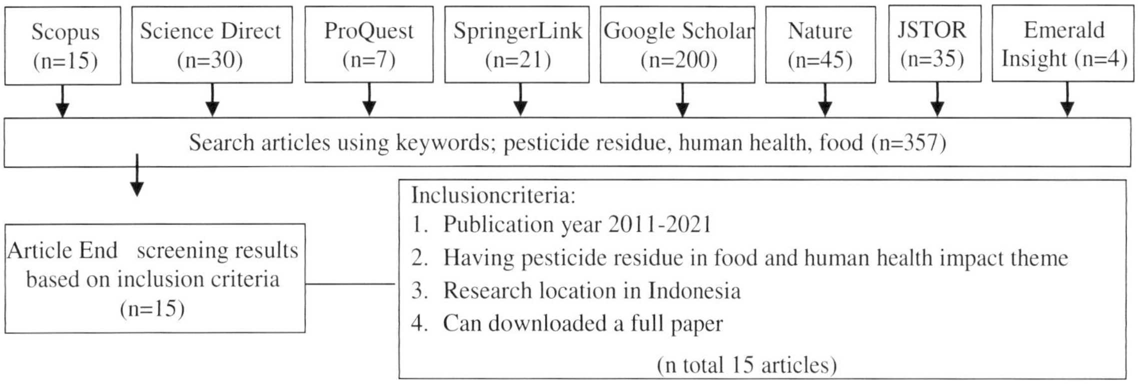

Figure 1. Article screening chart.

3. Result and discussion

Table 1. Result synthesis of article.

No	Author	Findings	Impact on Human Health
1	Tri Joko[1], Nikie A. Y. Dewanti, Hanan L. Dangiran. [11]	1. Conduct health checks on 100 indonesian farmers who use pesticides, checks such as headaches, dizziness, burning sensation in the esophagus, body cramps, difficulty breathing, coughing, and easily tired. 2. Conducting Focus Group Discussions with farmers regarding excessive use of pesticides. 3. It was found that excessive pesticide use was associated with headaches, difficulty breathing, excessive salivation, blurred vision, body aches, and excessive dizziness. 4. The types of pesticides used are organophosphates and chlorphyropos.	Symptoms appear in farmers such as headaches, dizziness, sensation in the esophagus, body cramps, difficulty breathing, coughing, and easily tired
2	Hestin A[1], Amin Leksono, Zulfaidah Gama. [12]	Conducting open interviews with farmers using the pitfall trap method. The pitfall trap used is 7 cm in size and crammed with 50mL 70% alcohol. Placed 2 days before spraying and a couple of days after spraying. Research locstion in Malang Regency, East Java, Indonesia	The occurrence of resistance, secondary pest outbreaks, and potential to environmental pollution
3	Andria M C, Abee L B, Kristina A T, Melissa JP. [2]	OP pesticides are harmful to the environment, and lots of developed and developing countries have monitored their residue levels in food. Although OP pesticide residues in white rice pose a possible risk to certain groups of individuals, many countries currently don't implement national white rice safety standards. it's necessary to conduct annual monitoring of OP pesticide residues in white rice. It should even be noted that the present research is merely focused on white rice, which has led to underestimation of the entire exposure to surgical pesticides. the continued monitoring plan should be extended to other plants treated with OP pesticides. apart from children	Fatalities, Tetralogy of Fallot, Anencephaly, Autism Spectrum Disorder, Memory Loss, Finger Tremor

		aged 2-4 years (119% of ADI) and 4-7 years old (104.3% of ADI), the daily intake of P99.9 of surgical pesticides altogether age groups is less than that of ADI (52-94.5% of ADI). it's also necessary to show other crops that are treated with AChE inhibitory pesticides to know the role of dietary intake in total OP pesticides exposure in china.	
4	Chen C, Li Y, Chen M, Chen Z and Qian Y. [13]	This examine attempted to get the residues of precise pesticides (organophosphorus and pyrethroids) and fungicides (triazole and chloronitrile) withinside the amassed culmination and vegetables. The Electron Capture Detector Gas Chromatograph (ECDGC) is employed to make a decision the notice of twenty-two pesticide residues utilized in pest management. Among the utmost regularly detected pesticide residues, cypermethrin became detected in 18.7% of the analyzed samples.	Immaturity, Fatigue, Fingerpain
5	Anna H, Jane, James, etc. [5]	We discovered that corporations of farmers are at better threat of reporting signs of poisoning. This locating became unexpected, however it might be thanks to the very fact participants of the agricultural network are greater aware about the power dangers of insecticides and consequently file greater signs. However, withinside the absence of farmers' expertise and practice, citizenry have discovered the usage of WHO Class II insecticides and insecticides with decrease toxicity. they're susceptible to acute poisoning, persistent fitness troubles and environmental pollution.	Headache, Fatigue, Vomit, Skin Irritation, Excessive Sweating, Blurred Vision, Dizzines
6	Pornpimol Kongtip, Noppanum Nankongab, Mathuros Tipayamongkholgul, Ariya Bunngamchairat, Jutharak Yimsabai, Aranya P, Susan Wokie. [7]	The Thai foodstuff has found high levels of pesticide residues in fruits and vegetables, which has become a current public ill health. Therefore, many Thais are increasingly choosing organic foods. Thailand Pestiicide Early Warning Network (ThaiPAN) tested 296 fruit and vegetable samples from five regions in Thailand in 2016, and located that 51.4% of the fruit and vegetable had obtained the "Q (Quality) Mark" standard issued by the National Agricultural and Food Administration of Thailand, and therefore the contamination by pesticide residues exceeded the utmost. Residue Limit (MRL). Most organic farmers during this study (86.9%) had used pesticides before they realized the toxicity of the pesticides they used and switched to organic farming. Therefore, they're older than traditional farmers on the average and have a better level of education. About 35% of traditional and organic farmers are in debt, and 57% of organic farmers have part-time jobs additionally to agricultural work. In 2013, among the 4.8 million people living in rural areas, most of them were engaged in agricultural work, of which 10.9% lived in poverty. 4,444. Because there are greater guys in the conventional farmer organization and male muscular tissues has a	Low Body Mass Index, Stunting, Heavy Breathing

		tendency to be better, your BMI might also additionally even His frame fats percent or waist circumference isn't , however better and falls into the "abnormal" range. Another reason behind the distinction in BMI may also be that natural farming is greater labor-intensive, at the same time as chemical farmers use greater agricultural equipment of their operations.	
7	Damalas C A and Eleftherohorinos I G. [10]	Pesticides have performed a key function in ensuring dependable deliver of agricultural merchandise at lower priced prices, enhancing product high-satisfactory and ensuring excessive income for farmers. Although insecticides are designed to paintings with affordable protection and minimum dangers to human fitness and therefore the environment, many research are involved approximately the fitness dangers resulting from non-occupational publicity of farmers (or different quit customers of insecticides) and therefore the public. However, their use shows decreased protection, which shows that chance signs want to be evolved to reinforce the accuracy and reliability of pesticide hazard assessment, thereby assisting to reduce the viable dangerous outcomes of insecticides on human health and therefore the environment.	Skin irritation, Eye Irritation, Oral Irritation, Dizzines, Vomit, Headache
8	Vu N Huyen, Nguyen S, Nguyen Thuy, Le Dung, Luong Khanh. [14]	This study conducted interviews with 300 farmer households randomly selected from 3 studies villages. This observe extensively utilized the fee of disease (COI) approach to degree the effect of crop safety merchandise on farmers' fitness costs. Formula for estimating fitness care costs.	Otorhinolaryngologic diseases, Eye diseases, Nervous system diseases, Dermatological diseases, Musculoskeletal diseases, Digestive diseases, Urologic diseases, Teeth - jaw – face diseases, cancer
9	Oluwatoyin T. Fatunsin, Aderonke O. Oyeyiola, Muyideen O. Moshood, etc.[15]	Analyze organophosphate and carbamate pesticide residues on a fueloline chromatograph (Agilent 7890A) including a spectrometer (GC-MS). The sensitivity of the tool is likewise decided with the help of using calculating the restrict of detection (LOD) (signal-to-noise ratio of 3:1) and therefore the restrict of quantification (LOQ) (awareness with a signal-to-noise of 10:1).	Does not pose a health threat
10	G Bhandari, Kishor Atreya, Paul T.J. Scheepers, Violette G. [16]	The soil samples had been gathered according with the concepts of EU guidelines. that they had been marked as easy soil samples through Wageningen Food Safety research facility and did now not include any of the 23 insecticides tested. Chemical evaluation and fine manage had been according with the ecu Commission Conduct Guide (SANTE / 11813/2017)	Cancer and non-cancer health problems Skin disease disorders due to direct contact
11	Michiel A. Mònica	Concentration of six dialkyl phosphate in urine	not explained, only measuring the

	Guxensc,d,e,f, Suzanne Spaanb, Trudy Voortmana,g, Vincent W. Jaddoea,g,h, Todd A. Juskoi, Matthew P. Longneckerj,Henning Tiemeierd.[17]		concentration of pesticide residues in the urine
12	Haji Mwevura, Henrik Kylin, Tash Vogt, Hindrik Bouwman. [18]	Use the Agilent GC 6890 and Varian 3400 series for fuel chromatographic (GC) quantification of pesticide residues. The fuel oil chromatograph is equipped with thermionic detectors for the selective detection of nitrogen and phosphorus compounds for the investigation of degradation products; Organochlorine pesticides (OCP) contain hexachlorobenzene (HCB), hexachlorocyclohexane (HCH), DDT and its decomposition products, cyclodiene (aldrine, dieldrin, heptachlor, heptachlorobenzene, gammachloride, endrin, and Ken's agent); and organophosphate (OP) pesticides (chlorpyrifos, diazinon, fenthion, oxon and malathion). The herbicide thiobendazim was also tested.	Only describes biota impacts where toxins accumulate biologically and possible food webs may be affected.
13	Norsita Agustina, Norfai. [19]	1. Conducted research on 50 horticultural farmers 2.The variable that features a significant relationship with pesticide exposure to anemia in horticultural farmers 3.There are 11 respondents with mild poisoning thanks to exposure to pesticides 4.There are 4 respondents who experience anemia	Pesticides decrease the production or increase the destruction of red blood cells. This causes hemoglobin to become abnormal resulting in hemolytic anemia
14	Siti A K, Onny S, Sri A N. [20]	1. Conducted on 40 respondents 2. The results of the Binary Logistics Regression test show that exposure to pesticides features a tendency of 5,333 times greater influence on the incidence of anemia compared to respondents who aren't exposed to pesticides. 3. Types of pesticides used by organophosphates	Causes anemia with symptoms of dizziness, weakness, if you get up from sitting, your eyes feel dizzy

3.1. The presence of food pesticide residues

The use of pesticides basically serves to control plant pests. Pesticides such as insecticides for insect pest control, fungicides for fungal pest control, and herbicides for weed pest control [21]. The active ingredients of pesticides that are often found are the organophosphate group (chlorpyrifos, paration, profenofos, diazinon, phenitrotion, metidation and malation), pesticide group IA and IB, insecticide (deltamethrin), organochlorine group consisting of lindane, heptachlor, DDT, aldrin, dieldrin, endrin and endosulfan [12][2].

Excessive use of pesticides, inappropriate doses, frequency of spraying will cause pesticides to leave residues. Pesticide residues can be found in plant products such as chili, rice, shallot, tomato, eggplant, lettuce, apple, orange, corn, papaya, banana, pumpkin, mustard greens, cauliflower, beans, dragon fruit, bitter melon, pear, grapes, longan, strawberries, cabbage and leeks [22][10]. It is also possible for pesticide residues to be present on the soil, due to the spraying process carried out the residue will fall to the surface of the soil. Pesticide residues found in groundwater occur because residues that start on the surface of the soil will seep and enter the pores of the soil to contaminate resistant water. Pesticide residues found in the air occur due to the spraying process that is not in accordance with the direction of the wind, the suitability of the plant height when spraying, and wind speed [14][15].

Measurement of food residues can be identified using gas chromatography is a type of chromatography that is commonly used in chemical analysis for the separation and analysis of compounds that can evaporate without undergoing decompos [17][18] Common uses of GC include testing the purity of a particular compound, or separating the different components in a mixture. Measurement of other pesticide residues can use the G9 Fast Pesticide Test Kit, which is a pesticide test tool to detect pesticide residues in the Organophosphate, Carbamate and Cholinesterase inhibitor sample groups which are very sensitive with qualitative results. QuEChERS Method (Quick, Easy, Cheap, Effective, Rugged and Safe) can be used in multi-analysis pesticide residues on various sample matrices [19] . The pesticide residue left on pepper can be harmful to human health. Therefore, the FAO/WHO Codex The Alimentarious Commission has established Maximum Residue Limit (BMR) of pesticides at pepper is 1 mg/kg for DDT; 0.5 mg/kg for -cyhalothrin, fenvalerate, deltamethrin; 0.1 mg/kg for permethrin, cyfluthrin, cypermethrin, heptachlor and 5 mg/kg for endosulfan [23].

3.2. The pesticide exposure on human health

Table 2. Exposure of active pesticide ingredients to human health.

No	Active Pesticide Ingredients	Symptom on Human Health
1	Organophosphates and chlorphyropos	Symptoms appear in farmers such as headaches, eye irritation, itchy skin, excessive saliva, dizziness, vomiting, burning sensation in the esophagus, body cramps, difficulty breathing, coughing, and easily tired
2	Organochlorine pesticides (OCPs) include hexachlorobenzene (HCB), hexachlorocyclohexanes (HCHs), DDT and its decomposition products, cyclodiene (Aldrin, Dieldrin, heptachlor, heptachlor epoxide, Gamma Chlordane, Endrin, and Ken's Agent); and organophosphorus pesticides (OPP) (chlorpyrifos, diazinon, fenthion, oxon, and malathion).	Dermatological diseases, Musculoskeletal diseases, Otorhinolaryngologic diseases, Eye diseases, Nervous system diseases, Digestive diseases, Urologic diseases, Teeth - jaw – face diseases, cancer
3	OP (Organophosphates) pesticides	Fatigue, headache, dizziness, loss of appetite, nausea, stomach cramps, diarrhea, blurred vision, tearing, sweating, excessive salivation, tremor, pupil constriction, slow heart rate, muscle spasms (twitching), inability to walk, discomfort and tightness, uncontrolled defecation and urination, incontinence, unconsciousness and convulsions. Complications that can occur include pulmonary edema, respiratory arrest, atrioventricular block, and convulsions.
4	Residues of selected insecticides (organophosphorus and pyrethroids) and fungicides (triazoles and chloronitrile) in the collected fruits and vegetables. Most frequently detected pesticide residues, cypermethrin was found.	Symptoms that arise are fatigue, headache, dizziness, loss of appetite, nausea, skin stomach cramps, asthma, allergies, irritation to hearing and taste where the effects of pyrethroids pesticides generally appear 1-2 hours after exposure and disappear within 24 hours.[23]
5	Organophosphate and carbamate pesticide residues	Nervous, diaphoresis (cold sweat),

		shortness of breath, rapid processive circulatory collapse or shock which will lead to anemia Symptoms of carbamate poisoning appear quickly but disappear quickly when compared to organophosphates [23]

4. Conlusion

Pesticide residues in food have an effect on human health. The health risks arising from exposure to these pesticides can vary from each type of pesticide. where some pesticides cause acute symptoms and prolonged illness such as organophosphates (OPPs), chlorphyropos, organochlorines (OCPs), cyclodienes, triazoles, and chloronitrile pesticides but some only cause temporary and quickly disappearing effects such as pyrethroids and carbamates.

References

[1] Okoffo E D, Mensah M and Fosu-Mensah B Y 2016 *Environ Syst Res* **5(1)**
[2] Cimino A M, Boyles A L, Thayer K A and Perry M J 2017 *Environ Health Perspect* **125(2)** p 155–162
[3] Teló G M, Marchesan E, Zanella R, Peixoto S C, Prestes O D and de Oliveira M L 2017 *Acta Sci - Agron* **39(1)** p 9–15
[4] Chen C, Li Y, Chen M, Chen Z and Qian Y 2009 *Food Addit Contam - Part A Chem Anal Control Expo Risk Assess* **26(3)** 340–347
[5] Oesterlund A H, Thomsen J F, Sekimpi D K, Maziina J, Racheal A and Jørs E 2014 *Afr Health Sci* **14(2)** 420–433
[6] and Jikun Huang S R, Fangbin Qiao, Linxiu Zhang 2001 *Farm Pesticides, Rice Production and Human Health in China* **53(9)**
[7] Kongtip P, Nankongnab N, Tipayamongkholgul M, Bunngamchairat A, Yimsabai J, Pataitiemthong A and Woskie S 2018 *Int J Environ Res Public Health* **15(11)** p 1–12
[8] Eqani S A M A S, Malik R N, Katsoyiannis A, Zhang G, Chakraborty P, Mohammad A and Jones K C 2012 *J Environ Monit* **14(6)** p 1645–1654
[9] Rahman S and Chima C D 2018 *Agric* 8(3) p 1–14
[10] Damalas C A and Eleftherohorinos I G 2011 *Int J Environ Res Public Health* 8(5) p 1402–1419
[11] Joko T, Dewanti N A Y and Dangiran H L 2020 *J Environ Public Health* **2020**
[12] Asih H A, Leksono A S and Gama Z P 2019 *Int J Sci Res Publ* **9(12)** p p9672
[13] Chen C, Qian Y, Chen Q, Tao C, Li C and Li Y 2011 *Food Control* **22(7)** p 1114–1120
[14] Huyen V N, Van Song N, Thuy N T, Dung, L T P and Hoan L K 2020 *Sustain Futur* **2** p 100026
[15] Fatunsin O T, Oyeyiola A O, Moshood M O, Akanbi L M and Fadahunsi D E 2020 *Sci African* **8** p e00442
[16] Bhandari G, Atreya K, Scheepers P T J and Geissen V 2020 *Chemosphere* **253** p 126594
[17] van den Dries M A, Pronk A, Guxens M, Spaan S, Voortman T, Jaddoe V W, Jusko T A, Longnecker M P and Tiemeier H 2018 *Int J Hyg Environ Health* **221(3)** p 489–501
[18] Mwevura H, Kylin H, Vogt T and Bouwman H 2021 *Reg Stud Mar Sci* **41** p 101607
[19] Agustina N and Norfai N 2018 2013 *Maj Kedokt Bandung* **50(4)** p 215–221
[20] Kurniasih S A, Setiani O and Nugraheni S A 2013 *Jurnal Kesehatan Lingkungan Indonesia* **12(2)**
[21] Louisa M, Joko T, Lingkungan B K, Masyarakat F K, and Diponegoro U 2018 *J Kesehat Masy* **6(1)** p 654–661
[22] Mergia M T, Weldemariam E D, Eklo O M and Yimer G T 2021 *J Heal Pollut* **11(30)** p 1–19
[23] Kurniasih S A, Setiani O, Nugraheni S A 2013 *Jurnal Kesehatan Lingkungan Indonesia* **12(2)** 132-137

IOP Conf. Series: Earth and Environmental Science **894** (2021) 012026 doi:10.1088/1755-1315/894/1/012026

Management aspects of indigenous lands in environmental conservation

H Thamrin

Department Islamic Economic, Post Graduate Programme, State Islamic University of Sultan Syarif Kasim Riau JL HR Subrantas 17, Pekanbaru, Indonesia

husni2017husni@gmail.com

Abstract. This study is intended to analyze the management aspects of indigenous lands in environmental conservation. This research applies qualitative grounded research methods from the sociology-anthropology-ecology—research shop in Riau Province. The results of the study found that many indigenous peoples had lost their indigenous lands and local wisdom. Therefore, to maintain the sustainability of indigenous land conservation, it is necessary to carry out eco-culture management by considering the following points: First, the right to self-determination regarding the cultural identity one has. The second is territorial rights and indigenous land. The third is collective human rights. Fourth is cultural rights. Fifth is the right to adhere to their own religious and moral belief system and values. Sixth is the right not to be discriminated. Seventh is the right to participate fully in the political process. Eighth is the right to obtain compensation for any activities that have a detrimental impact on the environment and social, cultural, spiritual and moral values. This eco-cultural management perspective needs to be implemented in the government's socio-political policies, people's economic policies and ecological sustainability policies.

1. Introduction

The management of indigenous lands in Indonesia is currently experiencing erosion or criticality, some of which are degraded due to several factors. One of these factors is the land ownership policy which adopts colonialism policies so that the government does not recognize many indigenous lands. In fact, before the colonial era, indigenous land and indigenous law in the archipelago were included in the ownership of indigenous land in Malay Land in Riau Province. Likewise, the existence of indigenous peoples has been marginalized. Indigenous peoples are distinguished based on kinship (genealogical) and territory (territorial). Based on the perspective of its form, indigenous peoples who stand alone become part of a higher indigenous law and consist of several lower indigenous law communities and an association of several equal indigenous law communities. Furthermore, it can be said that each of these forms of indigenous law communities can be dubbed as indigenous peoples who live, are stratified, and in series"[1]. The existence of indigenous peoples has been going on for a long time in Indonesia. However, its current existence has long been eliminated and degraded. The non-recognition of the existence of this indigenous land has caused much harm to indigenous communities and the sustainability of the environment. In the regulation of indigenous land, local wisdom is regulated regarding preserving the environment and its contents.

Land management today does not pay attention to elements of local wisdom but tends to be more concerned with economic-capitalistic profits that exploit nature and its contents and resources. It is what causes the degradation of local communities and gives birth to marginalization, where is what is called the Anthropocene perspective of environmental tricks. It is dominated by anthropocentric-capitalistic capitalistic economic activities and modern technology [2]. Nature, which is understood as a sacred place by the Malay indigenous people and holds a million values of local wisdom that has long been rooted in people's lives, can be explained by using reason to evoke an attitude of admiration full of respect. The sacredness and mystery of nature have been eliminated in the anthropocentric-capitalistic paradigm and modern technology. In the positivistic-capitalistic anthropocentric paradigm, nature is understood as merely a profane fact without any sacred-magical-religious vibrations and no moral values and content. Nature is reduced as objects and facts that can be sorted out, analyzed and explained scientifically-rationally.

The ecological crisis is also a phenomenon caused by the mechanistic-reductionistic-dualistic perspective based on Cartesian science. The anthropocentric perspective is corrected by the ethics of bio-centrism and ecocentrism, especially Deep Ecology, to review nature as an ethical community [3]. This anthropocentric-capitalism perspective has destroyed much of the Malays' environmental, economic, cultural, political, social, infrastructure, and technological lives. However, some of the Malay local wisdom in preserving the environment, among others, still survive amid the currents of shifting by the insistence of an anthropocentric perspective. Some are experiencing a crisis due to the pressure of the influence of modernization. Meanwhile, others have been swept away by modernization and an anthropocentric perspective. The concept of local wisdom and ulayat land of the Riau Malays has been practised by the Malay indigenous peoples or traditional communities in other places. The perspective of humans, indigenous lands and the environment as an integral part of a harmonious relationship between humans and nature, as well as behaviour that is full of responsibility, full of respect and care for the continuity of all life in the universe, has become the perspective and behaviour of various Malay communities. In this case, the perspective of the Malay community has a harmonious relationship between humans and the environment, which is then called the eco-culture perspective [4].

Summarizing the explanation above, an accurate solution is needed to answer the challenges of the times, especially related to preserving the environment and function of indigenous lands that are increasingly being degraded and marginalizing human life due to an eco-centric perspective.

2. Methodology

This study used a qualitative grounded research method. The research location is in Riau Regency, Indragiri Hilir Regency, Siak Regency, Dumai City, Pekanbaru City.

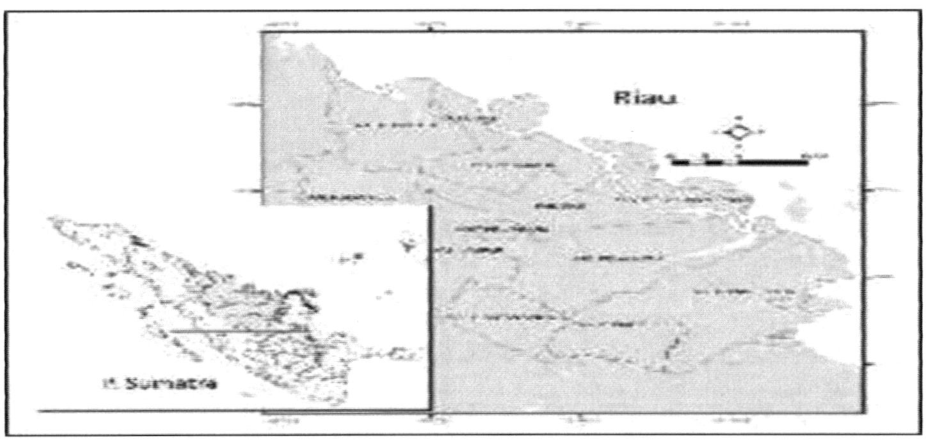

Figure 1. Map of the research site.

2.1. Data collection
Techniques This research was conducted through in-depth interviews with informants as follows; The research informants included: traditional figures = 17. Company leaders = 5 academics = 11, environmental Non Governance Organization = 12, Friends of the Earth Indonesia = 7, Ministry of Environment = 13 , Environmental Agency = 15, Religious Figures = 25, Ordinary People = 35, community leaders = 17 Data analysis activities started from classification, categorization, interpretation, and discussion.In addition to conducting in-depth interviews, data was also collected through observation and data on Indigenous Lands' location. extinct, cultural rights and territories and indigenous land rights losses, the right to self-determination and cultural identity of Indigenous Lands activities and facilities, right to profess a belief system and values and socio-politics, as well as local culture losses.

Data processing or descriptive analysis contains the meaning of an attempt to simplify and at the same time explain part of the entire data through classification and Indigenous Lands. Indigenous Lands categorization steps to compile a series of systematic and accurate descriptions in the Sociology-Anthropology Ecological approach. The process of classifying and categorizing data was carried out since they were in the field, along with the data collection process and continued in more detail and systematically after all data were collected. Classification and categorization of data are carried out in part but still maintain the position of Malay culture as a whole. In categorizing the data, there is an attempt at a qualitative interpretation that is carried out inductively based on the emic approach.

3. Results and discussion

3.1. Scrape anthropocentric value indigenous peoples
Vandana Shiva noted that modernization has led to traditional communities being uprooted from their cultural roots. Religiosity and spirituality of indigenous peoples who ranged in nature have been replaced by modernization as a new religion and spirituality [5]. No casualties were no longer in the form of offerings but in human form, individuals or groups that had to be sacrificed for the sake of modernization and the construction of a new religion. “Dams, mines, power plants, military bases all of this is a religious temple called the new development. What is sacrificed at the altar of religion are natural life and human life. Sacrament development consisted of destruction and desecration of sacred things more. When natural sacred has lost its strength and is seen merely as a resource for development, nature was destroyed without fear magical.

In the current of modernization, traditional knowledge is eroded and undermined by the dominance of the anthropocentric view. In addition to the modernization and anthropocentric domination of indigenous peoples, marginalized indigenous philosophy of value is very high, and there is also the invasion of the perspective and lifestyle of the anthropocentric and pragmatic people. Because everything is placed within the framework of human economic interests and a rational explanation-positivistic framework, local knowledge that does not meet both criteria were then crushed extinct anthropocentric.[6] The wisdom way of thinking and knowledge of indigenous people Malay are not recognized as knowledge even denounced as the occult-irrational. Thus, all the wisdom and practices of life and its values are so rich, particularly concerning nature, removed from modern society

The dominance of positivistic philosophy and ethics derived from Aristotle and reinforced by the paradigm of anthropocentric ethical buried indigenous Malays. With a view of humans only as social beings and by limiting ethics as only applicable to the human community, the ethics of indigenous Malays have primarily been forgotten altogether by modern society. Man's relationship with nature in ecological communities like those of indigenous peoples is denied altogether. Ethics indigenous peoples Indigenous understood as applicable to all relations of life in nature were also removed from the minds and practices of indigenous peoples live. Because of the dominance of anthropocentric philosophy, which disappeared Malay heritage of indigenous peoples ethics of public discourse in today's society.

The loss of biodiversity due to modernization with "development" as the religion of modern society, devastation and extinction of biodiversity is so rich in traditional Malay society. The impact of trade-off is the extinction of biodiversity. It is increasingly extinct and eroded Malay traditional wisdom with all its value because traditional knowledge is closely linked to biodiversity. Traditional knowledge may only be maintained if nature and all its wealth are still maintained when nature with all its richness is endangered, extinct throughout this traditional wisdom.

The loss of most of the indigenous land rights of indigenous peoples of Riau, including the right to live and survive by the identity and the uniqueness of their cultural traditions and the right to self-determination. During the invasion and domination of modern society because of the influence of anthropocentric Malay indigenous peoples of different cultural traditions and the loss of indigenous land rights. It happens either because of colonialism and imperialism, and modernization of the Western impact was eliminated and the extinction of indigenous peoples worldwide. What do they know life has no place in a modern legal and ethical framework. Along with the loss of traditional land rights of indigenous peoples of Riau, lost as local knowledge they so determined how indigenous Malay it. Instead, the loss of the land had it, then threatened the existence of indigenous Malays [7].

It is attributable to several factors: First, the ideology of developmentalism did not include indigenous Malay, especially regarding indigenous land in environmental management policies and socio-cultural wealth preservation as an integral part of all development programs. Second, misperceptions are people who think that the indigenous people of Malay are a destroyer of the environment that must be removed or relocated to save the environment. Nevertheless, indigenous peoples are the guardians of the Malay environment of invasion and destruction by outside communities, immigrant communities. Third, the nature only in terms of economic value, so that the release of the entire ecological value, the local political, social, cultural, spiritual and moral associated with indigenous land and the lives of indigenous peoples in the surrounding of Riau, Fourth, modernization and progress of civilization seen and measured primarily by the quality of the physical-economic-capitalistic. indigenous land and wealth and cultural values, social, spiritual, and moral attached to and owned by indigenous people considered to have no economic value to the modernization and advancement of civilization [8].

3.2. Eco-culture: A management aspects of indigenous lands troubleshooting ecologies

In the context of culture, especially Deep Ecology, support for leaving the anthropocentric perspective. When the life of a holistic perspective invites to abandon an anthropocentric perspective, real human beings are invited to return to local wisdom, the wisdom of the old indigenous Malays.

On that basis, we need to review the wisdom or knowledge of indigenous peoples in various regions of the human, natural, and man's relationship with nature. The review will focus on three things. First, how indigenous Malay peoples about themselves, nature and the relationship between humans and nature. Second, the peculiarities of the traditional knowledge of indigenous peoples of Malay as well as determine the lifestyle and behaviour of the indigenous nature. Thirdly, the rights of indigenous Malays need to be protected because protecting their rights, not only the existence of indigenous people is protected, but also their ethics and nature that were targeted.

Occurred desecration of nature, making nature no longer attractive to be respected, worshipped, and maintained in amazement. Thus, men had seen him so excellent and superior in dealing with nature that does not mean anything. Therefore, respect, caring behaviour, love, and harmony or rapport with nature are irrelevant and do not have a place anymore. "Desecration (nature) takes place through a change of meaning (space). Space is sacred, the centre of all meaning and life, the source of ecological survival, transformed into another place (site), the location because of anthropocentric. Whereas because of the Malays human relations -Natural-God is a sacred religious, metaphysical relationship, many of abstinence forbids, and advice is essentially derived from the Koran can be seen in relation with nature in a letter al-Baqarah (2): 164

Legal regulations and policies just for the sake of capitalistic economic value have marginalized the existence of the Malays' indigenous land as an example of a case where the Land of the Malay comprehensive Malay 8881.59 km2 or 888 159 hectares. However, the land of indigenous people in Malay in the area Bangko are at the border of Lengadai-slaughter Hilir area of 500 hectares, in Bangko Pusako, an area of 200 hectares. Land indigenous in Kubu are in the Gulf of Nilap, an area of 200

IOP Conf. Series: Earth and Environmental Science **894** (2021) 012026 doi:10.1088/1755-1315/894/1/012026

hectares. Sei-sei Panji-panji, measuring 200 Ha and on the island of drying an area of 200 hectares. While in the Land. The existence of Indigenous Lands in Malay in the Sultanate of Siak (1772-1942) is in crisis at the moment of the situation [9];. The Efforts to eliminate the existence of indigenous land have already existed since the Dutch colonial era. It can be seen in policies at the exit by the GOI Colonial and laws issued during the reign of Indonesia such as (1) Agrarische Wet stablat 1870 No. 55; (2) Contract Dutch politics in 1885; (3) Besluit 1915 (4) Act No. 13 of 1948 and Act 1950; (5) Basic Agrarian Law No. 1960; and (6) of Law No. 5 of 1979. The regulation is issued this law has been marginalized the existence of indigenous land

Thus, all the traditional knowledge of high economic value is retained and transformed into an anthropocentric view, closing the traces of its origin entirely on local wisdom. The most tangible evidence is the botanical knowledge of indigenous peoples, and traditional healing Malay foreclosed by little, leaving a trail of traditional, as is the original discovery of the modern scientific community. [10].

To save the Malay Malay indigenous land should be returned to the ethical values of the Malay community. The rights of indigenous Malays should be recognized and guaranteed by the government. There must be a political commitment at the global and national level to protect the indigenous land rights of indigenous peoples and all the wisdom of traditional Malay. Through eco-culture paradigm approach is to save the existence of indigenous land and all the wealth and traditional knowledge, but also to save the ecological crisis is mainly caused by faulty worldview and behavioural paradigm of anthropocentric society.

In the case of indigenous land in Riau, for example, there has been no significant political steps to protect the rights of indigenous peoples' indigenous land legally. Indigenous lands and indigenous peoples are essential to protect the cultural identity and a symbol of Riau. In other words, the translation of political and moral commitment at the global level to the national level is still not satisfactory. As stated by Darrell Addison Posey, "Many countries do not even recognize the fundamental rights of the indigenous tribes to live, let alone a guarantee for their right to self-determination, the right to belong to indigenous land, or the right to master traditional of power resources [11].

To protect the existence of indigenous peoples and their lands and all the wealth of their cultural traditions, including traditional wisdom and in order to protect biodiversity, few indigenous rights following should be recognized, guaranteed and protected by management aspects of indigenous lands in environmental conservation eco-culture approach as follow :

First, the right to self-determination and cultural identity of Riau. Moral and legal rights are attached to the existence of indigenous people Malay as civilized human beings. The main target of this right is that indigenous peoples have the legal and moral position equivalent to other civilizations community groups to be heard and be involved in all the political processes that determine their fate. Indigenous peoples in Malay should not be ignored, and their fate should not be determined unilaterally by outsiders.

Second, territories and indigenous land rights. It is essential because indigenous land is directly linked to their existence. Indigenous land can not be separated from its existence and all its cultural traditions. Therefore, seizing territories and their traditional lands, in the name of anything, is a denial and annihilation of the existence of indigenous people. Thus, the right to territory and land is the most fundamental right for indigenous peoples.

The Declaration of Rights of the tribes of Native of the United Nations, Section VI, acknowledged that "tribes indigenous have the right to maintain and strengthen the relationship spiritual and material typical to traditional lands, territories, water, and sea and the resources at their disposal, inhabit, or use for generations, and is responsible for the future generations in this matter. " Similarly, the "tribes native has the right to own, develop, control and use the lands and territories, including the entire environment of earth, air, water, ocean, sea ice, flora and fauna, sources and other resources that they have, live or use for generations. It includes the right to full recognition of the laws, traditions, customs, territorial system of land and their institutions for the development and management of resources, and the right to effective measures the state to prevent the interference of any kind, deprivation or violation of their rights[12]. Thirdly, collective rights. It is a reaffirmation that the rights agreed there should be a formal legal law for indigenous peoples Riau. However, in contrast to the regime under the United Nations human rights that tend individualistically, indigenous peoples demand that these rights are guaranteed to every human being and for groups of people, especially indigenous groups in Riau. So, in terms of the right to freedom, indigenous peoples Malay emphasize guarantees for the right to freedom to be themselves with all the unique cultural identity in a cultural group.

Fourth, cultural rights. Culture for indigenous peoples Malay include everything, including knowledge and traditional wisdom, wooded, resinous, berotan, fishing, give, farming, gardening, hunting, ensnaring, dancing, singing, language, sacred places, fairy tales, innovations and practices of life in traditional crafts. It is a wealth of invaluable, not just physically but also spiritual and moral wealth. Therefore, Malay indigenous people demand that their rights to maintain and preserve their culture even developed must be recognized and guaranteed. However, because it is such a close relationship between indigenous culture and nature, preserving it also means preserving the natural world around them. The whole culture is to be taken seriously in all development policies that impact the preservation of indigenous culture and ecology.

Fifth, Malay indigenous peoples have the right to profess a belief system and values of their religion and morals, which outsiders should not violate. They should not be forced to leave the religious practices they have hereditary. The freedom to live their religion must be guaranteed. Similarly, sacred places and objects of their worship, including animals and plants considered sacred, must be protected and kept their integrity. Guaranteeing this right will have a very positive contribution to the preservation of the environment.

Sixth, the right not to be discriminated against (the right of non-discrimination). It is a consequence of the recognition of human rights, with this want to say that the indigenous people of Malay's moral and legal position are the same and equal with the community and other human beings. Therefore, all forms of discrimination against them should be avoided. Countries need to develop a paradigm of eco-culture policy specifically for them, but this is a more specific policy within the framework of the protection of indigenous peoples from invasion and degradation of their cultural traditions, not in terms of treatment marginalizing demeaning them.

Seventh, their Malay community has the right to participate fully in the political process of mutual concern in all communities. It is essential within the framework of development that impacts the environment around them and the preservation of cultural traditions and their existence. The political approach that considers them merely an object that can be moved and sacrificed in every development is anthropocentric, highly unethical, and violates indigenous peoples' rights.

Eighth, the right to obtain compensation for any activity adversely affects the environment and the values of social, cultural, spiritual and moral indigenous peoples. This compensation shall be regulated in positive law to enable indigenous peoples of Malay to be not harmed economically. The riches and values of ecological, cultural, social, spiritual, and moral have not become extinct just like that, especially by the invasion of outsiders.

3.3. Alignments need the government to maintain the social-economic community

For the people of indigenous law, government recognition of indigenous land rights, including indigenous forests, is a substance essential to maintain the indigenous law system they run. Natural resources are a source of inspiration to increase their capacity for social and economic development with due regard to the wisdom of ecology has evolved during the life of indigenous peoples marginalization of indigenous happened throughout the history of the regime of the Indonesian government to be reorganized with the approach of eco-culture in restoring rights indigenous law communities in the management of forest resources.

The government, as the mandate of the forest management as stipulated in Law No. 41 of 1999 on Forestry, to put off scepticism about the ability of indigenous law community Malay in the management of their forest resources for the legal community Malay customs are groups of people who think comprehensively and holistic for the current generation and the future generation in the legal and institutional framework indigenous implementation. The government's focus is ensuring that indigenous rights with the laws that govern are justly earned by indigenous people.

IOP Conf. Series: Earth and Environmental Science **894** (2021) 012026 doi:10.1088/1755-1315/894/1/012026

Indigenous Land Rights in contrast to proprietary rights of the individuals obtained by open land or forest. Property rights to land may also be obtained from the land-use rights on social unity (indigenous law communities) where the right to become permanent and hereditary. Absolute property rights are not recognized in indigenous law [13]. Similarly, according to Ter Haar in the right to beschikken (an absolute master), the meaning of transfer can not be found in Indonesia's community of ethnic groups. Rule "seignorial area" (beschikkings gebeid) non-transferable fixed first of all apply, although there are some exceptions, such as war and pressure the central government [14]. Conditions that there are no absolute property rights in the Law of 1960 are referred to as soil's social function.

The government's commitment to encouraging community-based forest management (CBFM) in state forest areas is still weak. CBFM development is still inferior to the development of industrial timber estates (HTI). For example, until 2014, licensing Plantation Forest (HTR) and Community Forest (HKM) has reached about 165 thousand hectares and 44 thousand ha, respectively. At the same time, the total area of plantations increased sharply from about 30 thousand ha in 1990 to more than 10 million ha in 2011. As one of the schemes in the Community-Based Forest Management, indigenous forests should be seen in three essential aspects, namely social, economic, and ecological and the perspective of sustainable development capable of bridging the dynamics of the problem in a comprehensive and holistic. To achieve this, government policies that can be taken are as follows:

First, political and social policies, mainly through amendment of Act No. 41 of 1999 on Forestry to accommodate the interests of the essentials of indigenous people and generally encourage legislation regulating the indigenous law community itself in a broad sense. The government's political commitment (political will) must be realized by establishing norms, standards, procedures, and criteria that can interpret indigenous law communities as part of the national legal system.

Second, the economic policy of the people, by making the indigenous law community a significant driver of the region's economy and expanding their access to financial institutions that exist with legal systems based service that they profess. Economic institutions of indigenous peoples Malay and / or delegate tasks based services indigenous law system is in one of the national economic systems.

Third, ecological policy implementation through the optimization of Plantation Forest (HTR), Community Forest (HKM) and Village Forest (HD) within the framework of indigenous law system Malay and guarantee indigenous governance on lands outside the forest area [15]. The most important of these policies is to give communities control of Malay indigenous law in determining the spatial-based micro-ecological to achieve food self-sufficiency, preservation of resources, and capacity building. To achieve these objectives, the territories of indigenous law Malay that Management Aspects of Indigenous Lands foundation supported by indigenous law community Riau should be formed Indigenous Forest Conservation Institute as a unit-based management Malay indigenous law system based on eco-culture.

4. Conclusion

The management aspects of indigenous lands crisis caused by the ecological indigenous anthropocentric viewpoint that is different from the way the eco-culture saw indigenous community as a community of socio-economic, ecological, the social community are not only human as understood anthropocentric perspective. Because eco-culture of indigenous peoples is an integral part of the Management Aspects of Indigenous Lands community ecological, natural, and metaphysical-religious communities. Indigenous Malays develop into themselves, either individually or as a group, in the bond and relationship with the whole universe, with all beings in the universe, and integrate with substances that are metaphysical-religious in the socio-economic-environmental preservation.

To develop their Management Aspects of Indigenous Lands environment of indigenous peoples of Malay should be able to amend viewpoint anthropocentric -capitalistic to the view of the Eco-culture with due regard to the rights of indigenous peoples, either individually or in groups, which is the minimum standard for the survival of the Management Aspects of Indigenous Lands environment as well as respect for the dignity of the people indigenous.

IOP Conf. Series: Earth and Environmental Science **894** (2021) 012026 doi:10.1088/1755-1315/894/1/012026

Reference

[1] Fitrah A 2020 *Jurnal Hukum & Pembangunan* **(50)3** 586-602
[2] Naess A 1973 *The Deep Ecological Movement.Some Philosophical Aspect* (Boston: Shambala)
[3] Keraf A S 2010 *Etika Lingkungan* (Jakarta: Kompas)
[4] Husni T 2012 *Eco- Culture* (Pekanbaru: LPP UIN Suska Riau)
[5] Vandana S 1997 *Monoculture of The Mind* (London: TWD Press)
[6] Husni T 2013 *Jurnal Kutubkhanah* **16(2)** 61-72
[7] Husni T 2019 *Eco Islamic Cuture* (Yogyakarta: Magnum Press)
[8] Susilo R K D 2008 *Sosiologi Lingkungan* (Jakarta: Raja Grafindo Persada)
[9] Nijhoff M 1922 Adatrechbundels vol XX, s-Gravenhage
[10] Syahyuti 2006 *Jurnal Forum Penelitian Agro Ekonomi* **24(1)** 14-17
[11] VDarel A P 2002 *Resource management in Amazone : Indigenous and folk strategies* (New York)
[12] Vandana Sh 2014 *Making Peace with The Earth.Melbourne* (Spinifex Pess)
[13] Soerjono S 1984 Het Adatrecht van Naderlandsch- Indi, Dell III , Pokok-Pokok Hukum Adat Bandung
[14] Haar T 1931 *Het Adatrecht van Naderlandsch –Indie in Watenschap, Practij en Onderwijs*
[15] Lestuti Abu Bakar 2013 *Jurnal Dinamika Hukum UNPAD* **13(2)** : 320-331

IOP Conf. Series: Earth and Environmental Science **894** (2021) 012027 doi:10.1088/1755-1315/894/1/012027

Crack formation behavior of composite fly ash – bentonite (FAB) in landfill liner system

Syafrudin[1,2], M A Budihardjo[1,2], E Sutrisno[1], W D Nugraha[1], G Samudro[1]

[1]Environmental Sustainability Research Group, Department of Environmental Engineering, Diponegoro University, Semarang Indonesia 50275
[2]Department of Environmental Engineering, Diponegoro University, Semarang Indonesia 50275

syafrudin@lecturer.undip.ac.id

Abstract. This study aims to investigate the influence of different mixtures on the phenomenon of desiccation cracking in bentonite-fly ash mixtures as a landfill liner system. Fly ash is quite potential to be used as a landfill liner mixture because it has a low hydraulic permeability or conductivity value. This study uses class F fly ash from the Paiton power plant production process, Indonesia, which has been distributed commercially. Desiccation test was conducted in this study. The composition of fly ash and bentonite which is used are pure fly ash (FAB0), fly ash and bentonite 15% (FAB15), fly ash and bentonite 20% (FAB20) and fly ash and bentonite 25% (FAB25). The smallest CIF value is found in the pure fly ash layer. However, the pure fly ash cannot be used as landfill liner because the high permeability value. Therefore, the addition of bentonite will increase the possibility of crack formation. This study reveals that the addition bentonite in the fly ash composite will increase the crack. However, determining appropriate mixture composition is critical when working on the landfill liner system.

1. Introduction

Landfill liner system as a layer of retaining leachate contamination from waste on the surface of the soil is one of the important components in landfills that withstand the load of landfill and prevent leachate from leaking into the environment [1]. Liners used as base coatings in landfill systems generally have low hydraulic conductivity and are rich in clay material [2]. Clay consists of various soil minerals where bentonite is the mineral most often used both as a landfill liner and cover [3]. Bentonite is quite desirable to be used as a constituent of landfill liners because it has a low hydraulic conductivity value and is easily installed as a mixture of other materials. Montmorillonite in bentonite can bind enough water molecules during the hydration process. The ability of bentonite to bind water molecules is able to prevent bulk water from entering, causing the bentonite hydraulic conductivity to decrease. However, cracking or desiccation and leakage which can lead to failure of the containment system can occur at any time depending on the environment in which the material is placed and operational conditions such as chemical exposure, dry-wet cycle and increase in gradient temperature [4].

Optimization of the performance of the liner system is done in various ways one of them by using a mixture of waste material with binding material [5]. Fly ash is quite potential to be used as a landfill liner mixture because it has a low hydraulic permeability or conductivity value. The permeability value of pure fly ash is in the range of 10^{-4} to 10^{-7} cm/s [6]. Pozzolanic fly ash has a low shrinkage potential and is not susceptible to cracking [7]. The effectiveness of liner performance is influenced by several factors, such as initial water content and particle size distribution of the soil above and/or below the geosynthetic clay liner (GCL), the level of exposure to thermal cycles, sustained thermal gradients, and the stress on the GCL [8]. The increase in temperature due to the degradation process of the waste pile in the landfill can bring moisture downward which tends to reduce the temperature of the liner system. If the temperature increase gradient is higher than the possibility of desiccation in the liner layer will be even greater. Desiccation causes an increase in the value of hydraulic conductivity, compressibility and consolidation rate which causes a decrease in soil strength and affects self-healing capacity [9].

Nevertheless, studies that analyze the influence of bentonite mixtures to the phenomenon of desiccation cracking in the mixture of bentonite-fly ash landfill liner materials of the composite layer is lacking. In this study, the influence of the mixtures on the phenomenon of desiccation cracking in bentonite-fly ash mixtures as a landfill liner system was further investigated to overcome gaps in studies of landfill liner cracking. This research is expected to be able to contribute knowledge to the use of alternative composite materials for better landfill liner systems.

2. Methodology

This study uses class F fly ash from the Paiton power plant production process, Indonesia, which has been distributed commercially. The results of the XRF test show that fly ash is composed of mineral quartz (SiO_2), ferrous oxide (Fe_2O_3), periclase (MgO), alumina (Al) and titanium oxide (TiO_2). The permeability value of pure fly ash is 1.653 x 10^{-8} m/s. As a result of the high permeability value, it is necessary to add other materials that can reduce the permeability value of composite soils. Bentonite clay is taken from mineral and rock mining in Giriwoyo District, Wonogiri Regency, Central Java. To distinguish the effect of salinity on desiccation crack, NH_4Cl 1 N was used as a substitute for aquadest in one of the treatments [10].

This research was conducted at the Environmental Laboratory, Environmental Engineering and Soil Mechanics Laboratory, Civil Engineering, Diponegoro University in March - June 2019. The research was carried out by making a mixture of bentonite-fly ash [2]. The study was conducted with the same mixture and obtained optimal results of bentonite levels of 20%. Thus, in this study the variation in bentonite levels was 15%, 20% and 25%. In addition, researchers also made a mixture of 25% bentonite composite - fly ash and NH_4Cl 1 N solution with a concentration of 7.1 ppm following the salinity concentration at Jatibarang Landfill, Semarang. Bentonite-fly ash composites are put into a 2-inch hubcap and are loaded on them.

The desiccation test is carried out by placing a load on the bentonite-fly ash composite. Determination of the amount of pressure/load is based on the assumption that the area of the landfill is 1 m^2, the waste density is 800 kg/m^3 and the height of the landfill is 4 meters. The amount of load faced by the simulated landfill liner system is calculated using equation (1).

$$P = \frac{W}{A} = \frac{\rho \, x \, V \, x \, g}{A} \tag{1}$$

$$P = \frac{800 \; kgm^{-3} x \; 1 \; m^2 \; x \; 4 \; m}{1 \; m^2} = \frac{31{,}392 \; kgms^{-2}}{1 \; m^2} = 31{,}392 \; kgm^{-2}s^{-2}$$

The lab-scale liner that will be used in the study has a diameter of 0.06 m^2 so that according to equation (1), the load used (W) on a fly ash - bentonite composite dish is 9.043 kg or estimated to be ± 9 kg.

Crack observations on bentonite - fly ash composites were carried out by analyzing composite fracture photographs. Image pixel methods are used to measure the dimensions of cracks that occur [11]. Photo acquisition was carried out using a cellphone with a 13 megapixel camera specification. Image sizes vary from 3.077 x 2.304 pixels to 1.632 x 1.224 pixels which are converted into black and white images. The ratio of crack area to total surface area will describe the crack intensity factor (CIF). Crack intensity factor is defined as the ratio of crack area (Ac) to total surface area (At) of the mass of the soil experiencing drought [12]. The CIF value is determined using the 2018a Matlab® software. Analysis is carried out based on the contrast of the cracked soil which is characterized by darker colored segmentation of the gap compared to the area that has not experienced cracking or in other words, Matlab® will calculate the ratio of black and white in the image the liner layer [13]. Researchers analyzed the results of the software by comparing data taken every day for 39 days on a composite layer and pure fly ash.

3. Results and discussion

Water content testing was carried out on all samples of bentonite-fly ash, bentonite-fly ash with 7.100 ppm salinity solution and pure fly ash. The results of observations on changes in water content in the composite showed that during the 39 days of the study as a whole all mixes experiencing a decrease in water content until the end of the study time (See Figures 1). Likewise, the lowest moisture content values on the composite were observed in the pure fly ash layer. Fly ash + 25% bentonite composite have the highest water content compared to other mixtures and pure fly ash. The moisture content in the fly ash + 25% bentonite composite had increased on the 18th day and continued to decline until the last day of the study.

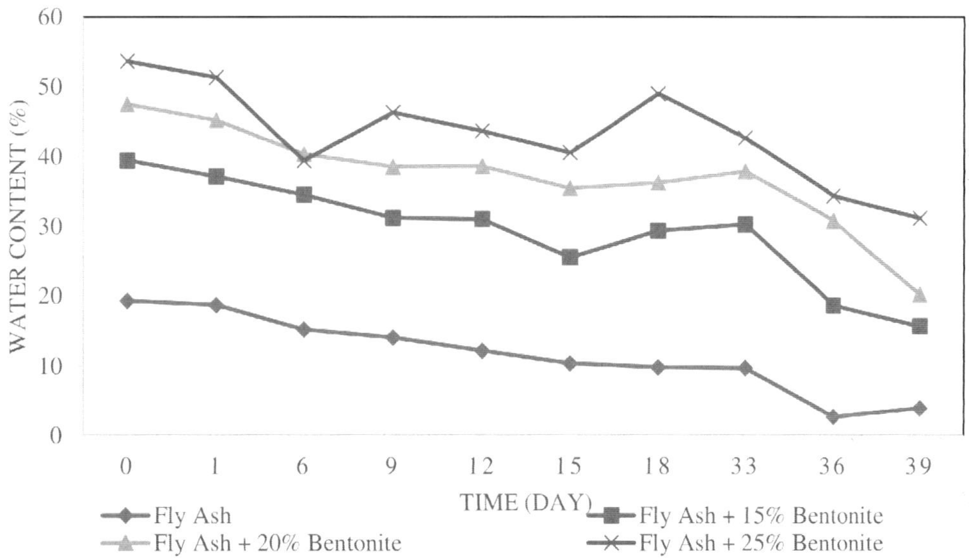

Figure 1. Fly Ash bentonite composite moisture with loading.

Researchers also observed cracks that occur in the fly ash bentonite composite layer and analyzed the crack intensity factor (CIF) value. The results obtained indicate that the CIF value is higher along with a decrease in water content. Based on the data presented in Table 1 it is known that the lowest CIF value was obtained on pure fly ash dish. Whereas the highest CIF value was found in dishes with a composite fly ash + 20% bentonite on the last day of the study of 2.439. The second highest value was found on the composite fly ash + 25% bentonite on the last day of the study amounting to 2.250. The results obtained are quite different in composite fly ash bentonite without loading. The lower the water content, the higher the CIF value. However, the CIF value at the end of the study time on the fly ash composite + 15% bentonite with a moisture content of 15,660 lower than the composite layer of fly ash + 20% bentonite with a moisture content of 20,219. Similarly, the pure fly ash composite layer with the lowest CIF value has a lower moisture content value compared to the fly ash composite + 15% bentonite at the end of the study.

Table 1. CIF value and moisture content in each composite coating cup by loading.

Day	Pure Fly Ash		Fly Ash + 15% Bentonite		Fly Ash + 20% Bentonite		Fly Ash + 25% Bentonite	
	W (%)	CIF	W (%)	CIF	W (%)	CIF	W (%)	CIF
0	19.300	0	29.428	0	47.481	0	53.655	0

Day	Pure Fly Ash		Fly Ash + 15% Bentonite		Fly Ash + 20% Bentonite		Fly Ash + 25% Bentonite	
	W (%)	CIF	W (%)	CIF	W (%)	CIF	W (%)	CIF
1	18.694	0	37.120	0.357	45.225	0.047	51.332	0.115
6	15.188	0	34.505	1.271	40.313	0.147	39.391	0.429
9	14.045	0	31.222	0.302	38.528	0.186	46.318	0.377
12	12.162	0.0285	31.018	1.989	38.600	0.661	46.677	0.655
15	10.354	0.0232	25.562	1.620	35.463	0.737	40.528	1.526
18	9.795	0.0406	29.371	1.942	36.235	1.132	48.986	2.827
33	9.687	0.0327	30.271	0.118	37.837	1.639	42.632	1.549
36	2.675	0.0522	18.648	0.695	30.823	1.707	34.357	2.175
39	3.945	0.0202	15.660	1.774	20.219	2.439	31.171	2.250

It is known that a decrease in water content and an increase in CIF values from day 1 to day 39 are illustrated by the appearance of cracks or desiccation cracks in the composite layer. The lowest CIF value in the pure fly ash layer can be observed in the absence of cracks which are represented by the presence of black in image analysis with MATLAB version 2018a. If these results are compared with studies conducted by Safari et al., the geotextile layer covering compacted clay soils was considered capable of reducing CIF by an average of 52% [14]. The burden on the composite layer is assumed to act like a geotextile layer that covers the layer of compacted clay soil so that the composite layer is protected from atmospheric conditions [15]. The reduced water content followed by cracks in the composite layer will certainly affect the permeability of the layer.

Fly ash can be used as a reactive liner material because some researchers say that fly ash is classified as pozzolanic material, a material that can do self-cementing due to the high content of silica or silica alumina so that it forms a chemical reaction with water to form an adhesive at room temperature [16]. On the other hand, bentonite clay soils are expansive when going through a wet-dry cycle, giving rise to the potential for cracking during the dry season. Bentonite is always negatively charged and isomorphic substitution of ions such as Al^{3+}, Mg^{2+}, and Fe^{2+} occurs. The gap between the bentonite layers can expand and be filled with water molecules and other cations [17]. The mixture of bentonite - fly ash is considered to be able to minimize the potential for crack desiccation because the liquid limit of the bentonite increases and the cation exchange capacity increases [18]. Both bentonite and fly ash have their advantages and disadvantages in their function as a leachate barrier. Therefore, determining the right mix composition is a critical factor in the decision to use this mixture.

4. Conclusion

Based on research that has been done, it is known that the phenomenon of desiccation which is marked by the value of CIF (crack intensity factor) appears along with the decreasing value of the water content. The smallest CIF value is found in the pure fly ash layer, but the permeability coefficient of pure fly ash exceeds the limits in force in Indonesia. So that the addition of bentonite to fly ash can reduce the value of the permeability coefficient of the composite landfill liner so that it complies with applicable regulations for use on pilot-scale. Loading on the composite can increase the value of the optimum moisture content thereby increasing the value of CIF.

Acknowledgements

This research was funded under research program *Penelitian Strategis Fakultas Teknik Universitas Diponegoro* number 195/UN7.5.3.2/HK/2021 appendix VIII.

References

[1] Akcanca F, Aytekin M 2012 *Environ Earth Sci* **66** 67–74
[2] Pandey M, Jain P R 2017 *Int Res J Eng Technol* **4** 2277–2280
[3] Yu B, El-Zein A 2019 *Geotext Geomembranes* **47** 142–153

[4] Azad F M, Rowe R K, El-Zein A, Airey D W 2011 *Geotext Geomembranes* **29** 534–543
[5] Cheriyan R, Chandrakaran S 2018 *Indian Geotechnical Conference 2018* 1–7
[6] Bhatt A, Priyadarshini S, Mohanakrishnan A A, Abri A, Sattler M, Techapaphawit S 2019 *Constr Mater* **11** e00263
[7] Yao Z T, Ji X S, Sarker P K, Tang J H, Ge L Q, Xia M S, Xi Y Q 2015 *Earth-Science Rev* **141** 105–121
[8] Rowe R K 2014 *Environ Geotech* **1** 3–21
[9] Mukherjee K, Mishra A K 2017 *Int J Geotech Eng* **11** 107–113
[10] Abd El-Salam M M, Abu-Zuid G I, 2015 *J Adv Res* **6** 579–586
[11] Harianto T, Hayashi S, Du Y.-J, Suetsugu D 2008 *Water Air Soil Pollut* **194** 141–149
[12] Chaduvula U, Viswanadham B V S S, Kodikara J 2017 *Appl Clay Sci* **142** 163–172
[13] Mukunoki T, Nakano T, Otani J, Gourc J P 2014 *Appl Clay Sci* **101** 558–566
[14] Safari E, Ghazizade M J, Abduli M A, Gatmiri B 2014 *Waste Manag* **34** 1408–1415
[15] Malizia J P, Shakoor A 2018 *Eng Geol* **244** 125–131
[16] Sivapullaiah P V, Baig M A A 2011 *Waste Manag* **31** 359–369
[17] Chen Y G, Liu X M, Lei H N, Ye W M, Cui Y J 2019 *Adv Civ Eng* **2019** 2879156
[18] Shariatmadari N, Salami M, Fard M K 2011 *Int J Civ Eng* **9** 103–110

Wave transformation around breakwater (case study: tourism harbour, Eastern Bali, Indonesia)

M N Arsyad[1], O Setyandito[1], L M Kesuma[2], H D Armono[3], M Anda[4], R A Kusumadewi[5]

[1)]Department of Civil Engineering, Faculty of Engineering, University of Bina Nusantara, Jakarta, Indonesia, 11480
[2)]Professional Engineer Program (Program Studi Program Profesi Insinyur PSPPI), Universitas Gadjah Mada, Yogyakarta, Indonesia, 55281
[3)] Department of Ocean Engineering, Institut Teknologi Sepuluh Nopember
[4)]Environmental Engineering, Murdoch University, Australia
[5)]Department of Environmental Engineering, Faculty of Engineering, University of Trisakti, Jakarta Barat, Indonesia, 11440

osetyandito@binus.edu

Abstract. An essential aspect in the sustainable design of breakwater is the determination of the design wave condition. It is predicted by utilizing severe wave conditions of the past 10 to 20 years. The tourism harbor at eastern Bali, Indonesia, is located where extreme wave condition occurs. Therefore, this research studies the wave height before and after constructing a breakwater in the harbor area. The wave height was simulated using numerical modeling. The methodology was performed by using the coastal modeling software of the SMS-CGWAVE model. The result shows the highest design wave height value of 3.9 m in the direction from the southeast. The design breakwater can reduce wave height up to 0.9 m or a 75.5% reduction. Further study is needed to simulate the extension of breakwater length to meet the criterion design of wave height in the harbor basin.

1. Introduction

When the waves approach the shoreline, a wave transformation occurs due to shoaling, refraction, and diffraction. Shoaling occurs when the water depths are less than half of the wavelength and causes a reduction of wave velocity. Refraction is a change of the direction of the wave that occurs when the wave velocity is reduced and the wave becomes more aligned with the depth contours. Diffraction occurs when a wave propagates into an obstacle such as breakwaters and travels to the area behind it [1,2]. The extreme wave condition at tourism harbor, eastern Bali, disrupts passengers' loading and unloading and has a safety concern. A breakwater facility can protect the harbor area by reducing the extreme wave height [3]. Many studies have been conducted about breakwater reducing a wave height by different location studies and different types of breakwater specifications. Each study has its limitations [4,5,6], while an analysis facing directly to the Indonesian Ocean with a double non-overtopping breakwater is not much studied earlier. This study aims to analyze the reduction of the extreme wave condition at the harbor area at eastern Bali by creating a wave simulation with the condition before and after the breakwater construction. A CGWAVE model can simulate the characteristics of waves coming towards the shoreline and breakwater [7]. Therefore, this study used a Surface Modelling System (SMS), CGWAVE model, to create wave simulation for analyzing the wave characteristic around the breakwater. The result of this study is the wave height that occurs at the harbor area and the percentage of wave height reduction after the construction of the breakwater.

2. Methodology

2.1. Research stages

The study starts with literature studies and research planning regarding wave characteristics and hydrodynamic references. The data used for this research were obtained from [8] and [9]. The wave characteristic was 50 years of predicted wave conditions with a significant wave height (Hs) of 3.9 m and a significant period (Tp) of 8.13 s. Other parameters being utilized are wave rose data as a reference for the direction of the dominant waves (east, northeast, and southeast), bathymetry, topography, and breakwater design layout. The bathymetry and the breakwater design layout can be seen in Figure 1. After the required data has been collected, the next step is to create a 2D wave modeling. Wave modeling was created with two conditions; the first condition was created in the existing condition or without the breakwater structure. Second, the model is created with the design conditions in the presence of a breakwater structure. Then after the running of the model is completed, data analysis can be done by comparing the existing condition data with the breakwater condition data by making an observation line profile. After collecting the wave height data from the model, the percentage of wave reduction that occurs in both conditions is analyzed.

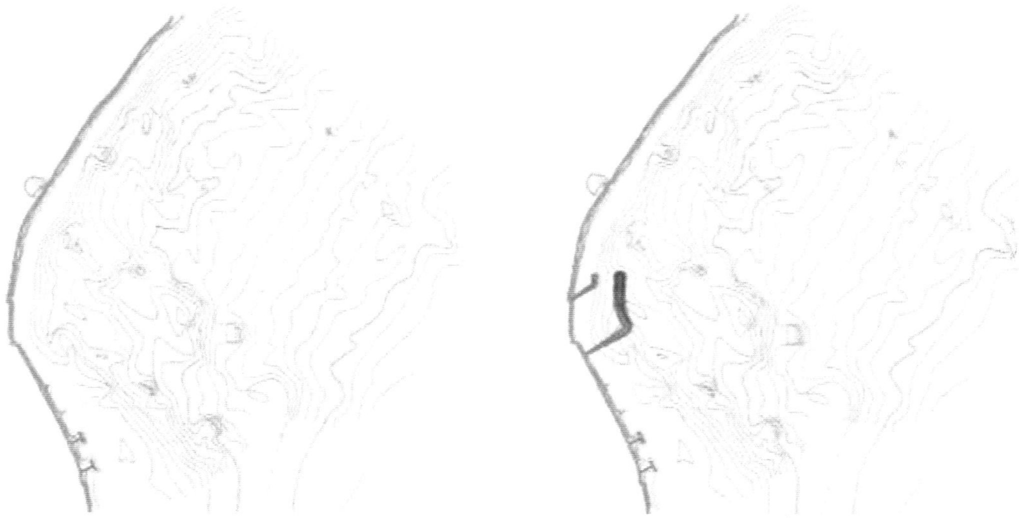

(a) (b)

Figure 1. Bathymetry layout of existing condition (a) and with breakwater structure (b).

2.2. Wave modeling

This study provides two-wave model conditions, existing and breakwater conditions with the incident wave from three directions, northeast (45°), east (90°), and southeast (135°) using a CGWAVE model. The analysis begins with the input geometry model, creating boundary area, creating a mesh, and solving the simulation. The bathymetry model is used to create a mesh model. The size of the mesh model depends on the wavelength data [10]. In this study, the significant period (Tp) is 8.13 s. A wavelength calculation was carried out to determine the size of the mesh [11].

$$L_0 = 1{,}56T^2$$

$$L_0 = 1{,}56(8{,}13)^2$$

L_0 = 103,11 m
Where:
L_0 = Wavelength at deep sea (m)
T = Wave period at deep sea (s)

The mesh size must have a minimum of 200 vertices along the shoreline to make the wave model output more accurate. The shoreline observed in this study is around 2 km. A 10 m mesh size is needed to create 200 vertices. Therefore, the equation that used to create a mesh in the CGWAVE modeling are:

- Wavelength = Period x Celerity
- Celerity = (gravity x depth)$^{0.5}$
- Mesh size = Wavelength/10

The following is a view of model layout at existing conditions for this study. Figure 2 shows the mesh layout that have over 200 mesh of elements with a 10 m size and a mesh model layout with a water depth information.

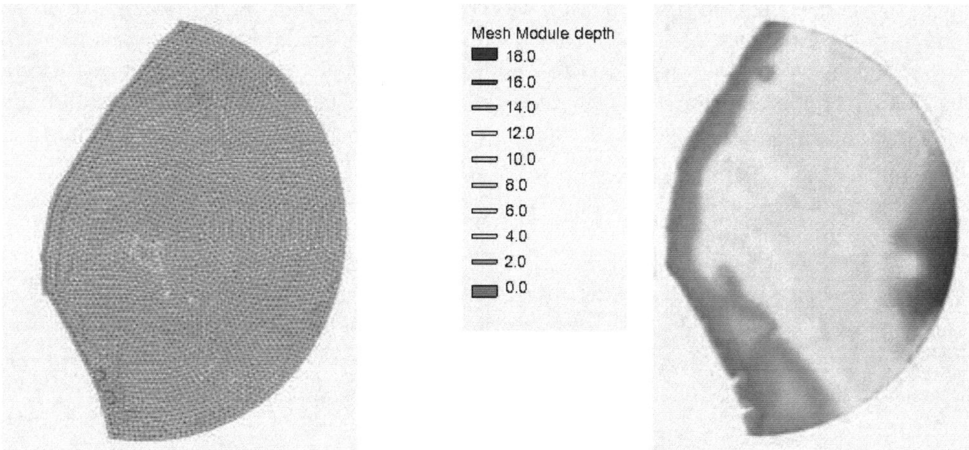

(a) (b)

Figure 2. Wave mesh model boundary, before (a) and after (b) the modeling process.

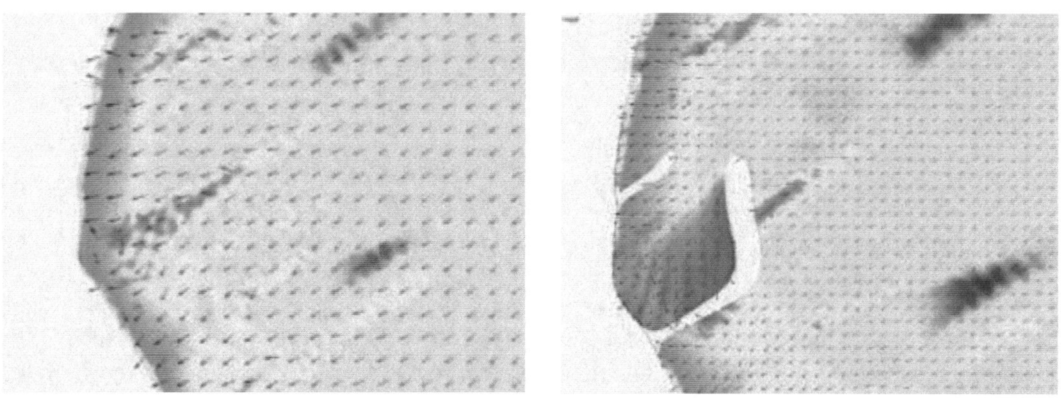

(a) (b)

Figure 3. Wave from northeast direction modeling results of existing condition (a) and with breakwater structure (b).

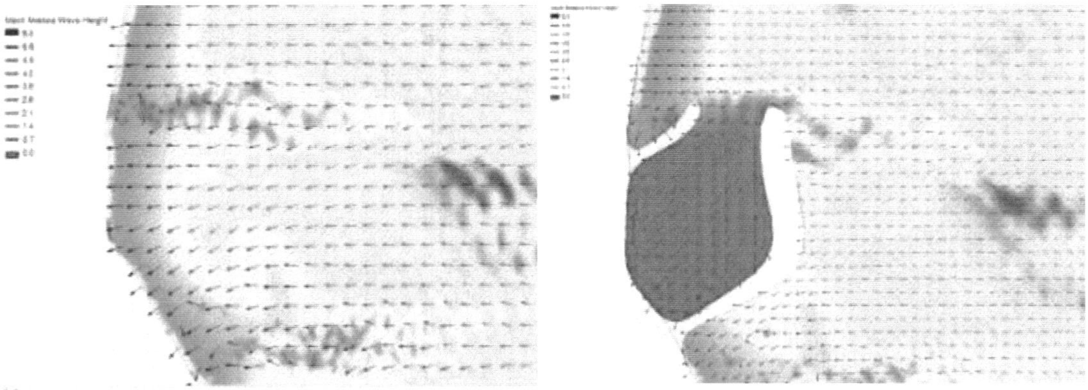

(a) (b)

Figure 4. Wave from east direction modeling results of existing condition and with breakwater structure (b).

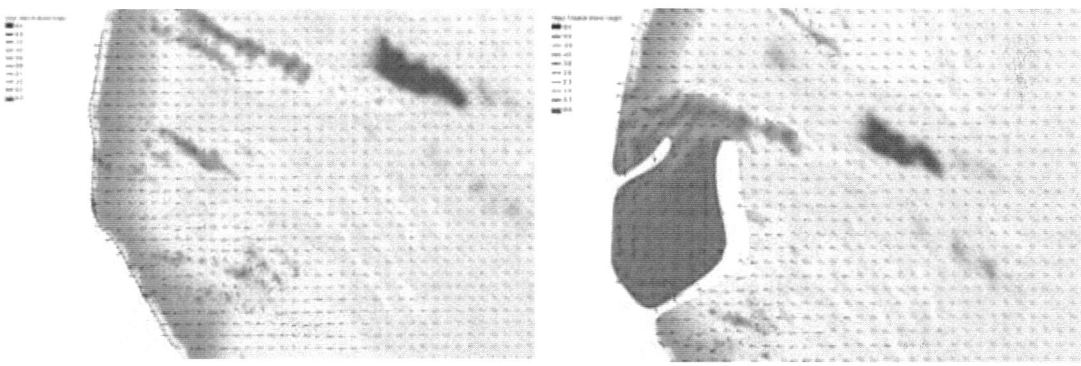

(a) (b)

Figure 5 Wave from southeast direction modeling results of existing condition (a) and with breakwater structure (b).

After a boundary mesh is created, the wave amplitude (wave height/2) of 1.95 m, wave period of 8.13 s, and dominant wave angle direction from the northeast, east, and southeast is entered at CGWAVE's model control. The wave mesh element needs to be adjusted to ensure there are no zero or minus at depth values before running the model. The comparison between the existing and breakwater condition model is shown in Figure 3 to Figure 5.

3. Results and discussion

3.1. Wave reduction result

The wave height reduction analysis is done by making an observation line to compare the results between the existing conditions model and the condition of the breakwater model. The layout of the observation sections layout to analyze the wave reduction can be seen in Figure 6.

Section A-A' is observing the harbor entrance channel to determine the wave reduction that occurs. The graph in Figure 7 shows that the wave height at the observation point of the existing condition A-A' section of the design harbor basin is about 2.4 m. Meanwhile, in conditions with a breakwater, the wave height drops to 1.8 m. From the east, the current wave height is around 1.1 m, and it drops to 0.1 m with a breakwater. From the southeast, in the existing condition, the wave height reaches 2.4 m, then it drops to 0.2 m with the breakwater. Section C-C' is observing the harbor entrance channel to determine the wave reduction that occurs.

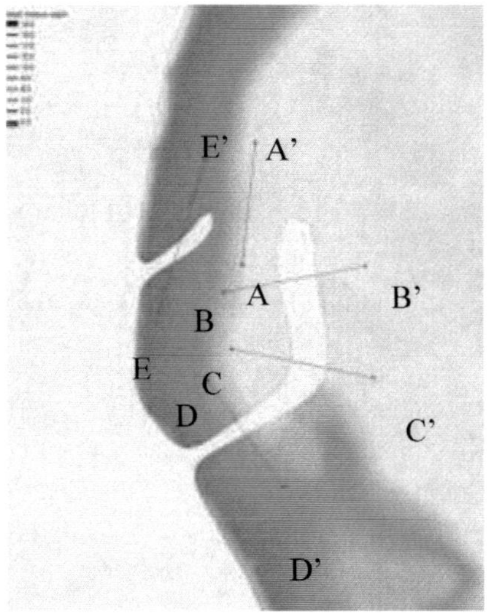

Figure 6. Observation wave height sections layout

The graph in Figure 8 shows that the wave height at the observation point of the existing condition A-A' section of the design harbor basin is about 2.4 m. Meanwhile, in conditions with a breakwater, the wave height drops to 1.8 m. From the east, the current wave height is around 1.1 m, and it drops to 0.1 m with a breakwater.

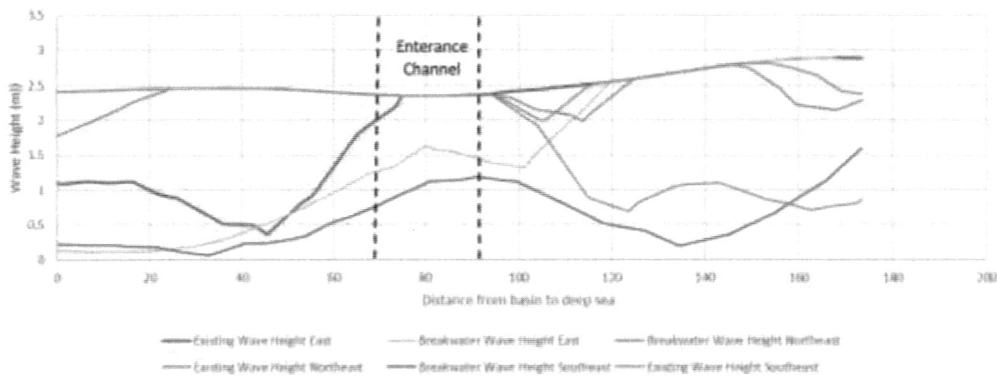

Figure 7. Wave reduction section A-A.

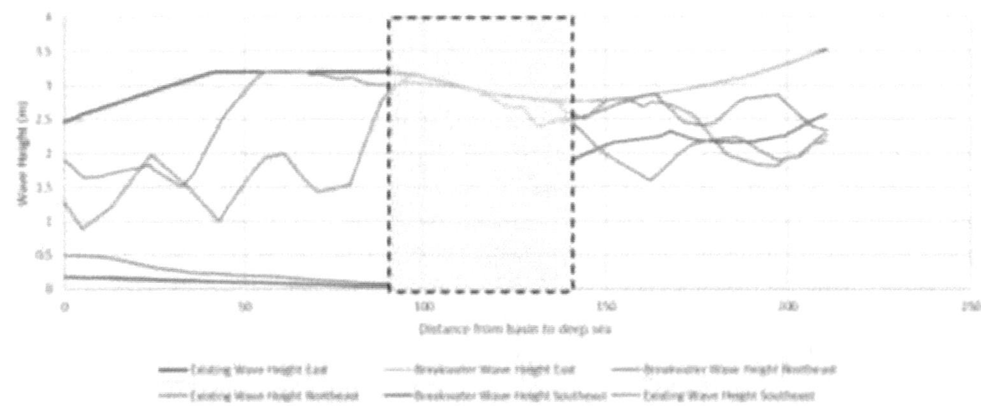

Figure 8. Wave reduction section C-C.

From the southeast, in the existing condition, the wave height reaches 2.4 m, then it drops to 0.2 m with the breakwater. The previous study shows the same pattern of wave reduction graph [12]. For more details, the percentage of wave height reduction is shown in Table 1.

Table 1. Percentage of wave height reduction between existing and breakwater conditions.

Wave Direction	Wave Height Existing Condition (m)	Wave Height Breakwater Condition (m)	Wave Reduction (%)
		Section A-A'	
Northeast	2.4	1.8	26.4
East	1.1	0.1	87.0
Southeast	2.4	0.2	90.6
		Section B-B'	
Northeast	1.9	1.3	32.8
East	1.9	0.2	90.2
Southeast	0.7	0.2	74.3
		Section C-C'	
Northeast	1.3	0.5	61.7
East	2.5	0.2	93.4
Southeast	1.9	0.2	90.7
		Section D-D'	
Northeast	1.6	0.4	75.5
East	1.6	0.1	90.8
Southeast	1.6	0.2	88.4
		Section E-E'	
Northeast	0.8	0.8	0.0
East	0.8	0.3	65.1
Southeast	0.8	0.3	55.1

It shows that the wave reduction due to the breakwater has the most significant percentage reduction in wave height in the A-A' sections coming from the southeast, reaching a value of 90.6%, and the minor percentage reduction in wave height coming from the northeast with a value of 26.4 %. It indicates that the breakwater design cannot protect the harbor basin from the northeast wave. The previous study shows that an alternative breakwater design is analyzed by extending the breakwater layout that can reduce the incoming wave height compared to the previous breakwater [13].

3.2. Analysis of alternative breakwater layout

In this section, an analysis of alternative breakwater layout is planned by modifying the addition of the length of the right side of the breakwater.

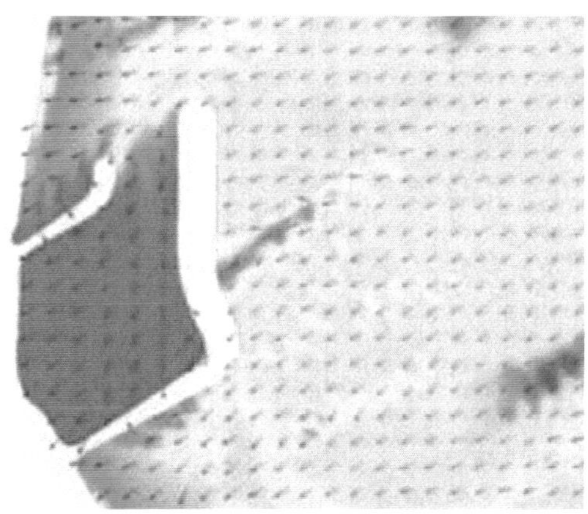

Figure 9. Modeling results of alternative breakwater conditions with northeast wave direction.

The alternative breakwater only modifies the south or right side of the breakwater by extending the breakwater length by a breakwater head diameter, which is 50 m, by still maintaining the minimum standard for two lanes for a ship, which is 86.6 m. To analyze the reduction by alternative breakwater layout, wave modeling was created the same way as before with a significant wave input of 3.9 m and a significant period of 8.13 s. The model result for the alternative breakwater with the wave coming from the northeast is shown in Figure 9. The model showed that the red color is dominant, which means have a low wave height at the harbor basin, unlike Figure 4, which shown that the orange color is more dominant at the harbor basin.

Table 2. Percentage of wave height reduction between existing and alternative breakwater conditions.

Wave Direction	**Existing Wave Height (m)**	**Alternative Breakwater Wave Height (m)**	**Maximum Wave Height for Vessel GT50 0,3 m**	**Wave Reduction (%)**
		Section A-A'		
Northeast	2.4	0.3	Ok	89.4
East	1.1	0.3	Ok	74.8
Southeast	2.4	0.3	Ok	88.9
		Section B-B'		
Northeast	1.9	0.2	Ok	91.2
East	1.9	0.3	Ok	86.4
Southeast	0.7	0.2	Ok	66.5
		Section C-C'		
Northeast	1.3	0.1	Ok	88.9
East	2.5	0.1	Ok	95.4
Southeast	1.9	0.1	Ok	92.7
		Section D-D'		
Northeast	1.6	0.1	Ok	92.1
East	1.6	0.1	Ok	94.2
Southeast	1.6	0.1	Ok	91.0
		Section E-E'		
Northeast	0.8	0.3	Ok	66.9
East	0.8	0.3	Ok	62.4
Southeast	0.8	0.2	Ok	74.3

From Table 2, it is shown that the highest percentage of wave reduction occurs in the east wave direction reaching 95.4% at the C-C' section, while the smallest wave reduction percentage occurs in the east wave direction at the E-E' section. The wave height from the northeast, which was previously reduced to only 26.4% in the initial A-A' breakwater section, can be reduced to 89.4%. Reduction of wave height tends to increase with alternative breakwaters. However, there is a decrease in minor wave height reduction, namely in the A-A' section from the east wave direction; in the initial breakwater layout, the percentage of wave reduction reaches 87%. However, in the alternative breakwater layout, the percentage of wave reduction drops to 74.8%. It can be caused by the influence of wave reflections that are getting bigger due to the addition of the breakwater length. The previous study shows the same percentage wave reduction pattern result by extending the breakwater layout [12]. However, further research is needed by considering cost and material factors in making alternative breakwater designs that can reduce wave height for east, northeast, and southeast wave directions.

4. Conclusion

4.1. Conclusion
Based on conducted study, it can be concluded that:

With a wave height in breakwater conditions with an observation point in the harbor pool from southeast reaching 2.4 m, from east reaching 2.4 m and from northeast reaching 2.5 m, the Sanur Harbor area is not protected from the threat of incoming waves;

With the breakwater structure, the wave height at the observation point at the Sanur Harbor breakwater is reduced from the east to 0.1 m with the largest reduction value of 90.8%, from the southeast it reaches 0.2 m with a reduction value of 90.7%, from the northeast. reached 0.4 m with a reduction of 75.5%;

In the A-A' section with the direction of the northeast wave with a breakwater condition, the wave reduction is still relatively small with a value of 26.4% with a wave height of 1.8 m;

The breakwater design cannot protect the harbor pool with the high waves coming from the northeast with the vessel's design requirements of a critical wave height of 0.3 m, but the wave height is still high with a magnitude of 1.8 m from the northeast;

To optimize the protection of the harbor from northeastern waves, a new breakwater layout is needed by extending the south side of the breakwater by 50 m which is the diameter of the breakwater head;

Based on the alternative breakwater layout modeling with the addition of the right breakwater length of 50 m, the percentage of wave reduction in the northeast direction reaches 0.1 m with the largest reduction value of 92.1%, from the east it reaches 0.1 m with a reduction value of 95.4%, from the southeast it reached 0.1 m with a reduction of 92.7%. The alternative breakwater layout can protect the harbor pool from the threat of wave height coming from the northeast, east and southeast.

4.2. Suggestion
Based on results of this study, it is suggested to do further research by considering cost and material factors in making alternative breakwater designs that can reduce wave height for east, northeast and southeast wave directions.

Acknowledgement
This work is supported by Research and Technology Transfer Office, Bina Nusantara University as a part of Bina Nusantara University's International Research Grant, with contract number: No.017/VR.RTT/Ill/2021.

References
[1] Sukhinov A I, Chistyakov A E, Protsenko E A, Protsenko S V 2021 *J Phys: Conf Ser* **1902** 012133
[2] Setyandito O, Purnama A C, Yuwono N, Juliastuti, Wijayanti Y 2020 *ComTech* **11(1)** 19-28
[3] Li M S, Hsu C J, Hsu H C, Tsai L H 2020 *J Mar Sci Eng* **8** 986
[4] Setyandito O, Yuwono N, Triatmodjo b, Bakti T E, Kesuma L M 2014 *World Applied Sciences Journal* **32(3)** 415-421
[5] Karathanasi F, Karperaki A, Gerostathis T, Belibassakis K 2020 *Atmosphere* **11** 280
[6] Rifka P A R, Baskoro R, Elis and Indrayanti 2020 *J Mar Res* **9(4)** 485-494
[7] Mohamed E, Paleologos E K and Mohamed E A *IOP Conf Ser: Mater Sci Eng* **471**
[8] Benoa Bali Port Authority 2020 *Basic Design Document: Development of Sanur Harbour* (Kantor Kesyahbandaran dan Otoritas Pelabuhan Kelas II Pelabuhan Benoa)
[9] Ministry of Transportation 2019 *Master Plan of Sanur Harbour Denpasar City Bali Province*
[10] Demirbilek Z and Panchang V 1998 *Semantic Scholar*
[11] Coastal Engineering Research Center, Department of the Army Corps of Engineers 1984 *Shore Protection Manual Volume 1* (Washington DC: Superintendent of Documents, U.S. Goverment Printing Office)
[12] Putra T W L, Sugianto D N and Siahaan H 2020 *IOP Conf Ser: Earth Environ Sci* **530** no 012033

[13] Fitri S and Hasan M 2020 *RekaRacana: Jurnal Teknik Sipil Jurnal Online Institut Teknologi Nasional* **6(3)**

IOP Conf. Series: Earth and Environmental Science **894** (2021) 012029 doi:10.1088/1755-1315/894/1/012029

Gas monitoring station in hazardous environment with gases containing

S Sendari[1]*, Y Rahmawati[1], H Rahmawati[2], D Arengga[1], F M Ramadhan[1], F Al-qodri[1], T Matsumoto[3], A Fujiyama[3], and I Rachman[3]

[1]Department of Electrical Engineering, Faculty of Engineering, Universitas Negeri Malang, Malang, Indonesia
[2]Department of Pshycology, Faculty of Psycology, Universitas Negeri Malang, Malang, Indonesia
[3]Department of Life and Environment Engineering, Faculty of Environmental Engineering, University of Kitakyushu, Fukuoka, Japan

siti.sendari.ft@um.ac.id

Abstract. Landfill sites collect tons of municipal solid waste (MSW) using an open dump mechanism, causing gases to emerge, which may cause disease and the greenhouse effect. Mainly, landfill environments are observed using a portable system that does not continuously monitor and measure emitted gas levels. It is also difficult to evaluate changes in landfill emissions over the long term unless they are monitored at regular intervals according to a detailed plan. This paper presents a new monitoring method to measure gas levels in landfill sites, which documents dynamic changes in gas composition concentrations over the long term. The system was placed in the middle area of the landfill and was charged using solar panels for convenience and greater efficiency during monitoring. While the instruments that are currently available are used for a specific parameter, this system can measure eight parameters, i.e., ambient concentration of methane (CH_4), carbon dioxide (CO_2), carbon monoxide (CO), temperature, humidity, wind direction, wind speed, and voltage level. The system was evaluated regarding its ability to monitor gas parameters continuously.

1. Introduction

A hazardous environment that contains poisoning gases could be harmful to people who reside nearby. One of the places which could be considered as a hazardous environment is a waste landfill site. It could also be a significant source of land, air, ground, and surface water pollution and harmful for nearby communities [1]. In Indonesia, municipal solid waste (MSW) has reached 64 million tons per year in 2017 [2]. It could be a potential problem because the open dump landfill system is mainly used. Furthermore, the statistic of the Central Bureau showed that 84.51% of MSW was disposed of without shorting it [3].

The landfill exposes various poisoning gases, such as methane (CH_4), carbon dioxide (CO_2), carbon monoxide (CO), and other gases depending on the waste mix. The potential problem of landfills could be affected by emissions of non-collected gases, where the situations depend on climate conditions. These conditions cannot be controlled [4]. Since a landfill covers a large area, and there is no gas collection system, the number of emission gases in the air should be predicted to prevent the negative effect of nearby people [5]. It is difficult to identify the effect of each individual caused by environmental contaminant exposure. Emitted gases can be considered the main factor of negative effect on health in sort and long-term effect. Some people worked at the landfill are potentially inhaling these gases and contracting diseases. Most of them can have coughs, chest tightness, breathlessness, and they suffer from an extraordinary range of illnesses that may be attributed to volatile organic compounds [6]. A study showed a relationship between the distance of houses with gas levels to the final spot of landfill [7].

Furthermore, waste landfill sites can also be a source of greenhouse gas (GHG), which is generally known as the responsible for increasing global warming [8][9]. Even this condition has been known, there is still a lacks actions to measure or monitor the environmental condition. Thus, how the landfill sites could contribute to gas emissions or diseases is difficult to observe. Regarding this issue, introducing a control and monitor system as preventive measures should develop a warning system [9].

Some studies have demonstrated the ability of gas detectors to measure the methane gas in-point using a scanning method [10][11]. The measurement process involves an open area mechanism, where the gas concentration is influenced by wind speed. The landfill environment is mainly observed using a portable system that continuously monitors and measures the emitted gases to inform the concentration. So, evaluating changes in landfills over the long term is difficult to determine unless it is conducted at regular intervals according to a detailed plan. In contrast, a continuous monitoring method, which could measure the gas level, is needed. This method needs a stationary monitoring system that should be developed and arranged according to some parameters, such as the selected gases, sampling method, sampling equipment, how well construction, and depth of screened intervals [12]. This method requires an expensive system, which combines mass balance, stationary plume, and dynamic plume [13]. Considering this situation, this paper describes how developing an internet-based monitoring system to inform the air condition of waste landfills continuously.

The developed system measures some parameters to find out how to monitor hazardous gases. Here, CH_4, CO, CO_2 are measured with temperature, humidity, wind direction, and wind speed. Since the developed system is placed in the landfill area, the electrical energy uses a solar system. A voltage sensor is attached to the system to monitor the battery used. The developed system is a prototype to connect the station at the final spot of the landfill site, which is connected to central information. Here, the working state of the station and how the data will be sent to the center will be discussed.

2. Methodology

2.1. Gas Emission in landfill site

Landfill sites produce gases by decomposing organic materials composed of 50% CH4 and 50% CO2 and a small amount of non-methane organic compounds. CH4 gas is the most potentially causing GHG, 28 to 36 times more effective than CO2 trapping heat in the atmosphere, as the information of the Intergovernmental Panel on Climate Change (IPCC) assessment report (AR5).

Estimated CH4 emissions were carried out on a lab-scale as the prior measurement of ambient CH4 concentrations, measured vertically using a portable gas detector. The gas detector was hanged at the surface to measure variation conditions, such as CH4 flow rates, wind speed, the position of the gas detector, and cover utilization [14]. From that study, the correlation between ambient CH4 concentration (x) and CH4 flux (y) was obtained, which uses significant ($p<0.01$). The correlation equations can be determined as follows.

$$R^2 = 0.7109 \text{ and } y = 0.1544x \tag{1}$$

2.2. Situation of landfill

This research was carried out in a waste landfill named TPA Supit Urang, i.e., the waste disposal center in Malang city, Indonesia, in 1993. Figure 1 shows the TPA Supit Urang landfill area from a satellite taken by Google Maps. The landfill is divided into six areas, covering 31 ha, where 190,053 people live surround it [15]. The volume of waste disposal each day can reach 1,500 m3/day or 375 tons/day.

86.67% of it can be collected at the waste disposal center. The composition of waste disposal mainly consists of organic garbage, i.e., 65%, while the remains are plastic, paper, metal, rubber, glass, and others [16].

The Supit Urang landfill receives garbage in the open dumped and produces CH4 gas. It was estimated that 5828.90 tons had been distributed to 319 households surrounding the landfill in 2018 [17]. However, it has no report about GHG emissions emitted into the air since this landfill uses the open dumped system.

In this research, the monitoring system uses a station placed at the center of the landfill with GPS positions of -7.982614, 112.577876. Then, the system will monitor, record, and give information on environmental conditions. Regarding the broad area of the landfill, with no electrical system, the station collects data and sends them wireless to the internet cloud. Then, the data will be received by the central host to analyze the data.

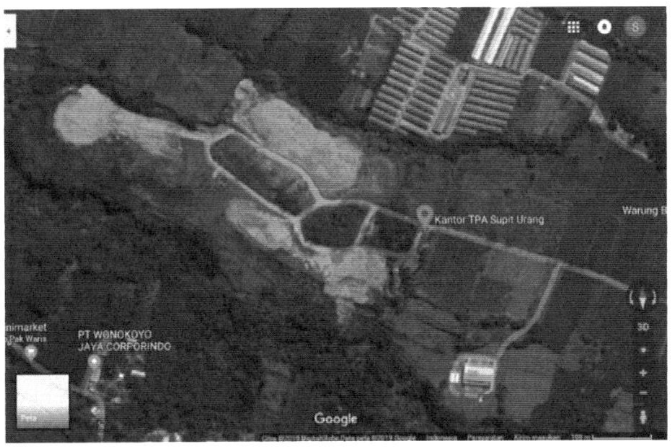

Figure 1. Satelite view of the TPA Supit Urang.

2.3. Methodology

The design method explains the Architecture of the Monitoring System of the Waste Disposal Center. The monitoring system is designed to observe the air condition of the final spot of the waste disposal center (TPA Supit Urang) using the internet-based system called Monitoring Station. The monitoring station takes the environmental data surrounding TPA Supit Urang. The data consist of eight parameters, i.e., CH_4, CO, CO_2, wind direction, wind speed, temperature, humidity, and battery voltage level, which are collected and send to the webserver at the Central Host in Universitas Negeri Malang via internet connection. The data can be accessed by three users, i.e., admin, government, and public. Admin has the privilege to access and change the system structure, while the government can access the data in real-time and historical data needed, and the public can only see the report of historical data and the current condition. The developed system is shown in figure 2.

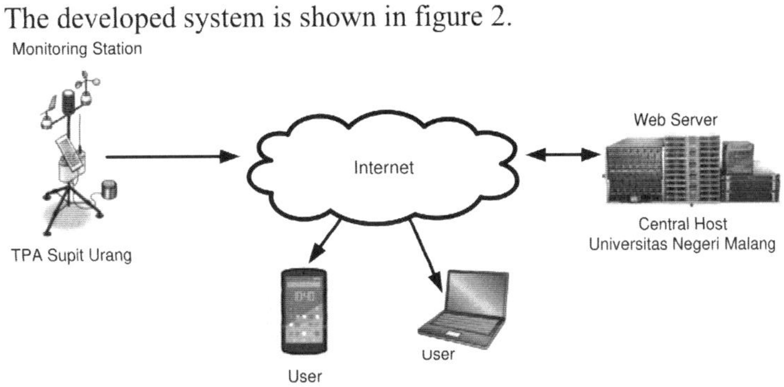

Figure 2. Diagram block of the developed system.

IOP Conf. Series: Earth and Environmental Science **894** (2021) 012029 doi:10.1088/1755-1315/894/1/012029

The system can use four layers, i.e., device, network, cloud management, and application layers [18], described as follows.

2.4. Device layer

This developed system's device layer is considered the monitoring station, which has functioned as the sensing and monitoring devices. This system uses eight sensors, i.e., temperature, humidity, wind speed, wind direction, carbon monoxide (CO), carbon dioxide (CO_2), methane (CH_4), and battery's level. The function of the sensor attached to the monitoring system has specifications explained as follows.

The primary gas which is emitted in landfills is CH_4. Thus, the CH_4 sensor has the primary function of monitoring the gases sensor, which monitors ambient CH4 exposure from landfills [19]. Besides CH_4 as the dominated gas produced, which is about 55%, CO2 also has a high concentration emitted from landfills, about 45% [20]; then, a CO_2 sensor is attached to this station. CO gas can be measured in emission gases in landfills because of solid waste degradation, where the gas concentration will increase when CH_4 concentration is low [21].

The concentration of gases is significantly affected by wind direction, and wind speed since gas concentration measured by the monitoring station is gas emitted large landfill area with open dumping system, so wind direction and wind speed sensors are added to the system. Furthermore, the monitoring system needs an electrical power source since the station is placed in a center landfill far from the electrical source. Then the electrical power source for the station is supplied using a solar panel system. This system is influenced by climate since Malang city has the characteristic that rain and cloud often occur. This situation affects the charging process for the battery used in the solar panel. Then the voltage level sensor is attached to the monitoring system to maintain the solar panel system. If the battery is insufficient, the electrical power will be switched off; while the battery is charged sufficiently, the electrical power will be switched on automatically, while the system usually starts work to capture data.

The sensors are integrated into the Arduino microcontroller to process the data, while the output is shown at liquid crystal display (LCD), while monitoring data are saved at the memory card. The data taken by the sensor are submitted to the Central Host using GSM module SIM800L via internet connection. The electrical system is shown in figure 3. Furthermore, the system is accomplished with a warning indicator that will blink on when the situation in the spot is harmful to the system; for example, the temperature is too high, or the battery's level is too low.

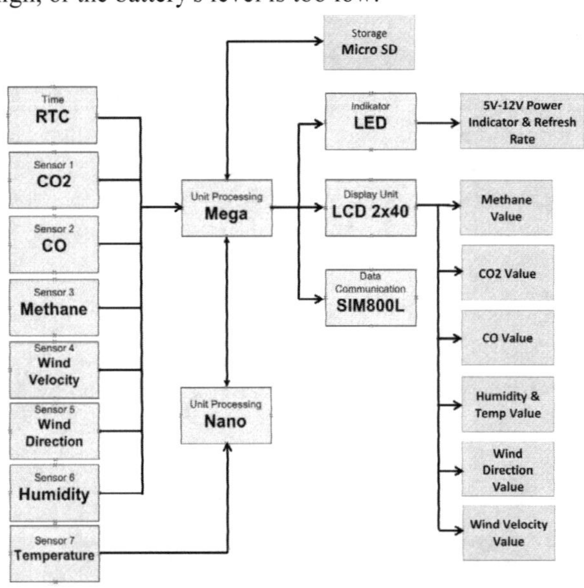

Figure 3. Block diagram of the electrical system.

2.5. *Network layer*

The proposed system is connected to the host using the GSM module via internet connection. GSM800L module [22] is used here, which supports quad-band GSM/GPRS network [23]. The module is connected to the microcontroller via Universal Asynchronous Receiver-Transmitter (UART) port, which supports AT command interface with auto baud detection.

To recognize a huge number of data from various monitoring stations, the ID station is determined firstly as STx representing station number "x." It is followed by sequentially data from the real-time clock (RTC) with the format of year-month-date (yyyy-mm-dd), hour:minute:second (hh:mm:ss), temperature (T), wind speed (S), wind direction (D), CO, CH_4, CO_2, humidity (H), voltage (V), which separated by punctuation comma. A format is used to manage the data collected from the sensor, as shown in table 1. Thus, the format data is represented as follows.

STx, yy-mm-dd,hh:mm:ss,xx.xx,xxx.xx,xx.xx,xxx.xx,xxx,x.xx,xx,xx.xx (2)

Table 1. Format data from sensor.

Sequence	Data Type	Format Data	Unit
ID Station	String	STxxx	-
RTC	Integer	yy-mm-dd,hh:mm:ss	-
Temperature	Float	xx.xx	C
CH_4	Float	xxxxx.xx	ppm
CO	Float	xxxx.xx	ppm
CO_2	Float	xxx.xx	ppm
Wind Direction	Integer	xxx	deg
Sequence	**Data Type**	**format data**	**Unit**
Wind Speed	Float	xx.xx	m/s
Humidity	Integer	xx	%RH
Voltage / VIT	Float	xx.xx	V
Current	Float	xxxx.xx	mA

Sequence	Data Type	Format Data	Unit
Duration (t)	Integer	xxxxxx	Second
GPS	Longitude F	xxxx.xxxxxx	-
	Laatittude F	xxxx.xxxxxx	-

2.6. *Cloud management layer*

The data from the monitoring station are sent to the webserver using two methods. The first method is named a normal situation, which collects the data from the monitoring station in each minute and saves them in a memory card; after 15 minutes, the collected data are sent to the webserver. This method is used to avoid data losses during data traffic in internet clouds. The second method is named an emergency, where besides saving data in each minute, the data are also pushed to be sent to the webserver directly. This mechanism will be done when the situation in the landfill is in an emergency, such as the temperature increase daily, which may alert that the landfill is burnt, or the gases concentration increase normally, which may inform that sensors work inappropriately or in hazardous situations occur.

The data from both mechanisms update the database in a web server cloud, where the data are used for data management, controlling, and reporting. When the emergency occurs, the system will send a notification using Google firebase. The detailed information is shown in figure 4.

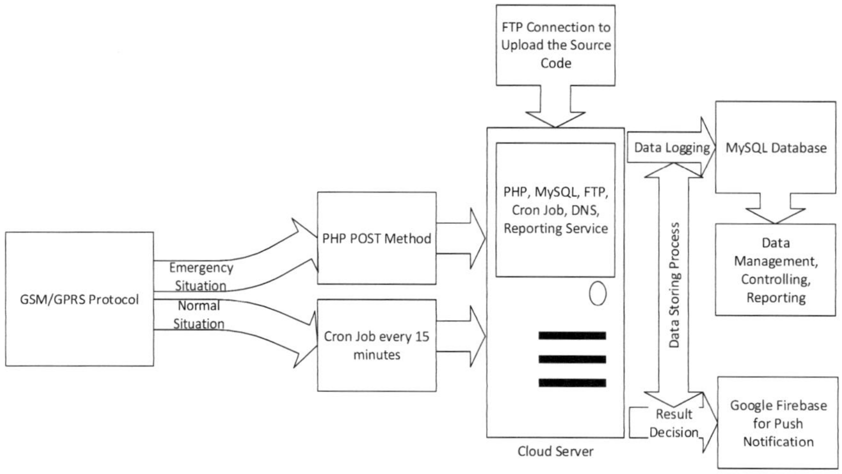

Figure 4. Cloud management layer schema for internet send, process, and the result of data.

The developed system works as shown in figure 5, while figure 6 shows the flowchart of the monitoring system at the network server.

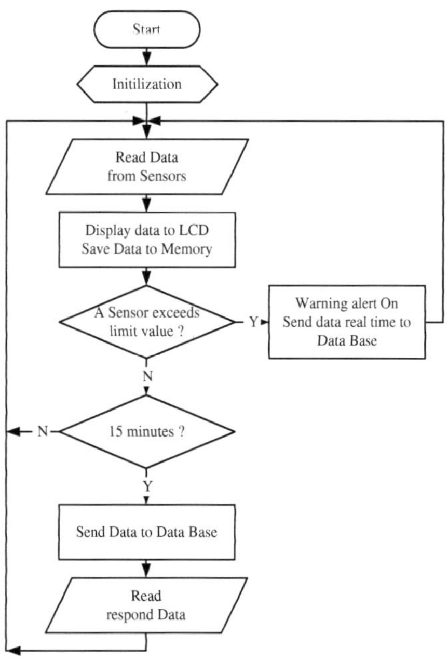

Figure 5. Flowchart of the monitoring station.

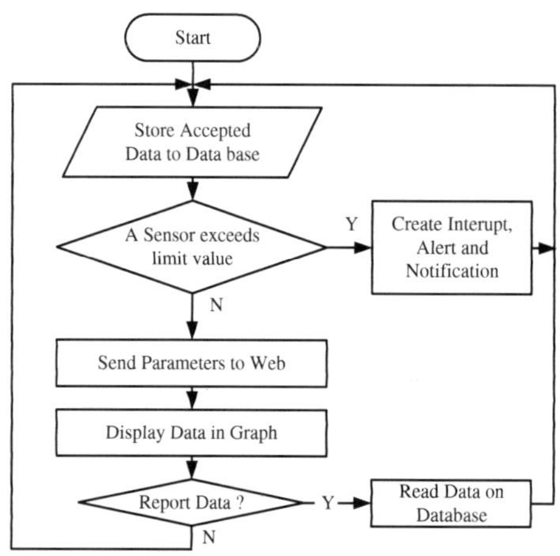

Figure 6. The Flowchart of the monitoring system at the webserver.

In this research, the monitoring station is placed at a waste disposal center in a city as a measurement node. In this research, the sample node is Malang city in Indonesia. Each waste disposal center could have one or more nodes representing monitoring stations. The database of the proposed system is developed to capture the data sent by the monitoring station, as shown in figure 7. Here, the database consists of tables, i.e., tbl_node, tbl_station, tbl_humidity, tbl_co, tbl_temperature, tbl_ch4, tbl_direction, tbl_speed, tbl_co2, tbl_scala, and tbl_user represent tables to store data of node, station, humidity, CO, temperature, CH_4, wind direction, wind speed, and CO_2, respectively. The table of tbl_scala is used to store ambient data for limit values of healthy environment determined by the regulation of the Indonesian government about occupational health and safety. The table of tbl_user is used to store data of the user, which could access data of the monitoring system. User data are categorized into three privileges, i.e., admin, government, and public. The public can access it if they register to the system. These tables are connected user Entity-Relationship (ER) diagram, as shown in figure 7.

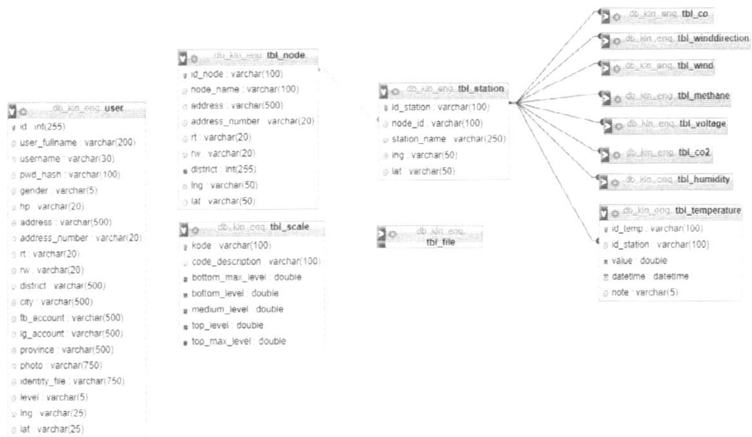

Figure 7. Database of a web server as the Central Host.

2.7. Application Layer

In the developed system, the web-based computer and Android applications have some functions, i.e., (1) show the data in real-time; (2) report the data like document, and (3) give the notification if the gas level exceeds the limit. To connect the data and user, the application layer shows the interface between system and user. The user is divided into three kinds in this system, i.e., admin, government, and public. Admin has the privilege to access all data; the government can access monitoring data in real-time and also access the historical data, while the public can access if they register to the system; then they will access to monitor real-time data. The dashboard of the user interface system is shown in figure 8.

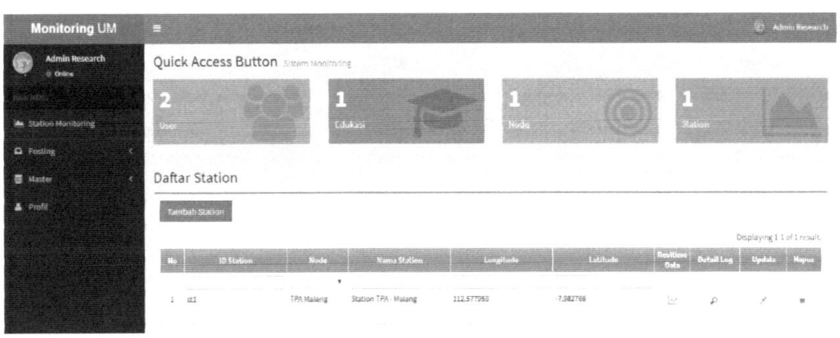

Figure 8. Administrator dashboard for manage logging process.

3. Results and discussion

The experiment measures gas in landfill area; the station is placed at longitude -7.982614, latitude 112.577876. The high of the station is 2 m over the waste surface. The method for experimenting is used to evaluate fluctuation in concentration. Here, we observe changing gas concentration daily.

3.1. Monitoring Results of CH_4

The average results of CH_4 when it was running in ten days are shown in figure 9. This stage aims to find the daily changing of gas concentration in the morning, afternoon, and night. From the result, CH_4 gas concentrations show decreasing at daily around 00.00 a.m. to 24.00 p.m. It shows that new disposal waste is going to decompose at night.

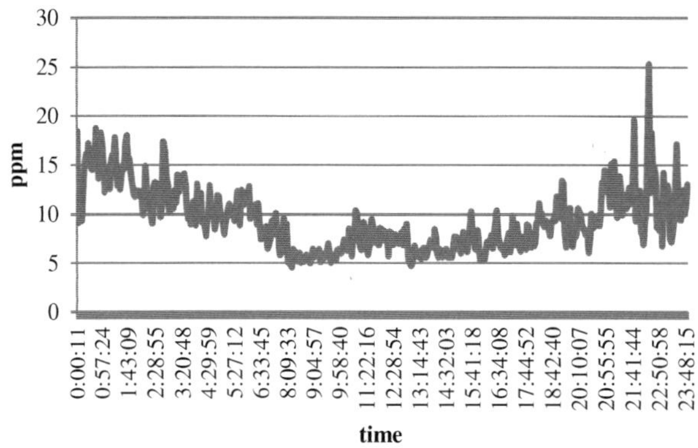

Figure 9. Averages of 10 days CH_4 gas concentration.

3.2. Monitoring Results of CO_2

In the case of the running results of CO_2 sensors for ten days, the daily changing of gas concentrations is shown in figure 10. From the result, CO_2 gas concentrations show relatively constant at daily around 24 hours.

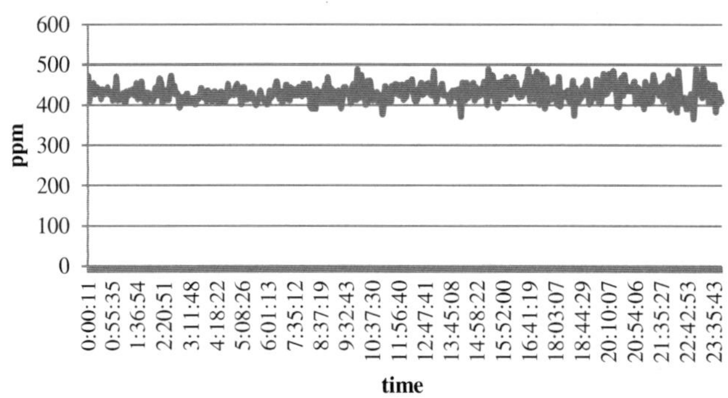

Figure 10. Averages of 10 days CO_2 gas concentration.

3.3. Monitoring Results of CO

Hereafter, the running results of CO sensors for ten days, the daily changing of gas concentrations are shown in figure 11. From the result, CO gas concentrations need to be observed regarding the situation in a landfill whether there are activities, which can affect the gas concentration, such as waste burned.

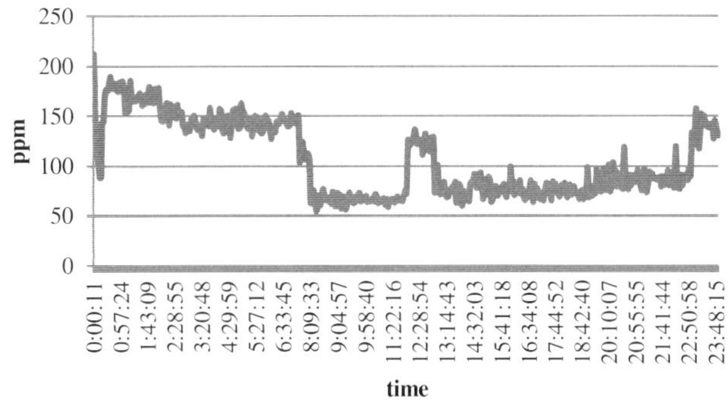

Figure 11. Averages of 10 days CO gas concentration.

After all, the situation of TPA Supit Urang should be observed furthermore, whether it is still safe for people whose life and work are surrounding it.

4. Conclusion

The propose system was placed in the middle area of the landfill. The result shows that the monitoring system can show the changing of gas concentration, where the instruments are used to measure the ambient concentration of methane (CH_4), carbon dioxide (CO_2), and carbon monoxide (CO). The system was evaluated regarding its ability to monitor gas parameters continuously. The concentrations are affected by the decomposing process of waste at the landfill, especially at the night stage. Furthermore, the system should be analyzed in more detail and was charged using solar panels for convenience and greater efficiency during monitoring.

Acknowledgment

We want to thank the Directorate of Higher Education of the Indonesia Government for supporting this research. This research is a collaboration result between the State University of Malang and the University of Kitakyushu.

References

[1] Murase N, Murayama T, Nishikizawa S, and Sato Y 2017 *Waste Manag Res* **35(10)** 1013-1022
[2] Sendari S, Agustin W, Fardyza AS, Rahmawati Y, Rahmawati H, Matsumoto T, and Rachman I 2019 *IOP Conf Ser: Earth Enviro Sci* **245(1)** 012034
[3] Sendari S, Rahmawati Y, Kamdi W, Elmunsyah H, Agustin W, Hadi N C, Wibowo FS, Matsumoto T and Rahman I 2018 *IOP Conf Ser: Earth and Environ Sci* **105** 1 012078
[4] Lee U, Han J, and Wang M 2017 *J Clean Prod* **166** 335–342
[5] Paraskaki I, and Lazaridis M 2005 *Waste Manag Res* **23(3)** 199–208
[6] Maheshwari R, Gupta S, and Das K 2015 *J Environ Sci Toxicol and Food Technol* **1(4)** 17–23
[7] Saepudin M and Amalia D 2016 *Buletin Penelitian Sistem Kesehatan* **19(4)** 243-249
[8] Singh SK, Anunay G, Rohit G, Shivangi G, and Vipul V 2016 *Journal of Climatology & Weather Forecasting* **4** 1
[9] Izumoto S, Hamamoto S, Kawamoto K, Nagamori M, and Nishimura T 2018 *Environ Sci Pollut Res* **25(24)** 24500-24506
[10] Lando AT 2015 *Memoirs of The Faculty of Engineering, Kyushu University* **75(1)** 29–46
[11] Thomas S and Haider NS 2014 *Journal of Engineering Research and Applications* **4(5)** 137-143
[12] Jovanov D, Vujić B, and Vujić G 2018 *Journal of Environmental Management* 216 32-40
[13] Scharff H, Martha A, Van Rijn DM, Hensen A, Flechard C, Oonk H, Vroon R, de Visscher A, and Boeckx P 2003 *Afvalzorg Deponie BV, Dutch Ministry of Housing, Spatial Planning and Environment*

[14] Lando AT, Nakayama H, and Shimaoka T 2017 *Waste Manag* **59** 255–266
[15] Neoza DK 2012 *Studi pengembangan Tempat Pengolahan Akhir (TPA) Supit Urang di Kelurahan Mulyorejo Kecamatan Sukun Kota Malang* (Doctoral Dissertation, University of Muhammadiyah Malang)
[16] Ministry of environment and Forestry Republic of Indonesia 2017 Sistem Informasi Pengelolaan Sampah Nasional *Information System of National Waste Management* [Online]. Available: http://sipsn.menlhk.go.id/?q=3a-tsph.
[17] Rahajeng A S, Meidiana C and Anggraeni M 2015 *KnE Energy* **2** 2 30
[18] Viswanath S K, Yuen C, Tushar W, Li W T, Wen C K, Hu K, Chen C and Liu X 2016 S *IEEE Wireless Communication* **23(5)** 90–98
[19] Bogner J and Spokas K 1993 *Chemosphere* **26(1)** 369–386
[20] Manfredi S, Tonini D, Christensen T H, and Scharff H 2009 *Waste Manag Res* **27(8)** 825–836
[21] Haarstad K, Bergersen O, and Sørheim R 2006 *J Air Waste Manag Assoc* **56(5)** 575–580
[22] Vimal P V and Shivaprakasha K S 2017 *International Conference on Intelligent Computing, Instrumentation and Control Technologies (ICICICT)* 1514-1519
[23] Pawar S S, Yadav A, Marathe D A, and Vyavhare N 2018 *Int J Elect Elect Comput Syst* **7(3)** 428–435

IOP Conf. Series: Earth and Environmental Science **894** (2021) 012030 doi:10.1088/1755-1315/894/1/012030

The effect of organic fertilizer chicken on the growth of papaya plants in palm oil plantation: a case study of Naga Sari Village, Mestong, Muaro Jambi District

A T Maryani[1], N Mirna[1], F F Bahar[2]

[1]Faculty of Agriculture Science, Universitas Jambi, Jambi, Indonesia
[2]Faculty of Science and Technology, Universitas Jambi, Jambi, Indonesia

atatikmaryani@gmail.com

Abstract. The use of chemical fertilizers in agriculture has caused soil degradation and is harmful to human health. Organic fertilizers made from animal waste have received worldwide attention because they are not detrimental to health or the environment. This paper describes the effect of using organic fertilizers from chicken manure on the growth of papaya plants. A completely randomized design was used in this study using two factors and three repetitions; the variations of the fertilizers given were 0, 50, 100, 150, and 200 g. This study informed that organic chicken manure had a significant effect on the growth of oil palm and papaya plants. The best growth is oil palm and papaya plants given 200 g of organic fertilizer from chicken manure.

1. Introduction

Soil contamination caused by agricultural chemical fertilizers has been widely reported around the world. The use of chemical fertilizers has resulted in a decrease in soil quality which causes the soil to be contaminated by heavy metals and other hazardous chemicals [1]. The content of chemical fertilizers such as sodium and potassium reduces soil quality [2]. The use of chemical fertilizers also affects the quality of fruit, plants and the environment [3]. The chemical content in chemical fertilizers is reported to have caused an increase in health costs in China from 1990 to 2012 [4].

One of the efforts to prevent pollution and maintain health is to use organic fertilizers. Various studies have reported the potential of organic materials to be converted into fertilizers. One of the potential materials used as organic fertilizer is animal (Chicken) manure [5]. Chicken manure was chosen because it is easier to obtain and has a complete content as organic fertilizer [6]. Chicken manure contains high N, P, K and has lower P and K content than cow and goat manure. Chicken manure is also beneficial for fertilizer on seasonal crops such as papaya [7].

The use of chicken manure as fertilizer for papaya plants is highly recommended, but animal manure is still limited to a monoculture planting pattern. There have not been any studies that reported the effect of using animal manure on the intercropping pattern with oil palms. This planting technique is intended to minimize environmental damage and the minimal benefits of oil palm plantations for the surrounding community. This paper will discuss the effect of using fertilizer from chicken manure and the intercropping pattern of papaya and oil palm on the development and diameter of the resulting papaya fruit.

IOP Conf. Series: Earth and Environmental Science **894** (2021) 012030 doi:10.1088/1755-1315/894/1/012030

2. Methodology

2.1. Materials

The research was conducted at the People's Garden in Naga Sari Village, Mestong District, Muaro Jambi Regency. The altitude is about 30 masl, with a climate type C2 (Oldeman) and red-yellow podsolic soil type. This research was started from March to August 2020. The plant materials used were 6-month-old oil palm (TBM 1) Dumpy variety and California and Pomegranate red papaya seeds, chicken manure organic fertilizer, NPK Mutiara, agricultural tools, namely tractors, hoes, machetes, plastic hoses, tape measure, raffia rope, cutting scissors, sprayer and others.

2.2. Method

The completely randomized design method was used in this study. The study was arranged in 2 factorials with three repetitions. The first factor that was tried was the variety (V)

V1 = California

V2 = Pomegranate red

The second factor that was tried was organic chicken manure (P)

P0 = without applying fertilizer P_1 = 50 g

P_2 = 100 g

P_3 = 150 g

P_4 = 200 g

Thus, there are ten treatment combinations, and each treatment combination is repeated four times so that there are 40 experimental units in total. Each experimental unit contained three papaya plants. The data obtained were analyzed statistically by variance; the analysis model used was:

$$Y_{ijk} = \mu + V_i + P_j + (VP)_{ij} + \sum_{ijk}$$

Observations were made on the growth of young oil palms, including stem circumference, plant height, number of leaves, leaf length, leaf width, number of leaflets and canopy area, which were observed descriptively. Whereas for papaya intercropping, growth components were observed, analysis of papaya plant income and papaya fruit production including plant height, number of leaves, fruit length, papaya fruit diameter and papaya fruit production. To determine the effect of treatment on experimental parameters, analysis of variance was carried out, followed by an analysis of the mean difference using the least significant difference at a 5% confidence interval, initial nutrient analysis and final nutrient analysis.

3. Result and discussion

3.1 Oil palm plant growth

The growth of oil palm and California varieties of papaya and pomegranate as intercrops during the immature plants did not show competition during the plant growth period. It is shown by the data in (Table 1). Plant height, number of leaves, and diameter of oil palms planted in poly-culture with papaya as an intercrop did not show further growth from oil palm plants planted in monoculture. Thus it can be neglected that papaya plants can be planted perceptually in oil palm plantations. Oil palm plants can also be planted in an intercropping system with corn plants without affecting the growth of oil palm plants in an immature plants phase [8].

Table 1. Plant height (cm) of oil palm aged 10 to 20 MST in chicken manure and intercropping on papaya plants of different varieties.

Varieties	Chicken manure					Average
	0 g	50 g	100 g	150 g	200 g	
10 MST						
California	120.25 a	148.75 bc	146.25 b	168.00 c	150.50 d	146.75

Varieties	Chicken manure					Average
	0 g	**50 g**	**100 g**	**150 g**	**200 g**	
	a	b	a	b	b	
Pomegranate red	138.75 b	135.00 a	153.75 c	155.00 c	140.00 c	144.50
	b	a	b	a	a	
Average	129.50 a	141.88 b	150 d	161.5 e	145.25 c	
Varietas	**Chicken manure**					**Average**
	0 g	**50 g**	**100 g**	**150 g**	**200 g**	
12 MST California	132.50 a	158.75 a	162.50 a	170.00 b	156.25 a	156.00 a
	a	b	c	d	b	
Pomegranate red	158.75 b	158.75 a	164.50 a	166.25 a	153.75 a	160.40 b
	b	b	c	c	a	
Average	145.63 a	158.75 c	163.50 d	168.13 e	155.00 b	
Varietas	**Chicken manure**					**Average**
	0 g	**50 g**	**100 g**	**150 g**	**200 g**	
14 MST California	162.00 a	167.50 b	170.00 b	175.00 a	150.00 a	164.90 b
	b	c	c	d	a	
Pomegranate red	162.50 a	146.25 a	166.25 a	173.75 a	162.00 b	162.15 a
	b	a	c	d	b	
Average	162.25 b	156.88 a	168.13 c	174.38 d	156.00 a	
Varietas	**Chicken manure**					**Average**
	0 g	**50 g**	**100 g**	**150 g**	**200 g**	
16 MST California	171.75 a	177.00 b	178.25 a	186.25 a	156.75 a	174.00 a
	b	c	c	d	a	
Pomegranate red	181.25 b	173.75 a	182.50 b	186.25 a	173.75 b	179.50 b
	b	a	b	c	a	
Average	176.50 b	175.38 b	180.38 c	186.25 d	165.25 a	
Varietas	**Chicken manure**					**Average**
	0 g	**50 g**	**100 g**	**150 g**	**200 g**	
18 MST California	178.00 a	186.25 b	190.25 a	192.50 a	170.00 a	183.40 a
	b	c	d	d	a	
Pomegranate red	183.75 b	181.25 a	188.75 a	197.00 b	177.50 b	185.65 b
	b	b	c	d	a	
Average	180.88 b	183.75 c	189.5 d	194.75 e	173.75 a	

Varieties	Chicken manure					Average
	0 g	50 g	100 g	150 g	200 g	
Varietas	**Chicken manure**					**Average**
	0 g	**50 g**	**100 g**	**150 g**	**200 g**	
20 MST						
California	182. 50 a	192.50 b	196.75 a	197.50 a	178.75 a	189.60 a
	b	c	d	d	a	
Pomegranate						
red	190.00 b	186.25 a	192.00 b	208.75 b	183.75 b	192.15 b
	b	a	b	c	a	
Average	186.25 b	189.38 c	194.38 d	203.13 e	181.25 a	

3.2. Papaya plant growth

Plant height, plant diameter, and the number of papaya leaves experienced a significant increase in the provision of 150 g Chicken manure/plant for all varieties (Table 2-4). This vegetative growth response proves that it is necessary to add nutrients through fertilizing chicken manure to increase the availability of low nutrients in the soil in areas under palm oil stands (IMMATURE PLANTS) to increase the growth of papaya plants. Recent research also informs that fertilizers play an essential role in regulating biophysical and biochemical functions in plants, the microbes contained in chicken manure can function to increase the elements available to plants [9]. The results of the analysis of soil nutrients (Table 5) show that by giving Chicken manure to the papaya growing media, there is an increase in nutrients N, P and K. These three nutrients are essential for plant growth so that if these nutrients are deficient, they can interfere with plant growth.

Variety analysis showed an interaction of pomegranate papaya varieties with 150 g of Chicken manure significantly affected all vegetative growth parameters at 18 MST and 20 MST. The addition of Chicken manure at a dose of 200 g does not show a significant difference to the provision of Chicken manure 150 g. It shows that although the role of organic fertilizers is vital in supporting plant growth, the excessive provision also harms vegetative growth, as seen in the provision of Chicken manure 200 g, in this case with the addition of the dose to 200 g/plant showed a decrease in growth, although not significant.

Nutrient N plays a vital role in supporting plant growth in the vegetative growth phase. The availability of very high N in the soil is sufficient to support papaya vegetative growth. P elements play an essential role in increasing the work of chloroplasts as absorbers of sunlight in photosynthesis. In addition, P elements also play an active role in transferring energy in cells. The K+ ion facilitates several plant physiological responses, including opening and closing leaf stomata and membrane regulation [10]. Potassium is an activator of many enzymes required to form starch and protein [11]. The opening and closing of the stomata are controlled by K+ through turgor pressure [12]. The osmotic pressure generated by the accumulation of K+ in the cells is also used to encourage leaf expansion [13]. Thus, the physiological process is optimal. The low nutrient content in the soil causes the photosynthetic activity to below, and assimilates are not optimal to support plant growth as experienced by low plant growth without giving Chicken manure.

Table 2. Plant height (cm) of papaya aged 10 to 20 MST in Chicken manure and intercropping on oil palm plants.

Varietas	**Chicken manure**					**Average**
	0 g	**50 g**	**100 g**	**150 g**	**200 g**	
10 MST California	14.00	16.75	17.00	17.75	15.00	16.10

Pomegranate red	15.75	15.50	15.75	16.75	13.75	15.50
Average	14.88	16.13	16.38	17.25	14.38	

Varietas	**Chicken manure**					**Average**
	0 g	**50 g**	**100 g**	**150 g**	**200 g**	
12 MST California	15.25	18.25	17.00	15.50	14.00	16.00
Pomegranate red	17.00	15.50	16.00	16.00	16.00	16.10
Average	16.13	16.88	15.75	15.75	15.00	

Varietas	**Chicken manure**					**Average**
	0 g	**50 g**	**100 g**	**150 g**	**200 g**	
14 MST California	17.50	20.00	18.75	18.25	16.50	18.20
Pomegranate red	18.25	17.500	17.50	19.50	17.50	18.05
Average	17.88	18.75	18.13	18.88	17.00	

Varietas	**Chicken manure**					**Average**
	0 g	**50 g**	**100 g**	**150 g**	**200 g**	
16 MST California	19.75	21.25	20.50	19.75	18.00	19.85
Pomegranate red	19.50	19.000	19.50	22.00	19.00	19.80
Average	19.63	20.13	20.00	20.88	18.50	

Varietas	**Chicken manure**					**Average**
	0 g	**50 g**	**100 g**	**150 g**	**200 g**	
18 MST California	21.75 a	22.75 a	21.75 a	23.25 a	20.25 a	21.95
	a	a	a	a	a	
Pomegranate red	19.25 a	21.00 a	21.00 a	25.00 a	20.50 a	21.35
	a	a	a	b	a	
Average	20.50 a	21.88 ab	21.38 ab	24.13 b	20.38 a	

Varietas	**Chicken manure**					**Average**
	0 g	**50 g**	**100 g**	**150 g**	**200 g**	
20 MST California	21.50 a	21.50 a	21.50 a	22.75 a	21.75 a	21.80
	a	a	a	a	a	
Pomegranate red	21.00 a	21.00 a	22.00 a	27.50 b	22.25 a	22.75

Varietas	Chicken manure					Average
	0 g	**50 g**	**100 g**	**150 g**	**200 g**	
	a	a	a	b	a	
Average	21.25 a	21.25 a	21.75 a	25.13 b	22.00 ab	

Note: numbers followed by different letters in the same row and column and the time of observation that is significantly different according to Duncan's test at the 5% level

Table 3. Plant diameter (cm) of papaya aged 10 to 20 MST in Chicken manure and intercropping on oil palm plants.

Varietas	Chicken manure					Average
	0 g	**50 g**	**100 g**	**150 g**	**200 g**	
10 MST California	15.50	16.50	17.75	16.25	16.00	16.40
Pomegranate red	16.25	16.50	16.25	16.75	15.00	16.15
Average	15.88	16.50	17.38	16.50	16.13	
Varietas	**Chicken manure**					**Average**
	0 g	**50 g**	**100 g**	**150 g**	**200 g**	
12 MST California	21.75	21.75	21.00	18.75	19.75	20.60
Pomegranate red	19.75	19.25	19.00	19.00	19.00	19.20
Average	20.75	20.50	20.00	18.88	19.38	
Varietas	**Chicken manure**					**Average**
	0 g	**50 g**	**100 g**	**150 g**	**200 g**	
14 MST California	23.00	23.50	22.50	22.50	21.25	22.55
Pomegranate red	21.00	21.00	21.25	22.25	20.50	21.20
Average	22.00	22.25	21.88	22.38	20.88	
Varietas	**Chicken manure**					**Average**
	0 g	**50 g**	**100 g**	**150 g**	**200 g**	
16 MST California	25.25	26.25	24.75	26.25	23.75	25.25 b
Pomegranate red	22.75	22.50	23.50	24.75	22.00	23.10 a
Average	24.00	24.38	24.13	25.50	22.88	

Varietas	Chicken manure					Average
	0 g	**50 g**	**100 g**	**150 g**	**200 g**	

18 MST California	27.00 a	27.25 a	27.75 a	26.75 a	27.75 a	27.30
	a	a	a	a	a	
Pomegranate red	24.50 a	24.25 a	26.00 a	28.75 a	23.75 b	25.45
	a	a	ab	b	a	
Average	26.00	25.75	26.38	28.25	25.75	
Varietas	**Chicken manure**					**Average**
	0 g	**50 g**	**100 g**	**150 g**	**200 g**	
20 MST California	30.25 b	30.50 b	29.75 a	31.25 a	28.25 a	30.00 b
	a	a	a	a	a	
Pomegranate red	26.00 a	26.25 a	28.50 a	32.25 a	26.50 a	27.90 a
	a	a	a	b	a	
Average	28.13 a	28.38 a	29.13 ab	31.75 b	27..38 a	

Note: numbers followed by different letters in the same row and column and the time of observation that is significantly different according to Duncan's test at the 5% level

Table 4. The number of leaves (cm) of papaya aged 10 to 20 MST in Chicken manure and intercropping on oil palm plants.

Varietas	**Chicken manure**					**Average**
	0 g	**50 g**	**100 g**	**150 g**	**200 g**	
10 MST California	19.00	18.75	18.00	16.50	16.00	17.65
Pomegranate red	16.50	16.25	16.75	16.50	16.25	16.45
Average	17.75	16.25	17.38	16.50	16.25	
Varietas	**Chicken manure**					**Average**
	0 g	**50 g**	**100 g**	**150 g**	**200 g**	
12 MST California	21.75	21.75	21.00	18.75	19.75	26.60
Pomegranate red	19.75	19.25	19.00	19.00	19.00	19.20
Average	20.75	20.50	20.00	18.88	19.38	
Varietas	**Chicken manure**					**Average**
	0 g	**50 g**	**100 g**	**150 g**	**200 g**	
14 MST California	23.00	23.50	22.50	22.50	21.25	22.55
Pomegranate red	21.00	21.00	21.25	22.25	20.50	21.20

Varietas	Chicken manure					Average
	0 g	50 g	100 g	150 g	200 g	
Average	22.00	22.25	21.88	22.38	20.88	

Varietas	Chicken manure					Average
	0 g	50 g	100 g	150 g	200 g	
16 MST California	25.25	26.25	24.75	26.25	23.75	22.55
Pomegranate red	22.75	22.50	23.50	24.75	22.00	21.20
Average	24.00	24.38	24.13	25.50	22.88	

Varietas	Chicken manure					Average
	0 g	50 g	100 g	150 g	200 g	
18 MST California	27.00 a	27.25 a	27.25 a	26.75 a	27.75 b	27.30
	a	a	a	a	a	
Pomegranate red	24.50 a	24.25 a	26.00 a	28.75 a	23.75 a	25.45
	a	a	ab	b	a	
Average	25.75	25.75	26.88	27.75	25.75	

Varietas	Chicken manure					Average
	0 g	50 g	100 g	150 g	200 g	
20 MST California	29.75 a	30.25 b	30.25 a	32.75 a	30.50 b	30.70
	a	a	a	a	a	
Pomegranate red	27.00 a	26.25 a	28.25 a	32.25 a	26.25 a	28.00
	a	a	a	b	a	
Average	28.38	28.25	29.25	32.50	28.38	

Note: numbers followed by different letters in the same row and column and the time of observation that is significantly different according to Duncan's test at the 5% level

Table 5. Soil characteristic anaylisis result.

No	Sample	pH H_2O	C Organic	N Total	P HCl 25%	K HCL 25%	P Total	K Total
			%		(mg P_2O_5 100^{-1})	(mg K_2O100^{-1})	%	
1	Soil	6.00	0.16	1.48	72.75	16.50	-	-

3.3. Plant production

Variety analysis showed an interaction between California papaya varieties and 200 g Chicken manure, while there was no interaction for fruit circumference. In general, giving Chicken Manure can increase papaya fruit length and papaya fruit circumference compared to without Chicken manure; giving Chicken manure 200 g shows the best fruit length and circumference for all varieties. This phenomenon shows that papaya plants planted as oil palm intercrops require relatively large amounts of nutrients in the production phase. The nutrient requirements in papaya plants during the productive period are not much different from those during vegetative growth. It shows that papaya requires nutrients from vegetative growth to production.

The results showed that the treatment of Chicken manure had a significant effect on fruit length at 20 MST, the treatment of varieties had a significant effect on the length of the papaya fruit and the interaction between the two varieties of treatment with Chicken manure had a significant effect on the length of the papaya fruit at 20 MST.

Table 6. Papaya fruit length (cm) at 20 MST in chicken manure and intercropping on oil palm plants.

Varietas	Chicken manure					Average
	0 g	50 g	100 g	150 g	200 g	
10 MST California	24.00 a	27.75 a	27.75 a	33.35 a	35.00 a	29.55 b
	a	ab	ab	bc	c	
Pomegranate red	22.50 a	24.75 a	26.00 a	29.50 a	32.25 a	27.00 a
	a	ab	ab	bc	c	
Average	23.25 a	26.25 b	26.88 b	31.38 c	33.63 c	

Table 6 shows that the treatment of giving Chicken manure has a significant effect on the length of papaya fruit; giving Chicken manure with a dose of 200 g has the highest average at 20 MST, which is 33.63 cm, and the interaction between 200 g Chicken manure treatment with California has a significant effect on 20 MST, namely 35.00 cm. The length of the California papaya varieties is superior to the red pomegranate varieties. From the results, it was known that the treatment of giving Chicken manure had a significant effect on the circumference of the fruit at 20 MST, the treatment of varieties had a significant effect on the length of the papaya fruit, but the interaction between the two varieties of treatment with Chicken manure had no significant effect on the circumference of the papaya fruit.

Table 7. Papaya Fruit Circumference (cm) at 20 MST in chicken manure and intercropping on oil palm plants.

Varietas	Chicken manure					Average
	0 g	50 g	100 g	150 g	200 g	
10 MST California	28.25	30.50	30.25	32.50	34.00	30.90 a
Pomegranate red	30.75	34.00	34.00	35.50	36.75	34.20 b
Average	29.50 a	32.25 abc	32.12 ab	33.50 bc	35.38 c	

Table 7 shows that the treatment of giving Chicken manure has a significant effect on the circumference of the papaya fruit, giving Chicken manure with a dose of 200 g has the highest average at 20 MST, namely 35.38 cm and treatment of red pomegranate varieties has a significant effect on fruit circumference and is superior to papaya varieties. In California, the average pomegranate red variety was 34.20 cm, but the interaction between 200 g Chicken manure treatment and the two varieties had no significant effect on fruit circumference.

Ten untrained panellists carried out organoleptic testing on California papaya varieties and pomegranates. The parameters assessed in the organoleptic test of papaya were taste, colour and texture. Figure 1 histogram of organoleptic test results for papaya fruit giving on chicken manure and intercropping on oil palm plants. Based on the results of the colour test on the average papaya fruit, the panellists assessed a very orange colour, delightful taste and soft texture in the V2P1, V2P3 and V2P4 treatments. From the results of the tests carried out by the panellists, it is found that the organoleptic test can be seen in the following figure:

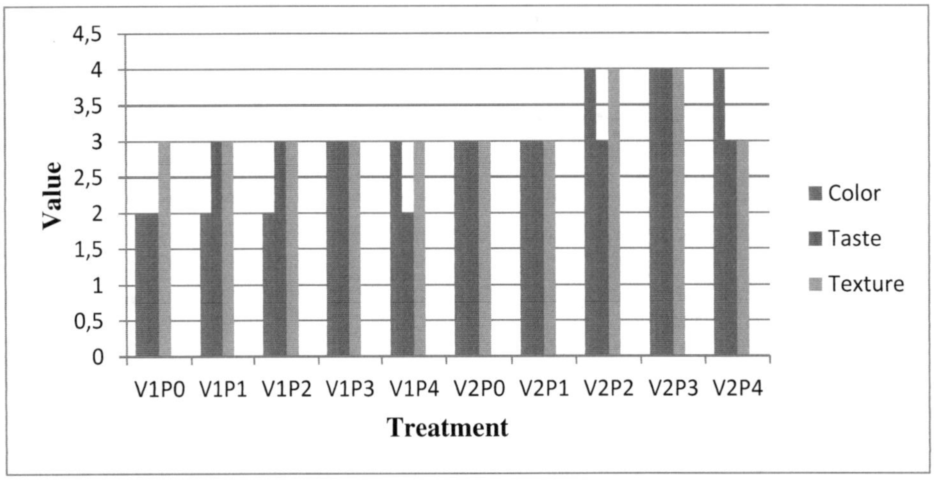

Figure 1. Organoleptic test of papaya fruit on both varieties.

4. Conclusion

The addition of Chicken manure affects the growth of California papaya and pomegranate papaya; this study also confirms that the more Chicken manure is given, the better fruit length and circumference will be. In addition, this study also informed that the growth of oil palm and papaya varieties of California and red pomegranate as intercrops during the immature plants period did not show competition during the plant growth period. Plant height, number of leaves, and diameter of oil palm plants planted in polyculture with papaya as an intercropping did not show further growth from those planted in monoculture.

Reference

[1] Savci, S 2012 *Int J Environ Sci Dev* **3(1)** 11–14
[2] Sanchez F, Barna R, Garrabrants A, Kosson D S and Moszkowicz P 2000 *Chem Eng Sci* **55** 113–128
[3] Spångberg J, Hansson P, Tidåker P and Jönsson H 2011 *Resour Conserv Recycl* **55** 1078–1086
[4] Yang H and Shen X 2012 *Sustainability 2012* 1–15
[5] Vandecasteele B, Reubens B, Willekens K and Neve S De 2014 *Waste Biomass Valor* **5** 491–503
[6] Ghosh P K, Bandyopadhyay K K, Manna M C, Mandal K G, Misra A K and Hati K M 2004 *Bioresour Technol* **95** 85–93
[7] Kirad K S, Barche S, Singh D B and Pradesh U 2010 *Proceeding IInd International Symposium on Papaya* 377–380.
[8] Darmawati N and Siregar A R 2014 *AGRIUM* **19(1)** 59–67
[9] Szczerba M W, Britto D T and Kronzucker H J 2009 *J Plant Physiol* **166** 447–466
[10] Elumalai R P, Nagpal P and Reed J W 2002 *The Plant Cell* **14** 119–131
[11] Haro R, Amtmann A, Ann T, Dreyer I., Biotecnología C De and Madrid U P De 2014 *J Plant Physiol* **171** 723–731
[12] Perez M E, Scherxer S and Al-Rashied K 2014 *Molecular Plant* **7(4)** 744–746
[13] Zivanovic B D, Cuin T A and Shabala S 2007 *Plant Cell Physiol* **48(4)** 598–605

IOP Conf. Series: Earth and Environmental Science **894** (2021) 012031 doi:10.1088/1755-1315/894/1/012031

Community sustainability and environmental quality improvement on studentified area in Malang City, Indonesia

R Situmorang[1*], A Sudikno[2], S Surjono[3], A D Wicaksono[3]

[1] Urban and Regional Planning Department, Trisakti University, Jakarta 11510, Indonesia
[2] Architecture Department, Brawijaya University, Malang 65145, Indonesia
[3] Urban and Regional Planning Department, Brawijaya University, Malang 65145, Indonesia

rahelsitumorang@trisakti.ac.id

Abstract. Studentification occurs in a college town so that the large number and activity of students and those related to universities affect the condition of the city. The impact of students consists of socio-cultural and economic changes that can be seen in physical changes. In the previous research, these changes often result in conflicts between students and residents, less service for residents, and a worsening physical environment. This study aims to find the contributions of the studentification model in Malang city as one of the college towns in Indonesia that occurs in the area adjacent to the campus Brawijaya University Malang City. The method used in this research is Structural Equation Model Partial Least Square. Data were taken through in-depth interviews and online surveys to 32 neighbourhood leaders as respondents. Some of the findings show additional thought contributions to studentification theory, namely: community sustainability and environmental quality improvement.

1. Introduction

Studentification is a process where students begin to live in parts of the city near the campus so that the previous population is reduced due to the internal migration process [1]. The influx of students into this part of the city also causes changes in the composition of the population and housing, increases land and rent prices [2], and causes damage to housing and the surrounding campus environment [3]. There was also a change in the structure of the house from a family residence to a boarding house, and segregation between students and non-students, so that the area became a student residential area, which causes spatial dysfunction because only the needs of students will be met [4]. It is also considered a demographic disorder because the transformation of the built environment accompanies the changes in life and the displacement of families from an area, besides being a reinvestment process, environmental landscape changes, class changes, and residents' income because students are a source of income [5]. In addition, studentification is a process of urban spatial planning that impacts socio-cultural, economic, and physical changes in urban spaces due to the number of students who live temporarily in the surrounding campus area.

This study aims to find the contributions of the studentification model in Malang city as one of the college towns in Indonesia [6] to community and environmental sustainability. The research was conducted in Ketawanggede Village, Malang City, adjacent to the Universitas Brawijaya campus. This campus is the second-largest university in Indonesia, with 72,721 students and 2,341 lecturers (in the 2019/2020 academic year), and is the most favourite university by prospective students since it has 178 study programs located on the same campus. The students come from all over Indonesia with various ethnicities and cultures as well as socioeconomic backgrounds. Student accommodation is provided by private boarding house owner located in Ketawanggede Village so that this area becomes dominated by boarding houses because very close by to campus, availability of rooms with different rates and facilities, easy access to obtain daily needs, and acceptance of students by the local population [7].

2. Methodology

The method used in this study is the Structural Equation Model Partial Least Square (SEM PLS) to examine the relationship between variables with indicators and sub-indicators and to obtain a comprehensive picture of the entire model. The model has higher flexibility for researchers to link theory and data [8]. Variables in this model are formative, and indicators of this model are reflective and have a different pattern of indicator relationship with other variables. The primary data used to construct the structural model was obtained from interviews with 32 neighbourhood leaders (Ketua Rukun Tetangga). Interviews were conducted using questionnaires directly by cell phone conversations and Google-Forms. A preliminary study was conducted on perceptions of students to support this research, owners/managers of boarding houses, residents, and business actors in the study area [7]. Another preliminary study was to determine 3 variables, 9 indicators, and 13 sub-indicators in the studentified area around Universitas Brawijaya [9].

3. Results and discussion

3.1. Result

The study area there has changed in the socio-demographic structure with students predominated to the population (87%) whose also temporary migrants since they only stay in Malang for four years and will move to another city after completing their studies. Changes in demographics can lead to social cohesion, and segregation is essential to avoid social conflicts between students and residents around the campus [10] because residents have different backgrounds and cultures. Social cohesion goes well, describing that students participate at community prayer and locals known students since they must report themselves as temporary residents in the area. Social segregation was identified by students' interest in choosing boarding house accommodation based on similar backgrounds (town of origin and beliefs) for the convenience of living together [7]. Economic changes on studentification seen from students as rental market [11] determined by 71.66 % of all houses are boarding houses and showed there is a balance between providing products and services to students without narrowing the supply of goods for residents. Another indicator in the physical variable is the quality of the environment [1][12][5] which can be seen through changes in environmental quality (cleaner and better) and security of the surrounding area (commotion or theft of student goods). The formulation model by SEM PLS describes those socio-cultural and economic variables that have a positive and significant effect on physical variables. The higher socio-cultural and economic impact can increase the pressure on physical condition area.

Change of studentified area begins with the influx of students into the area around the campus, thus causing an increase in population density. This situation causes a need for the availability of boarding houses room in the area around the campus. Locals see this as an opportunity to earn income by renting out rooms to students. The boarding house owner stated that renting a boarding house was initially done to increase income but became the primary income [9]. The increase of population density causing physical changes to the building (43.83%) due to increasing the number of rooms for rent, converting family rooms into rental-worthy spaces, optimizing the space that can be rented, to adding floors to the building. The increased student numbers affect the increase of the number of boarding rooms and the need for business stalls to serve students' needs, which causes a change in the space to become business stalls. The function changes of residential houses turned into the boarding house, and business stalls are 80.22%.

Physical condition changes in the study area are measured through form and function changes in houses and indicators of environmental changes in the area. Dimensions of houses form and function changes represented by sub-indicators of houses form changes showed by loading factor of 0.924. In contrast, the contribution of sub-indicators of house function changes from (residential into a boarding house or business stalls) in measuring the indicator of houses form and function changes seen from loading factor of 0.927%. Changes in houses are carried out by increasing the space area, the number of floors, and the form of space. The change of houses function is carried out by changing the residential houses into a new function (boarding house, business stall, or shop houses). Most respondents (65.6%) stated that more than 10 houses had changed form to multi-level buildings, added bedrooms, bathrooms, and more spaces to accommodate students and their activities. Furthermore, respondents (78.1%) stated that more than 16 houses in their neighbourhood changed function from residential houses to boarding houses/rental or places of business.

Indicators of physical change can be seen from the sub-indicators: environmental improvement and changes in environmental security. Environmental improvement is measured by improving the quality of the environment due to infrastructure and cleanliness, and changes in environmental security are seen from the level of security disturbances (theft, commotion, theft of laptops/cell phones, etc.). The respondents (50%) state that the environmental quality is better overall, while in terms of environmental security, 53.1% of respondents also stated that security disturbances have only occurred less than 5 times in 10 years.

3.2. Discussion

The previous community will change and tend to shift with the need for student boarding houses, strategic locations, and opportunities to get high property selling prices [13]. It shows that the area is not proposed inhabited by residents who are not students, so that the type and availability of property in the area are intended for students. Residents have difficulty selling their houses due to disorder, noise, and the declining physical quality of the area due to studentification [14]. Socio-cultural differences between students and residents are one of the causes of residents leaving the area around campus [14][15]. This study determined the community sustainability since those residents feel there is no urgent need to sell their property (70%) and still do not want to sell the house because they have not shared their inheritance yet (20%), and they know each other for a long time. Since most residents have lived in this area for more than 15 years stated that they want to keep the Javanese values, which cause the comfort of being familiar, and living in a strategic area are the strong motivation for survival. Several respondents also mentioned that investors would buy the property at whatever price the owner offers due to the limited number of properties offered on the market.

The influx of students around campus can cause environmental a decrease in cleanliness because students tend to disorganize and do not like to keep the environment clean [5][16], and students poses a security threat since they may not be involved in criminal, but become attractors for these criminal acts [5][17]. Students tend to have no habits or awareness in maintaining goods or the environment, so it is not easy to expect them to care for the surrounding area. This study shows that although it is recognized that students often do not care about the cleanliness of the environment; residents are trying to provide garbage collection. On the other hand, students try to maintain a better and cleaner environment because they are willing to follow the existing rules and regulations. The number of students living in the area around the campus also poses a security threat [5]. Students may not be involved in criminal acts but become attractors for these criminal acts [4][5]. Students tend to have no habits or awareness in maintaining goods or the environment, so it is not easy to expect them to care for the surrounding environment.

IOP Conf. Series: Earth and Environmental Science **894** (2021) 012031 doi:10.1088/1755-1315/894/1/012031

Figure 1. Garbage invites disease (mural by locals and students).

Figure 2. Beautify the world (mural by locals and students).

Figure 3. Green and clean entrance gate.

Figure 4. Green and clean alley.

Figure 5. Boarding house in the alley.

Figure 6. Boarding house on the main road.

The results of this study indicate that the students increase the income and welfare of the residents, causing the possibility of security disturbances. However, the study area is relatively safe from security disturbances and theft because most respondents are from neighbourhood leaders (53.1%). Stated that in the last 10 years, there have only been fewer than five security disturbances (theft, commotion, laptop/mobile theft, etc.) 43.8% of respondents stated that security disturbances (theft, commotion, theft of laptops/mobile phones, etc.) only occurred once in 12 months. It is also shown by efforts to use RT CCTV at 8 points monitored by residents and students, installing portals in the area's entrance area, and increasing regional rules and regulations with restrictions on night activities.

4. Conclusion

This research shows that socio-cultural changes directly affect the economic and physical changes, as well as economic changes, will affect the physical changes in the study area. The structural model showed that changes in socio-cultural variables could be reflected by social cohesion and social segregation indicators. In contrast, economic variables can be shown by indicators of income growth and new business opportunities, and physical variables can be described by indicators of l changes in houses form and changes in the environment. Some of the findings show additional thought contributions to studentification theory, namely: community sustainability, environmental quality improvement, and security of the surrounding area.

Acknowledgement

This work was supported in part by Universitas Trisakti scholarship program in Doctor of Civil Engineering specializing in Urban and Regional Planning, Universitas Brawijaya, Malang, Indonesia, based on Universitas Trisakti Chancellor's Decree No: 1194/Usakti/SKR/BSDM/DT/IX/2017.

References

[1] Allinson J 2006 *Plan Pract Res* **21(1)**
[2] He S 2015 *Urban Stud* **52(15)**
[3] Smith D P and Hubbard P 2014 *Area* **46(1)**
[4] Donaldson R Benn J Campbell M and De Jager A 2014 *Urban Izziv* **25**(Special Issue)
[5] Ackermann A and Visser G 2016 *Bulletin of Geography* **31(31)**
[6] Situmorang R Antariksa S and Wicaksono A D 2019 *Int J Sci Technol Res* **8(12)**
[7] Situmorang R Antariksa S and Wicaksono A D 2020 *Int J Sci Technol Res* **9(3)**
[8] Hair J Black W C Babin B J Anderson R E and Tatham R L 2006 *Multivariate Data Analysis Pearson* p 878-880
[9] Situmorang R Sudikno A Surjono S and Wicaksono A D 2020 *Int J Adv Sci Technol* **29(05)**
[10] Hubbard P 2009 *Environ Plan A: Economy and Space* **41(8)**
[11] Kinton C Smith D P Harrison J and Culora A 2018 *Geogr J* **184(3)**
[12] Avni N and Alfasi N 2018 *City and Community* **17(4)**
[13] Ramadhani A and Tribhuwaneswari A B 2018 *Seminar Nasional Kota Berkelanjutan* **1(1)**
[14] Revington N P 2021 *Urban Stud* July 2021
[15] Fincher R and Shaw K 2009 *Environ Plan A* **41**(8)
[16] Woldoff R A and Weiss K G 2018 *City and Community* **17(1)**
[17] Sabri S and Ludin A M 2009 *The 3rd (SEATUC) Symposium* (69-75) (Skudai Johor (SIT))

IOP Conf. Series: Earth and Environmental Science **894** (2021) 012032 doi:10.1088/1755-1315/894/1/012032

Domestic wastewater treatment with household-scale constructed wetland system using water fern plant (Azolla pinnata)

A Magfhira, P Kinasih, D Salsabila, E Marchella, M F Fachrul

Departement of Environmental Engineering, Trisakti University, Jakarta, Indonesia, 11440

melati@trisakti.ac.id

Abstract. The purpose of writing this scientific paper is to analyze more deeply the management of the quality of the aquatic environment, especially regarding the treatment of domestic wastewater on a household scale. Domestic wastewater treatment with a constructed land system using Air Fern (*Azolla pinnata*) is expected to reduce concentration parameters such as BOD (Biochemical Oxygen Demand), COD (Chemical Oxygen Demand), and Ammonia. The growth rate of Air Fern (*Azolla pinnata*) after acclimatization is quite good so that it can reduce BOD (Biochemical Oxygen Demand), COD (Chemical Oxygen Demand) and Ammonia, which are a source of nutrients for plant growth so that it can be used as a promising material for plant growth. It can be used as a promising material for plant growth domestic wastewater treatment. The analysis was carried out using the blended concept, a combination of virtual, online, and laboratory/field activities (offline/outside the network) while still paying attention to health protocols. The priority of this literature study research is to provide input for the government in the form of alternative solutions for managing the quality of the aquatic environment, especially regarding domestic wastewater treatment on a household scale. In addition, this research also supports the international Sustainable Development Goals (SDGs) program: Goal 6: Access to Clean Water and Sanitation" in point 3, namely "Improving water quality by reducing pollution, eliminating waste disposal, and minimizing the disposal of chemicals and hazardous materials, halving the proportion of untreated wastewater and substantially increasing recycling and safe reuse globally". This research can contribute to appropriate technology in environmental biotechnology, namely wastewater treatment with a system that has simple technology, low cost, energy-saving and is environmentally friendly, which can create a level of hygiene and comfort for the community and maintain environmental sustainability future.

1. Introduction

The volume of domestic wastewater increases by 5 million m^3 per year, with an average content increase of 50% [1]. The increase in the volume of domestic wastewater is not matched by the escalation of the receiving waterbody, both in terms of capacity and quality, which causes the amount of wastewater entering the water body to exceed its capacity and carrying capacity.

The amount of domestic wastewater is waste resulting from domestic activities, including housing (households), buildings, trade, offices, and similar facilities. Untreated wastewater is one of the causes of water ecosystem pollution because of the pollutants contained in wastewater. The domestic quality exceeds the quality standard and is not following the Minister of Environment and Forestry Regulation of the Republic of Indonesia No. P.68/Menlhk/Setjen/Kum.1/8/2016 concerning Domestic Wastewater Quality Standards.

This increase in the amount of domestic wastewater, if not appropriately managed, will cause various diseases and environmental damage because domestic wastewater contains materials both organic and inorganic pollutants which, if in large quantities, will have an impact on the quality of surface water and groundwater [2].

After knowing the characteristics of domestic wastewater, which contain much organic matter, an alternative biological waste treatment system can be used as the leading choice. With concentrations of contaminants that are not too high, the treatment system can be implemented with simple and practical technology in its maintenance. On this basis, a wastewater treatment system (WWTP) is needed that is simple, easy to operate, and inexpensive for its manufacturing and operational costs. One alternative to the wastewater treatment system is the Constructed Wetlands System, which uses aquatic plants as phytoremedians to remove pollutants in wastewater.

Constructed wetlands are divided into the surface flow (Surface Flow) and subsurface flow (Sub Surface Flow). However, knowing that the type of surface flow (Surface Flow) can increase the mosquito population around the WWTP location, then the subsurface flow (Sub Surface Flow) is more suitable for domestic wastewater treatment systems in Indonesia.

2. Methodology

2.1. Research method

The research method used is a blended concept, a combination of virtual, online, and laboratory/field activities (offline/outside the network) while still paying attention to health protocols.

The equipment used in this research are:

1. 2 units of containers which are domestic wastewater treatment reactors (grey water) made of glass with size: pxlxt= 90 cm x 30 cm x 50 cm
2. 1 Unit of wastewater container size: volume = 30 litres
3. Sample bottle
4. Faucet Stop, Speed Control / Dimmer, Timer Outlet
5. Aquarium Pump 1liter/second capacity, 2” pipe 2 rods

The materials used include:

1. The *Azolla pinnata* water fern is purchased online with a plant age of around two weeks in the form of seeds
2. Domestic wastewater comes from the Wastewater Treatment Plant (WWTP) Campus A Trisakti University
3. Gravel planting media size 15 cm, sand planting media size 10 cm

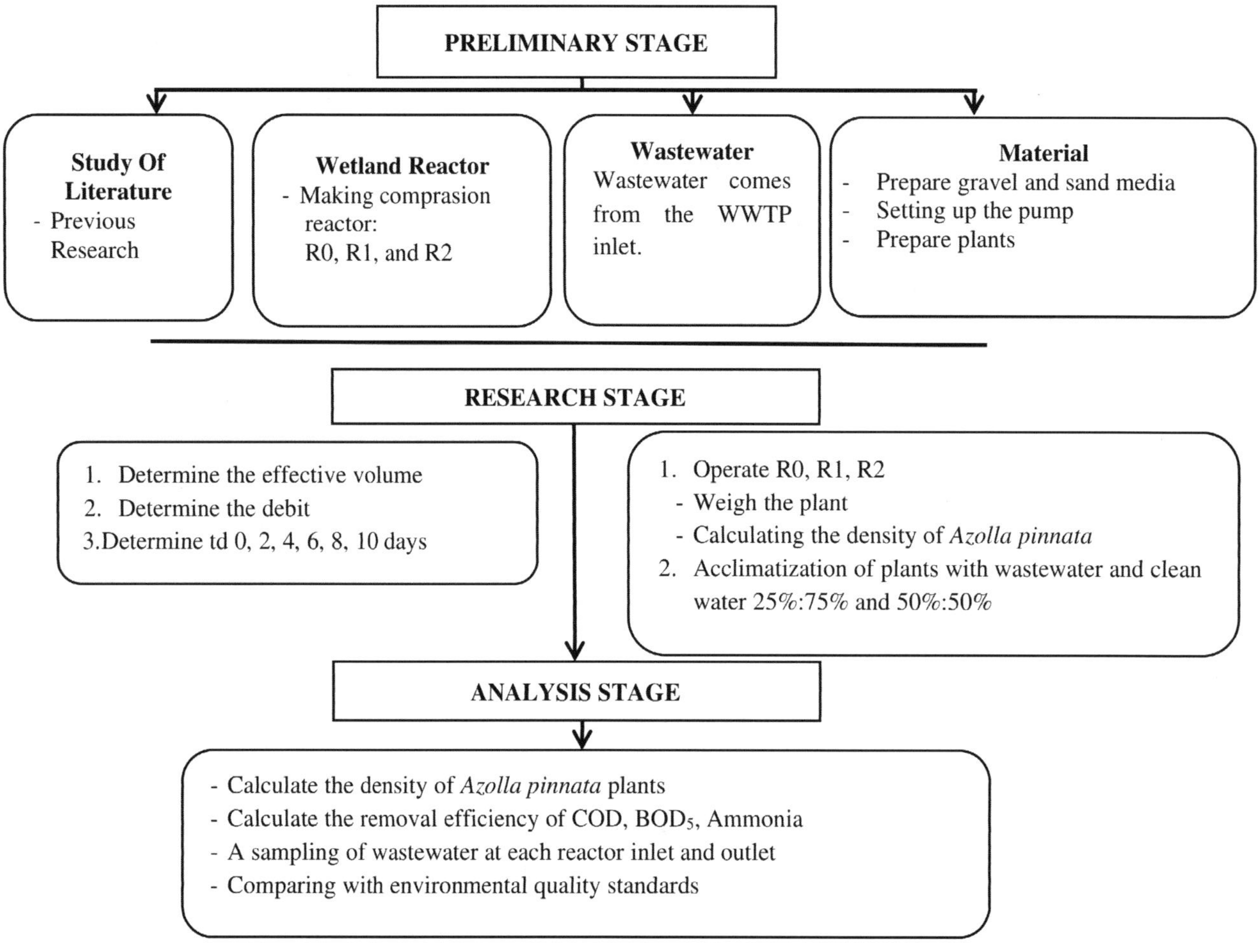

Figure 1. Research stage.

Preliminary Stage: Obtain laboratory test results regarding the quality characteristics of domestic wastewater (greywater) originating from the Wastewater Treatment Plant (WWTP) Campus A of Trisakti University and compare it with the Decree of the Minister of Environment and Forestry of the Republic of Indonesia No. P.68/Menlhk/Setjen/Kum.1/8/2016 concerning Domestic Wastewater Quality Standards.

Research Stage: Obtaining the percentage (%) of the concentration removal parameters of BOD (Biochemical Oxygen Demand), COD (Chemical Oxygen Demand) and Ammonia in domestic wastewater in a Constructed Wetland System using the Air Fern (Azolla pinnata) plant.

Analysis Stage: Get the calculation results regarding the optimal time needed to remove BOD (Biochemical Oxygen Demand), COD (Chemical Oxygen Demand) and Ammonia.

2.2. Data collection

The quality of wastewater can be determined by testing water samples with methods following SNI 06-6989.2-2004 for COD, SNI 6989.72-2009 for BOD and SNI 6989.3-2019 for TSS, which is shown in table 1.

Table 1. Water quality parameter analysis method.

Parameter	Unit	Method	
COD	mg/L	Reflux	$COD = \frac{(A-B) \times N\ FAS \times 8 \times 1000}{V\ sampel}$
BOD	mg/L	Winkler	$DO_0 = \frac{1000 \times A0 \times N \times 8 \times F}{298}$ $DO_5 = \frac{1000 \times A5 \times N \times 8 \times F}{298}$
Ammonia	mg/L	Spectrofotometer	

2.3. Data analysis method

1. Debit Calculation

Determine the flow rate using the following formula:
Where: Q = Flow rate (m/day); V = Volume (L/sec)

2. Removal Efficiency

According to [3], the calculation of the efficiency of the removal of wastewater is as follows:
Efficiency (%) =
Where: Cin: Domestic wastewater parameter values before treatment;
Cout: Domestic wastewater parameter values after treatment.

3. Determining Hydraulic Residence Time

Hydraulic residence time is done by calculating the detention time (Td) = ;
where: Q = Flow rate (m/day), V = Volume of constructed wetland (m^3), T = Time required (days)

4. Calculating The Relative Growth Rate

The growth rate is calculated using the formula for the relative growth rate of plants:
RGR =
where:
RGR= Relative growth rate (grams/day); W1= Initial weight (grams); W2= Weight on day - i (gram); T1= Initial observation time; T2= Observation time on day - i

5. Calculating Plant Density (ind/m2)

The plant density of Azolla pinnata was calculated using the formula:
D =
where: D = Density (ind/m^2); n = Number of individuals; A= Sampling area (m^2);

6. Mass Allowance Rate

Calculation of the mass removal rate using the formula:
R = q(Cin – Cout),
where: R = mass removal rate (g/m^2/sec); q=hydraulic loading rate (m/day)

3. Result and discussion

3.1. Domestic wastewater characteristics

Before treatment, wastewater has the following characteristics:

Table 2. Domestic wastewater initial characteristics.

No	Parameter	Unit	The Quality of the Wastewater	Quality Standard
1.	BOD	mg/L	67	30
2.	COD	mg/L	185	100
3.	Ammonia	mg/L	16.1	10

3.2. Laboratory wastewater analysis

The graph below depicts the examination of BOD parameters in wastewater from day to day.

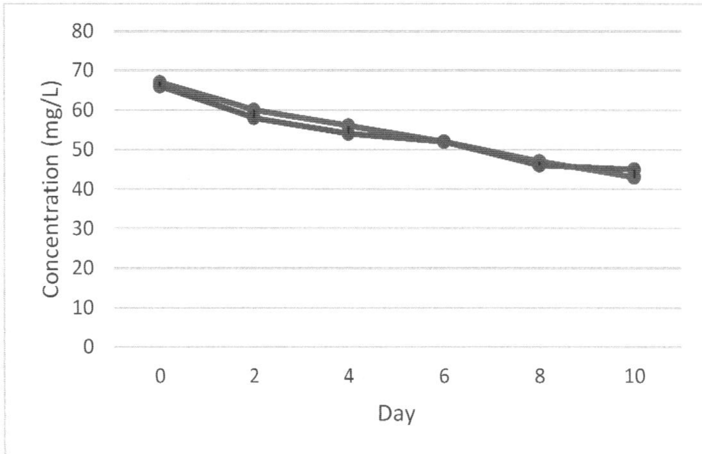

Figure 2. The value of BOD in wastewater.

The graph below depicts the examination of COD parameters in wastewater from day to day.

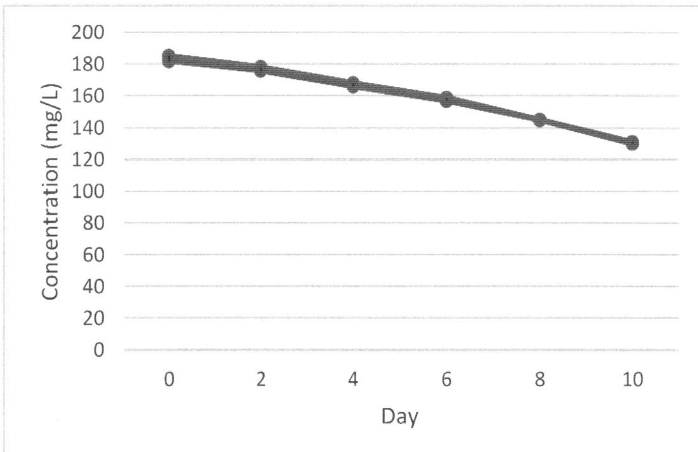

Figure 3. The value of COD in wastewater.

The graph below depicts the examination of amonnia parameters in wastewater from day to day.

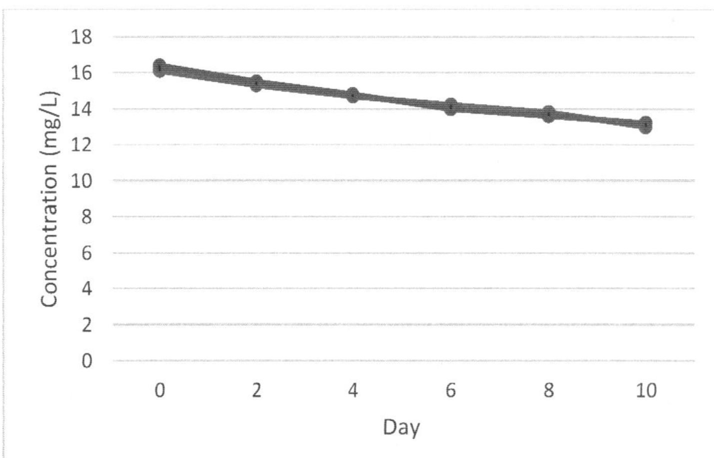

Figure 4. The value of ammonia in wastewater.

3.3. Removal efficiency of each parameter with plant

Table 3. Removal efficiency with plants.

No	Parameter	Removal Efficiency
1.	BOD	32.83%
2.	COD	29.72%
3.	Ammonia	18.01%

The highest removal efficiency is found in the BOD parameter, while the lowest is found in the Ammonia parameter, as shown in table 2. The effectiveness of the allowance is calculated using day 0 as the starting point for the computation. The amount of organic matter in the constructed wetland determines whether the allowance is high or low.

During ten days of domestic wastewater treatment for parameter values of BOD, COD and Ammonia are still not complying with the Quality Standards, so it needs to be treated again with a detention time of more than ten days, but the constructed wetland is effectively used as domestic wastewater treatment on a household scale.

4. Conclusion

The constructed wetland has removal efficiency for BOD 32.83%, COD 29.72% and Ammonia 18.01%. This research can contribute to appropriate technology in environmental biotechnology, namely wastewater treatment with a system that has simple technology, low cost, energy-saving and is environmentally friendly, which can create a level of hygiene and comfort for the community and maintain environmental sustainability future.

Acknowledgement

The authors would like to thank the Ministry of Education and Culture of the Republic of Indonesia, the Directorate General of Higher Education, for the overall funding of this research regarding the 2021 Field PKM Funding, especially the RE PKM. The authors also thank the Rector of Trisakti University, Vice Chancellor for Student Affairs, Dean of the Faculty of Landscape Architecture and Environmental Technology, Head of the Environmental Engineering Department, Ms Melati Ferianita Fachrul PKM supervisor, and friends who continually provide support and motivation in implementing PKM activities.

References

[1] Yaqin M A and Indriyati E R 2019 *Jurnal Sylva Scienteae* **2(1)** 80-92
[2] Farah H H 2019 *Fitoremediasi Limbah Domestik (Grey Water) Menggunakan Tanaman Kiambang (Salvinia molesta) Dengan Sistem Batch* (Universitas Islam Negeri Sunan Ampel)
[3] Alfiah T, Damatra R R and Kusuma M N 2015 *Jurnal ITATS* **1(1)** 1-10
[4] Ambarwati Y and Bahri S 2018 *Analit: Analytical and Environmental Chemistry* **3(02)**
[5] Astuti A D, Lindu M, Yanidar R and Kleden M M 2016 *Jurnal Penelitian Dan Karya Ilmiah* **1(2)** Juli: 91-108
[6] Arimbi A 2017 *Efektivitas Tanaman Melati Air (Echinodorus palaefolius) Dalam Menurunkan Kadar BOD (Biologycal Oxygen Demand) Dan COD (Chemical Oxygen Demand) Serta TSS (Total Suspended Solid) Pada Limbah Cair Tempat Pemotongan Ayam Di Kecamatan Delitua Kabupaten Deli Serdang Tahun 2017* (Universitas Sumatera Utara)
[7] Darmawan R 2017 Baku Mutu Air Limbah Domestik yang Diatur Oleh Peraturan Menteri LHK Republik Indonesia No.68 Tahun 2016
[8] Hidayat M S, Putra A and Diana S 2018 *Proceeding Seminar Nasional Politeknik Negeri Lhokseumawe* **2(1)** September
[9] Anggraeni M, Henna R S and Hadiyanto 2013 *Proceeding Biology Education Conference* **10(2)**

[10] Muhsinin N 2019 *Pengolahan Air Limbah Domestik Secara Fitoremediasi Sistem Constructed Wetland Dengan Tanaman Pandanus amaryllifolius dan Azolla microphylla* (Jogjakarta: Universitas Gadjah Mada)
[11] Pemerintah Pusat 2001 *Peraturan Pemerintah (PP) tentang Pengelolaan Kualitas Air Dan Pengendalian Pencemaran Air*
[12] Rohmah S N, I W, H R and Hilal N 2019 *Buletin Kesling*
[13] Salafiyah N 2016 *Pengaruh Lama Tanam dan Luas Penutupan Azolla microphylla Terhadap Kualitas Kimia dan Fisika Limbah Cair Laundry* (Malang: Univertitas Negeri Maulana Malik Ibrahim)
[14] Sisilia L 2016 *Pemanfaatan Tanaman Akar Wangi (Vetiveria zizanioides (L.) Nash) untuk Penyerapan Logam Berat Tembaga (Cu)* (Yogyakarta: Universitas Atma Jaya)
[15] Sudewo A R 2016 *Hybrid Constructed Wetland (HCW) Sebagai Upaya Pengolahan Air Limbah Domestik Pada Pemukiman Padat Penduduk Di Daerah Bantaran Sungai (Studi Kasus: Kejawan Gebang Kelurahan Gebang Putih Surabaya)* (Surabaya: Fakultas Teknik Sipil dan Perencanaan Institut Teknologi Sepuluh Nopember)
[16] Supradata 2005 *Pengolahan Limbah Domestik Menggunakan Tanaman Hias Cyperus altenifolius, L. Dalam Sistem Lahan Basah Buatan Aliran Bawah Permukaan (SSF-Wetlands)* (Semarang: Fakultas Teknik Universitas Diponegoro)

IOP Conf. Series: Earth and Environmental Science **894** (2021) 012033 doi:10.1088/1755-1315/894/1/012033

The effect of maximum iteration using 3 dimensional limit equilibrium method in open pit mine

F M Syahputra[1], M A Azizi[1*], I Marwanza[1]

[1]Mining Engineering Department, FTKE, Trisakti University, West Jakarta, 11440, Indonesia

masagus.azizi@trisakti.ac.id

Abstract. Nickel ore mines have a high potential of landslides due to their weak material base, which consists of soil. It is caused by the increase of soil density in rain conditions, leading to decreased soil shear strength (c) and internal friction angle (ϕ). This research aims to determine the optimum value of the maximum iteration number based on the Cuckoo Search and Particle Swarm Optimization search method. In this research, the slope is analyzed using the 3 Dimensional limit equilibrium method "Simplified Bishop," a slope stability analysis method that uses the principle of static equilibrium. Alongside this method, the Cuckoo Search and Particle Swarm Optimization is adopted. The Cuckoo Search and Particle Swarm Optimization are metaheuristic optimization techniques used as the slipped surface search method. Series of 3-dimensional limit equilibrium computation is performed using different amounts of nests in the cuckoo search method and different particle values and maximum iteration number. Cuckoo Search method to achieve optimal nest 100 and iteration of 80 with the fastest compute time of 3 minutes 49 seconds. While the Particle Swarm Optimization to achieve optimal on particles 60, iteration as much as 480 with a compute time of 6 minutes 46 second, with a factor of safety value of 1,12.

1. Introduction

In PT. X mining activity, 3-dimensional models of the pit design were made to ensure the stability of the slope for safe and productive mining practice. Slope stability is one of the essential aspects of mining practice because it determines the efficiency and productivity of production. Mine stability is highly governed by rock mass characteristics and other external factors such as groundwater and vibrations caused by the mining activities. The stability of a slope is represented by the Factor of Safety. Therefore, slope stability analysis plays an integral role in the mining process. There is various type of slope stability analysis method. The 3-dimensional Limit Equilibrium with simplified Bishop is one of the most applied in the field. This research utilizes this method to determine the FoS value and estimate the landslide volume. In the 3D LEM with simplified Bishop, the metaheuristic optimization method, cuckoo search, and PSO were applied to determine the slide surface.

The PSO and CS are algorithms that are used to determine the optimum and sub-optimum in numerical problems. The PSO is a population-based metaheuristic optimization method. Cuckoo search is a metaheuristic optimization method inspired by the breeding of cuckoo birds by Yang and Deb (2013) [2].

1.1. Limit equilibrium method

In this method, the calculation of slope stability analysis only uses static equilibrium conditions and ignores the stress-strain relationship on the slope. The geometry of the slip, the surface must be known or determined first.

There are two types of landslide analysis using the limit equilibrium method: the translational type and the rotational type. The analysis of the rotational type of landslide can use the slice method. The slice method itself can be divided into two, named the method that complete all conditions of forces balance and moments (Spencer, Morgenstern-Price, and General Limit Equilibrium) and which does not complete all conditions of forces equilibrium and moments (Ordinary Slice, Simplified Bishop, and Simplified Janbu).

1.2. Factor of Safety

The stability level can be determined with the safety factor. FoS is the value of the difference between the restraining force with the driving force.

$$FoS = \frac{resisting\ force}{driving\ force} \quad (1.1)$$

$$Shear\ Stress\ (\tau) = c + \sigma_n \tan\phi \quad (1.2)$$

Direction:
FoS = Factor of Safety
τ = Shear Stress
σ_n = Normal Stress
c = Cohesion
ϕ = Friction Angle

If the FoS value for a slope is > 1.0 (resisting force > driving force), the slope is in a stable condition. However, if the value of FoS < 1.0 (resisting force < driving force), the slope is unstable, and landslides may occur on the slope.

1.3. Search methods slide3

The slipped surface search method in Slide3 is highly automated and requires a little user input. The shape of the surface can be made and selected with two existing surface types, named Spherical and Ellipsoid; then, with the optimization of surface changes (Surface Altering Optimization (SAO)), the overall landslide area can be searched with a lower FoS. The 3D avalanche direction will be automatically determined during the analysis.
In Slide3, four slip surface search methods are provided, named Grid Search, CS, and PSO. Based on the development of optimization methods, two of the four mentioned above are the most modern methods and have a high level of effectiveness in solving engineering problems. The methods are CS and PSO.

1.4. Cuckoo search

Cuckoo Search is a metaheuristic optimization method [1] developed by Yang and Deb (2010). The method was inspired by cuckoo's breeding behavior. In this research, the CS is used as a slip surface search tool with the lowest SF. CS is coupled with Lévy Flights random walk. There are few rules to use this algorithm as stated in [2]:

- Each cuckoo lay one egg at a time and dumps it in a randomly chosen nest;
- The best nests with high quality of eggs (solutions) will carry over to the next generations;
- The number of available host nests is fixed, and a host can discover an alien egg with a probability pa∈ [0, 1]. In this case, the host bird can either throw the egg away or abandon the nest to build an entirely new nest in a new location.

The last rule can be approached using a fraction pa to determine the worst solutions of n nests that will randomly replace a new nest. In order to solve the problem, it can be simply illustrated that every egg in a nest represents one new solution. The purpose is to use the new and potentially better solution to replace the current solution in the nest. The nest may have two eggs (2 solutions) in a specific condition, but this problem is simplified, so one nest has only one solution [3,4].

1.5. Particle swarm optimization
Kennedy and Eberhart initialized PSO by simulating the behavior of a bird swarm with defined instructions for individual behaviors and intercommunications. These instructions help in the decision-making process of individuals, which is based on the following items [5]:

- Experience of the individual as its best results so far;
- An outlay of experience of a swarm is the best result among all individuals.

This process starts by randomly generating a certain number of individuals, namely particles, representing a possible solution for the problem. It is based on the individual attitude (Particle) and the influence of other individuals within the pack (Swarm). Every individual acts according to their intelligence and is influenced by the collective behavior of the pack. Therefore, if a particle finds a practical path to the food source, the pack will follow the path to the source where ever they are located.

2. Methodology
The slope stability analysis of this open-pit mine uses the 3D Bishop's limit equilibrium method with two parameters of the landslide field search method, named Particle Swarm Optimization (PSO) and Cuckoo Search. Research variables include the number of particles, the number of nests, and maximum iterations. The assumption of the type of landslide is circular because the limonite and saprolite materials have almost the same mechanical properties as the soil.

3. Result and discussion
In this study, the data was obtained in the final pit in 3D form and data on material properties of soil (limonite and saprolite) and rock (bedrock) such as tensile strength, cohesion, internal friction angle, and physical properties. Testing is carried out in the laboratory to obtain material properties. The material results can be seen in Table 1.

Table 1. Material properties.

NO.	Lithology	Unit Weight (kN/m^3)	Cohesion (Kpa)	Friction angle (°)
1.	Limonite	17,2	41,1	15,7
2.	Saprolite	15,8	38	12,6
3.	Bedrock	41,7	3.270	29

Slope stability analysis was carried out on the entire slope of the pit, which has limonite, saprolite, and bedrock lithology, along with the mechanical and physical parameters of the material entered into the Slide3 software, as shown in Figure 2.

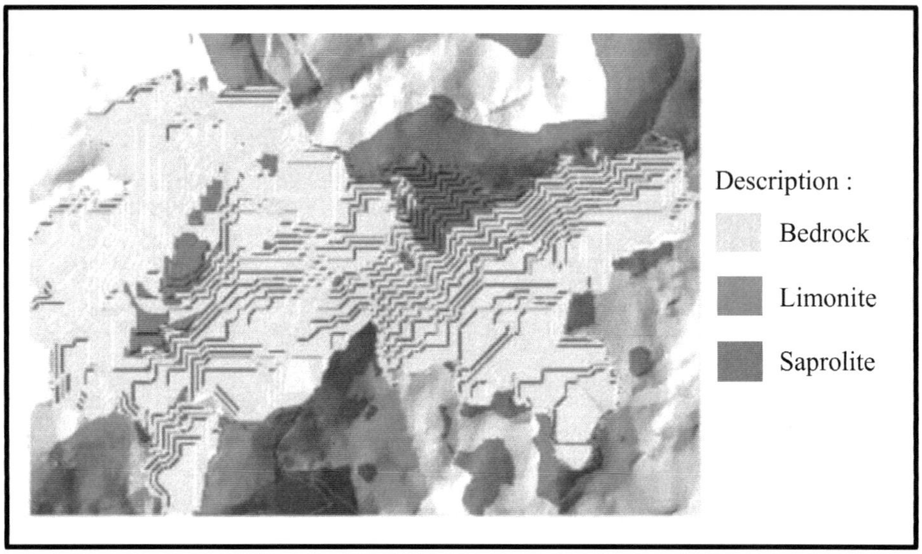

Figure 1. 3D Model Pit Design and Lithology.

Table 2. Computing result with 10.

Max Iteration	Nest	FOS	Volume (m^3)	Position	Time (Second)
20	100	1,63	34.708	North	65
40	100	1,16	37.182	North	114
60	100	1,19	46.280	North	179
80	100	1,12	45.423	North	229
100	100	1,14	42.073	North	291
120	100	1,12	10.103	North	345
140	100	1,14	41.384	North	403
160	100	1,12	43.591	North	447
180	100	1,14	38.988	North	509
200	100	1,12	50.680	North	564
220	100	1,12	43.599	North	636
240	100	1,13	8.951	North	696
260	100	1,12	8.732	North	773
280	100	1,13	43.322	North	838
300	100	1,13	41.299	North	919
320	100	1,12	7.767	North	947
340	100	1,12	40.812	North	1038
360	100	1,12	43.649	North	1060
380	100	1,12	9.097	North	1143
400	100	1,12	42.008	North	1207

Table 3. Computing result with 60 Particle.

Max Iteration	Particle	FOS	Volume (m3)	Position	Time (Second)
40	60	1,47	20.197	North	32
80	60	1,31	48.432	North	103
160	60	1,26	51.856	North	149
240	60	1,19	55.037	North	251
320	60	1,14	50.867	North	306
400	60	1,13	48.953	North	350
480	60	1,12	49.656	North	406
560	60	1,12	49.662	North	480
640	60	1,12	49.689	North	638
720	60	1,12	49.675	North	683
800	60	1,12	49.690	North	747

The table above shows the result of Particle Swarm Optimization dan Cuckoo Search consisting of Factor of Safety, critical position, potential failure volume, and computation time. Based on the results from both search methods, the nest, particle, and iteration number will influence the LEM analysis results. The high amount of particle, nest, and iteration computation needs more time to calculate. When the number of iteration is high, the analysis will have more repetition in the calculation and yielding an accurate FOS value, but it takes a longer computation time. In this research, the particle and nest number used varies from 60 to 100, with the maximum iteration of 20 to 400 in the CS search and 40 to 800 iterations for PSO.

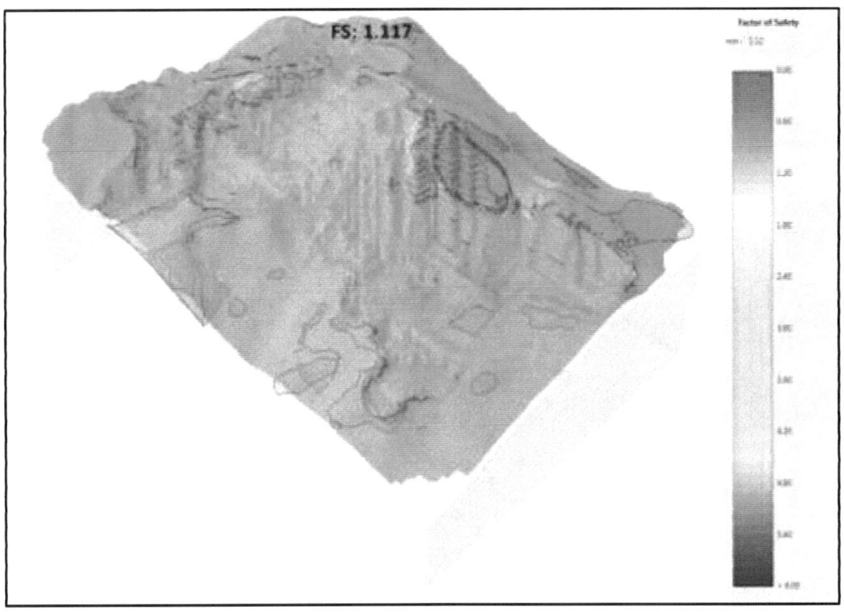

Figure 2. Critical position using CS with 100 nests and 80 iterations.

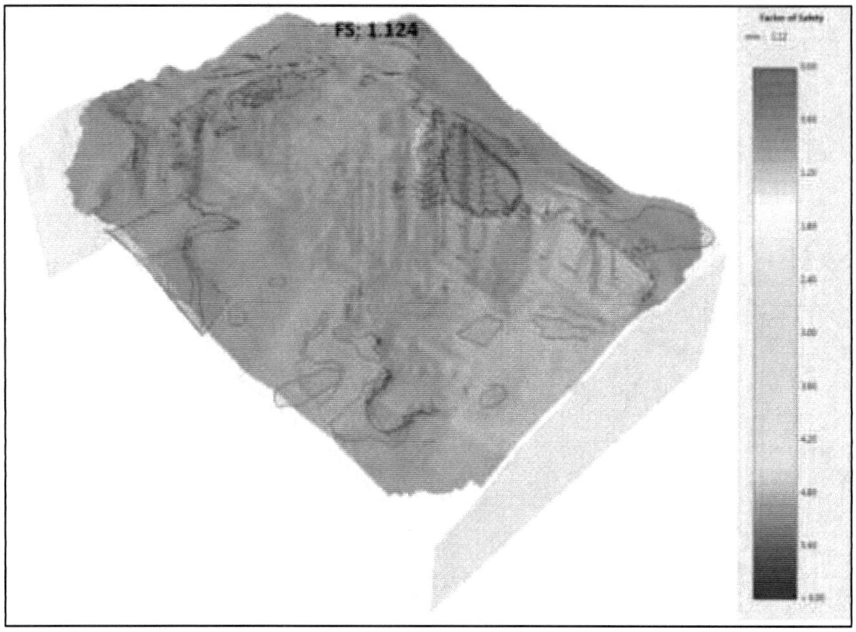

Figure 3. Critical position using PSO with 60 particle and 480 iterations.

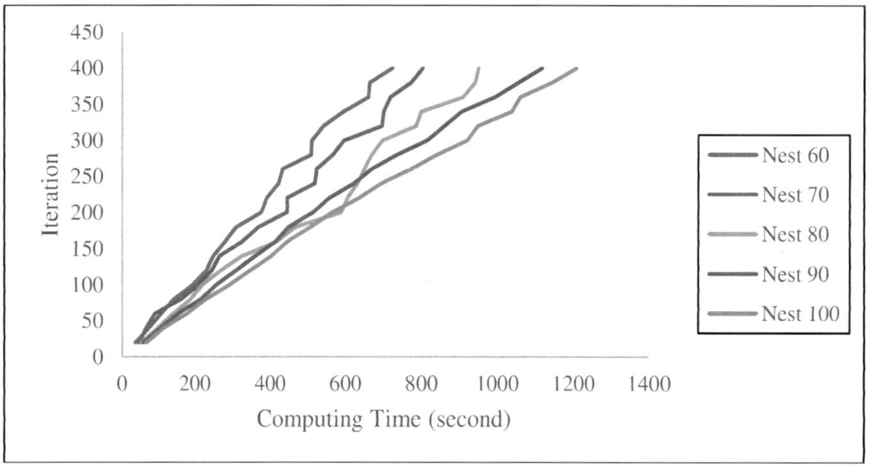

Figure 4. Computing time versus maximum iteration based on the number of the nest.

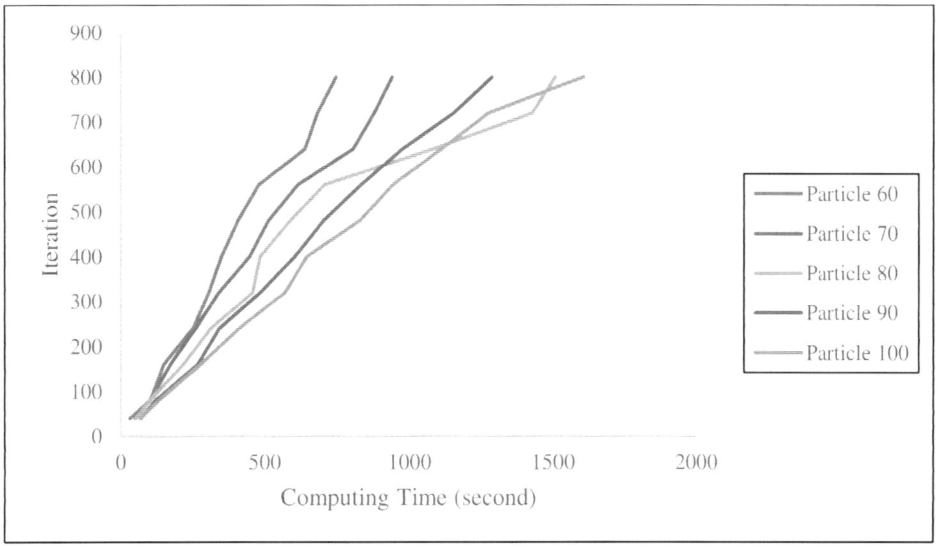

Figure 5. Computing time versus maximum iteration based on the number of particles.

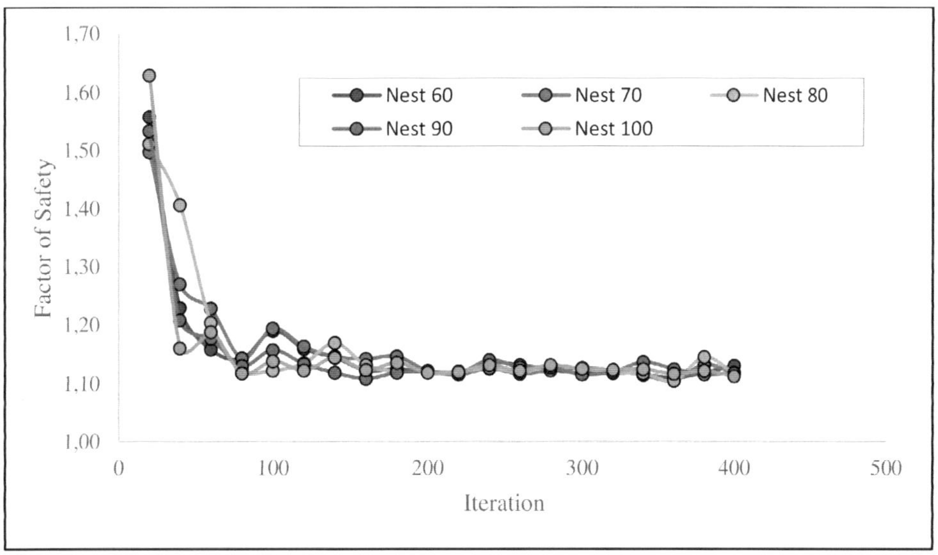

Figure 6. Minimum FOS versus iterations based on the number of nests.

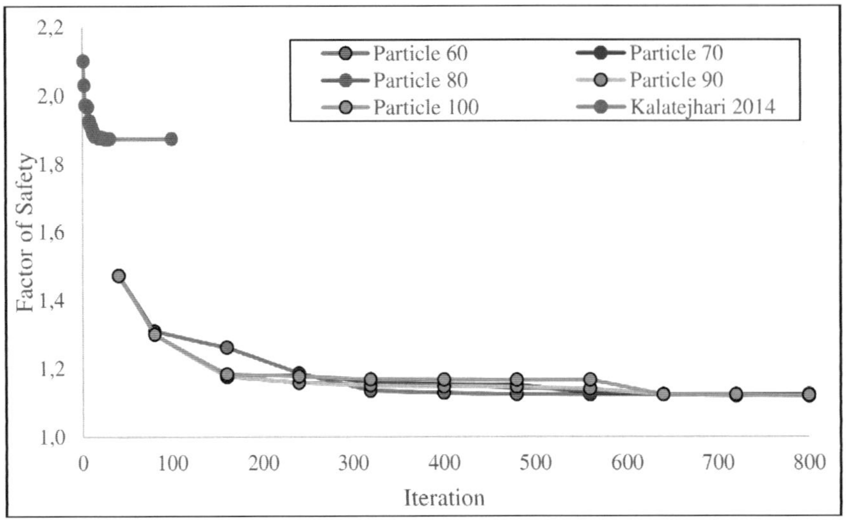

Figure 7. Minimum FOS versus iterations based on the number of particles.

From the result of figure 5, the limit equilibrium analysis using the Cuckoo Search slipped surface search method with the number of nest research variables and maximum iteration proves the more number of safety factor value iteration, the more critical it is to reach the optimal value. The result of the graph is in figure 4, which shows that after the iteration is 200, the changes of the safety factor are not too significant and stable. From the result of the CS graphic method, in the next iteration, the number of the nest is no longer influential. Based on the result of the CS graphic method can be concluded that in the next iteration, the number of nests is no longer influential, and the value has reached the optimum in the 1,12 safety factor. From the CS (figure 6.) method simulation, the optimal number of the nest and maximum iteration is in the 100 nests and 80 iterations with the fastest computation time of 3 minutes and 49 seconds.

The graph (figure 7.) of iteration number and the FoS compared to the one generated in the research by (Kalatejhari, 2014). From these graphics, it can be concluded that the FoS value will increase with the increase of particle, nest, and iteration number. The research was conducted on a single slope with slope forming material consisting of soil. From the graphic, the FoS value tends to stabilize with iteration above 680. The same trend is also seen in the cuckoo search graphic. From this point, it is safe to say that the particle number no longer influences the actual FoS value if 1.2 is reached. The PSO simulation method found that the optimum number of particles and iteration is when particle amount is 60 with the iteration number of 480 and 46 seconds computation time.

The graphic proves that the more maximum iteration is, nest or even the particle, the more accurate the safety factor value will be. This research could optimize the number of nests, particles, and maximum iteration based on the safety factor value result and the critical slipped surface position.

4. Conclusion

Based on the research conducted, it can be concluded as follows:
Computation using Slide3 software with Cuckoo and PSO as the slip surface search method generated an FoS value of 1.12 and a circular failure located on the north side of the pit. The analysis will yield an accurate FoS value at a higher amount of iteration. In the CS, the fastest computation time recorded is 3' 49" with iteration numbers of 80 and 60 particles. For the PSO method, the fastest computation time recorded is 6' 46" with the iteration number of 480 and 60 particles. In this location, the CS is more realistic because it generated the optimum value of FOS in a shorter computation time. The estimated volume of potential landslide for the Cuckoo search method is 42.286 m3 and 47.694 m3 for the PSO method.

References

[1] Yang X S and Deb S 2010 *International Journal of Mathematical Modelling and Numerical Optimisation* **1(4)** p 330

[2] Yang X S and Deb S 2013 *Computers & Operations Research* **40** p 1616–1624

[3] Anderson L R, Kiefer F W 1980 *Fundamentals of Geotechnical Analysis* (New York : John Wiley & Sons, Inc)

[4] Azizi M A, Marwanza I, Hartanti N A 2020 *Indonesian Mining Journal* **23(2)** p 57-65

[5] Azizi M A 2018 *Three dimensional slope stability analysis of open pit limestone mine in Rembang District, Central Java* (IOP Publishing Ltd)

[6] IOP Conference Series: Earth and Environmental Science, Volume 212, International Conference on Earth Science, Mineral, and Energy 11–12 October 2018, Yogyakarta, Indonesia

[7] Azizi M A, Karim R 2019 *Indonesian Mining Professionals Journal* **1(1)**

[8] Azizi M A 2019 *Proceedings in Earth and geosciences* **6** (CRC Press)

[9] Azizi M A 2021 *Three Dimensional Slope Stability Analysis of Open Pit Mine, in Slope Engineering Book* (IntechOpen Publishing)

[10] Muhammad Z A 2020 *Indonesian Mining Professionals Journal* **2(2)** 51-56

[11] Reyes A and Parra D 2014 *3D Slope Stability Analysis by The Using Limit Equilibrium Method Analysis of A Mine Waste Dump, Proceedings Tailings and Mine Waste, Colorado, USA, 127–139*

[12] Rocscience 2018 *Slide3 – 3D Limit Equilibrium Slope Stability Overview*

[13] Roohollah Kalatehjari et al 2014 *The Scientific World Jurnal* **2014** *2009 World Congress on Nature & Biologically Inspired Computing (NaBIC) IEEE* p 210–214

IOP Conf. Series: Earth and Environmental Science **894** (2021) 012034 doi:10.1088/1755-1315/894/1/012034

A bibliometric analysis of permeable reactive barrier enhanced electrokinetic treatment for sustainable polluted soil remediation

M A Budihardjo[1], R P Safitri[1], B S Ramadan[1], A J Effendi[2], S Hidayat[2], Y V Paramitadevi[3], B Ratnawati[3], A Karmilia[1]

[1]Environmental Engineering Department, Faculty of Engineering, Universitas Diponegoro, Indonesia
[2]Environmental Engineering Department, Faculty of Civil and Environmental Engineering, Institut Teknologi Bandung, Indonesia
[3]Environmental Management and Engineering Program, College of Vocational Studies of Bogor Agricultural University, Kumbang Street no 14, Bogor, West Java 16151, Indonesia

bimastyaji@live.undip.ac.id

Abstract. Research on soil remediation continues to develop, one of which is electrokinetic remediation combined with a permeable reactive barrier as a medium to prevent the migration of metals removed from the anode and cathode spaces. Thus, it is hoped that there is no need for reprocessing the residue resulting from electrokinetic remediation. This study aims to conduct a bibliographical analysis related to electrokinetic remediation coupled by permeable reactive barriers for heavy metal contaminated soil and to examine the effect of using various types of reactive barrier materials and their placement on the pollutants removal in the soil. Based on the results of bibliographic analysis, 26 relevant scientific articles were obtained, and the most publications in 2020 with 27% additional article publications are found. China and Environmental Science and Pollutant Research are the countries and journals that contribute the most to publications related to EK-PRB on heavy metal polluted soils.

1. Introduction

Rapid industrial development has a positive impact on human life and has a negative impact with losses that can even be greater and have a broad impact on the lives of other living things. These negative impacts are often caused by industrial process. If the waste is not managed correctly, it can cause water, soil, air pollution, and environmental damage and harm the human health. The nature of heavy metals is difficult to degrade and can accumulate in the body of living things, causing disease. Heavy metals or metalloids such as Pb, Hg, As, Cu, Zn, Cr, Cd, Sr, Fe, Mn, Sn, Ni, Cs, and U are pollutants that pollute the soil the most because of their mobility and solubility [1]. Although heavy metals pollute only a tiny part of the soil, it has a very high risk for surrounding environment [1], because it can enter the food chain and accumulate long term in the body. The accumulation of heavy metals is because they cannot be degraded both biologically and chemically [2].

There are various technologies for remediation of soil contaminated with heavy metals, including bioremediation, phytoremediation, permeable reactive barrier, electrokinetic separation, pump and treat system, and soil cover and coating [3]. Electrokinetic remediation is technology that can be a solution in removing metal content in polluted soil because it can be used on various types of soil, both fine-grained soil [2], mud, and ash from combustion. Remediation using electrokinetic is a process that involves an electric field on polluted soil produced by applying a direct current (DC) to the electrode [4]. Therefore, this study aims to conduct a bibliographical analysis of the publication of electrokinetic remediation with the permeable reactive barrier to find gaps from previous studies and to assess the effect of the type and location of reactive barrier placement on the removal of metal content in polluted soil.

2. Methodology

2.1. Systematic review

The bibliographic analysis consists of collecting and analyzing step to the information related to scientific articles [5]. In conducting the bibliographic analysis, PRISMA method was used to identify and analyze publications related to electrokinetic remediation in metal contaminated soils combined with permeable reactive barrier technology. The PRISMA method is an analytical method consisting of the stages of identification, screening, eligibility, and inclusion [6]. The article database was obtained from Scopus, which was accessed on June 30, 2021. Several keywords and limitations were used for elimination at the identification stage. The keyword "electrokinetic remediation" (1446 documents), followed by the keywords "metal" (1223 documents), "soil" (1209 documents), and "permeable reactive barrier" (214 documents). In addition, the limitation of the publication year from 2012 to 2021 (183 documents); types of article documents, review journals, and conference journals (180 documents); and writing in English (169 documents). At the screening stage, elimination is carried out based on the title and abstract. From the results of the elimination obtained 69 documents that are relevant to the research objectives. Furthermore, elimination is carried out at the eligibility stage based on the research method, and 40 relevant documents are obtained. In the inclusion stage, 26 selected documents, less relevant to research related to the types of pollutants in the polluted soil set aside, are eliminated.

2.2. Data analysis

The selected documents are then downloaded in RIS format from Scopus. Bibliometric analysis was carried out to find gaps from previous studies. This analysis can be done using VOSviewer 1.6.15, which can display keyword relationships in selected documents in the form of a complex and dynamic map [6]. The downloaded database in RIS format is inputted into VOSviewer and analyzed using co-occurrence analysis type with keyword analysis unit. The selected keywords are then automatically eliminated again to get the most relevant keywords to the objectives to be researched.

3. Result and discussion

3.1. Descriptive analysis

The trendl of publications on electrokinetic remediation of metal removal in contaminated soils with permeable reactive barriers fluctuated from 2012 to 2021. Publications on electrokinetic remediation with permeable reactive barriers in 2013 experienced an increase of 12% in 2016 and 2019. The addition of publications was 4% in 2014 and 2018, which was less than the previous year. An 8% increase in publications occurred in 2017 and a 15% increase in publications in 2021. The most prominent number of publications was in 2020, reaching 27% from the previous year and with the number of publications being 7 documents. The number of publications on electrokinetic remediation of metal removal in contaminated soil with permeable reactive barriers can be seen in Figure 1.

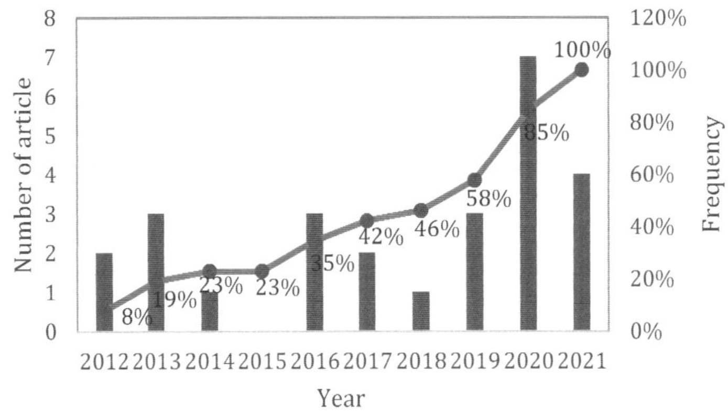

Figure 1. Number of publications on metal removal electrokinetic remediation in soil contaminated with permeable reactive barrier.

China, Iran, Australia, and Taiwan have the most publications on metal removal electrokinetic remediation in contaminated soils with the highest permeable reactive barrier. China has an outstanding contribution to research on remediation of metal-contaminated soils because the rapid progress of modern industry in China causes soil pollution [7], so a method is needed to restore the polluted soil. There are 16 different journals publish research on electrokinetic remediation to remove metal content in soil using a permeable reactive barrier. Journal of Environmental Science and Pollution Research has the highest contribution, which is 19% of the total publication found. Furthermore, other journals with the most contributions include Chemosphere and Chemical Engineering Journal as much as 12% and Journal of Hazardous Materials and Separation and Purification Technology as much as 8%.

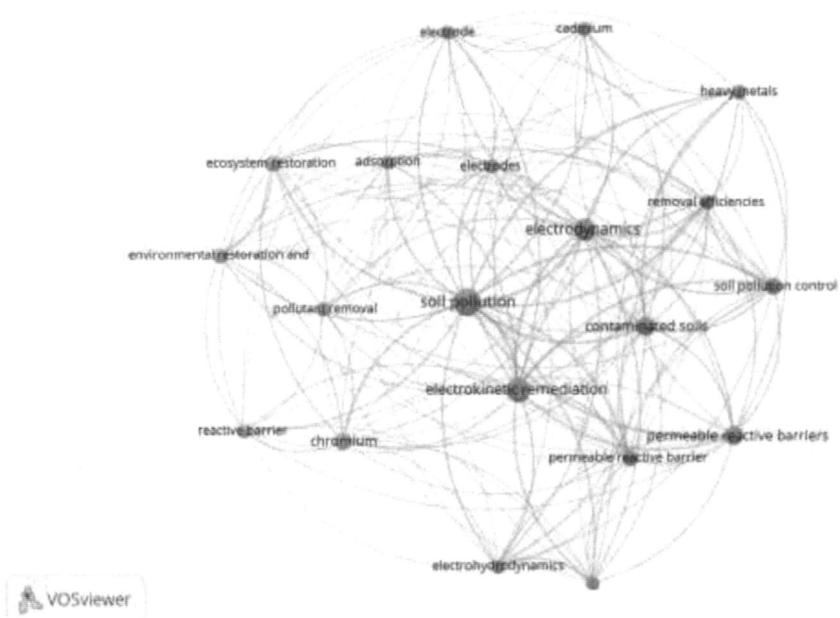

Figure 2. Co-occurence keyword map of publications related to electrokinetic remediation using permeable reactive barrier based on their relationship.

3.2. Bibliographic analysis

The relationship between keywords and the division of keyword clusters is shown in Figure 2. Cluster 1 with the highest number of keyword variations is shown in red. In general, this cluster grouped critical words related to electrokinetic remediation mechanisms. Cluster 2 is shown in green, where grouped keywords are related to the utility of electrokinetic remediation. Cluster 3 is blue, a group of keywords associated with factors affecting electrokinetic remediation.

Based on the year of publication, the keywords "chromium" and "electrohydrodynamics," indicated in purple, are primarily found in older articles, namely the year of publication in 2016 and below. Meanwhile, the keywords "heavy metals," "ecosystem restoration," "soil pollution control," and "cadmium," which are indicated by yellow, are often found in recent articles. This condition shows that electrokinetic remediation on polluted soil is currently being developed and is becoming a concern in the ecosystem and environmental restoration efforts. More fully, the analysis of keyword groups based on the year of publication is shown in Figure 3.

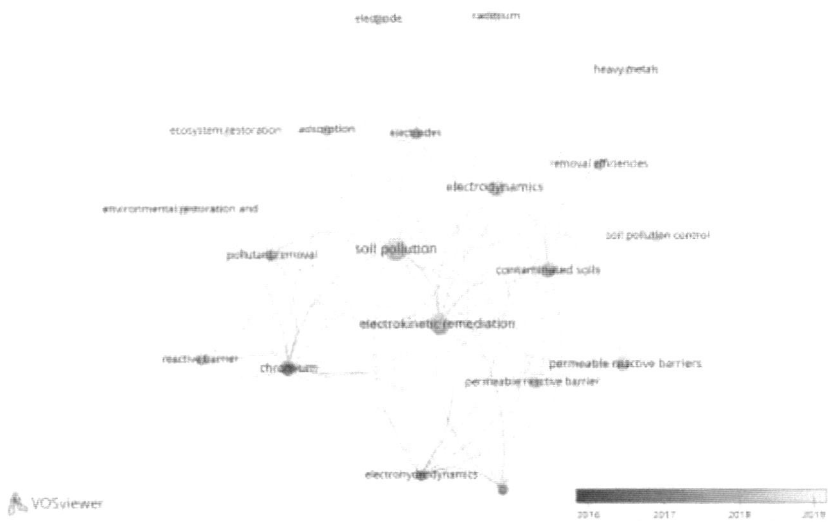

Figure 3. Keyword map of publications related to electrokinetic remediation using permeable reactive barrier based on the year of publication.

3.3. Contaminant type

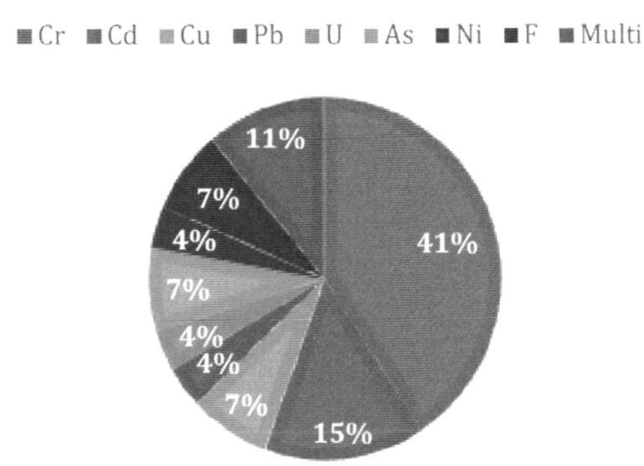

Figure 4. The number of metals used as targets of electrokinetic remediation with permeable reactive barrier.

Based on the analysis results, the removal of Cr content in contaminated soil using electrokinetic remediation with PRB was the most studied. This situation is because Cr(VI) can dissolve over a wide pH range. In addition, during migration to the anode, the species will change according to the existing pH conditions. Therefore their movement and mobilization need to be investigated [8]. Cr(VI) is carcinogenic which can be a soluble oxyanion [9]. Cr metal pollution occurs due to various human activities, such as industrial activities, littering, metal smelting activities, thus requiring fast and precise handling [9].

3.4. PRB material

Various kinds of materials have been developed as reactive barriers in electrokinetic remediation, both in composite materials and materials from waste recycling. In general, zero-valent iron is the most widely used material, either with a combination of other materials, zero-valent iron with activated carbon [11], zero-valent iron with quartz sand [12], zero-valent iron with zeolite [13] or without combination [4,14]. Based on the research of Saeedi et al. (2013) for the removal of Cr in kaolinite, the use of nano zero-valent iron as PRB was the most optimum when combined with 0.1 M EDTA as catholyte [4]. It was found that the reduction of Cr reached 63.75%, and the reduction reached 18.9%. The same study was also conducted by Zheng et al.(2020) using zero-valent iron grain to remediation soil contaminated with Cr [14]. This study used citric acid as an electrolyte and obtained a higher Cr removal of 82.86%. The combination of zero-valent iron with zeolite can remove Cr metal up to 97% and reduce Cr (VI) to Cr (III) by 98% [13]. Besides being effective in the removal of Cr, zero-valent iron is also effective in the removal of other heavy metals in contaminated soil, such as Xiao et al. (2020), who conducted a study on uranium (U) removal in soil samples obtained in the area near uranium tailings mining by combining zero-valent iron with activated carbon as a reactive barrier [11]. Pre-treatment of uranium-contaminated soil samples using a mixture of citric acid and $FeCl_2$ resulted in a U removal efficiency of 80.58 ± 0.99% and could even reduce U(VI) to U(IV).

Table 1. Elimination targets and reactive barriers used in electrokinetic remediation.

References	**Target Contaminant**	*Reactive Barrier*
Cappai et al., 2012	Cr, As	Red mud
Yu et al., 2019	Cr	Cetyl trimethyl ammonium bromide-zeolite (CTMAB-Z) and Fe(0) mixture
Wang et al., 2019	Cr	Polypyrrole-linen fabric (PPy-LF)
Zheng et al., 2020	Cr	Zero valent grain
Xu et al., 2016	Cr	Hydrocalumite (CaAl-LDH)
Xu et al., 2017	Cr	Hydrocalumite (CaAl-LDH)
Xu et al., 2019	As and Cr	Hydrocalumite (CaAl-LDH)
Saeedi et al., 2013	Cr	Nano zero-valent iron
Suzuki et al., 2014	Cr	Fe_3O_4
Wang et al., 2021	Cr	Hyperbranched polyethyleneimine (HPEI)-enriched polyacrylonitrile (PAN) electrospun nanofiber membrane (ENFM)
Nasiri et al., 2020	Cr	Fe_3O_4
Fu et al., 2013	Cr	Fe^0 and zeolite mixture
Hu, 2013	Cd	Nanometer powder Fe, granular activated carbon, and sand mixture
Xu et al., 2020	Cd	Porous graphene oxide (PGRO) carbon nanomaterial
Zhou et al., 2020	Cd	Zeolite and ZVI mixture
Ding et al., 2017	Cd	Humin

References	**Target Contaminant**	***Reactive Barrier***
He et al., 2021	Cd and Zn	SA/PVA/MgO/ATP composite hydrogel
Ghobadi et al., 2021	Cd, Cu, Pb, and Zn	Organic compost
Ghobadi et al., 2021[a]	Cu	Compost
Ghobadi et al., 2020	Cu	Biochar
Ribeiro et al., 2018	Pb	Eggshell inorganic fraction powder (EGGIF)
Xiao et al., 2020	U	ZVI and AC mixture
Yao et al., 2020	As	Activated carbon
Zanjani et al., 2012	Ni	Activated carbon
Zhou et al., 2016	F	Activated alumina and calcium chloride mixture
Zhu et al., 2016	F	Bamboo charcoal

However, Fe_3O_4 has a better ability than Fe0 in recovering Cr to Cr(III) [17]. Nasiri et al. (2020) used Fe3O4 as a reactive barrier and combined it with a chelating agent [19]. The results showed that the combination with the chelating agent EDTA had a more significant effect on Cr removal when compared to the combination using the chelating agent citric acid. In addition, activated carbon and hydrocalumite (CaAl-LDH) are also widely used as reactive barriers in metal-contaminated soil's electrokinetic remediation. Activated carbon can remove Ni from the soil without reverse osmosis at a voltage gradient of 1.25 V/cm [29]. However, in removing Cd from contaminated soil, according to Ghobadi et al. (2020), biochar can capture more Cd than activated carbon [27]. Meanwhile, research on reactive barrier hydrocalumite (CaAl-LDH) has been extensively studied by Xu et al. (2016) by comparing the efficiency of Cr metal removal using hydrocalumite (CaAl-LDH) as a reactive barrier with different variations of soil moisture [15]. The removal efficiency of Cr can reach 96.6% for Cr(IV) and 67.3% for total Cr at 40% soil moisture. Then Xu et al. (2017) using the same reactive barrier combining electrokinetic remediation with different voltage gradients [16]. The highest Cr(VI) and total Cr removal were obtained by applying a voltage gradient of 1 V/cm. Xu et al. (2019) investigated the pre-treatment using sodium citrate in electrokinetic remediation with hydrocalumite (CaAl-LDH) having a higher effect on As removal than Cr, while the use of EDTA-2Na pre-treatment had a higher effect on Cr removal than As [10]. As the research of Xu et al. (2017), in Cappai et al. (2012), red mud as a reactive barrier is more effective in removing Cr than As [8]. Thus it can be concluded that As metal is more difficult to clean than Cr metal. Therefore, it is necessary to develop research on the efficient reactive barrier in As removal in soil.

Reactive barriers can come from composite materials, such as polypyrrole-linen fabric (PPy-LF) and modified activated alumina. PPy-LF with a molar ratio of SDBS:CTAB = 1:2 has a Cr removal rate of 93.1%. Cr(VI) on the surface of PPy-LF will be reduced to Cr(III), some of which will settle on the surface of PPy-LF, and the rest will enter the soil [7]. Modifying activated alumina with the addition of calcium chloride as a reactive barrier can increase the defluorination ability of activated alumina [30] so that it is better at removing fluorine content in contaminated soil. Reactive barriers can also come from organic matter, for example, peat soil [23], compost, biochar [26], and bamboo charcoal in removing fluorine [31].

3.4. PRB placement

Several important reactions occur during electrokinetic remediation, including adsorption-desorption, precipitation-dissolution, crystallization, and corrosion of the electrodes at the anode [32]. Proper placement of reactive barriers can prevent the negative impact of the reactions during the electrokinetic remediation process. Replace reactive barriers in the electrokinetic remediation process also needs to consider the type of pollutant to be removed. Heavy metal cations will migrate towards the cathode [1], so the location of the reactive barrier should be placed close to the cathode. Meanwhile, heavy metal anions will migrate towards the anode [2]. Therefore a reactive barrier should be placed near the anode. The placement of the right reactive barrier effectively reduces and even eliminates pollutant pollution in the anode and cathode chambers, thereby reducing the processing of liquid waste in the anode and cathode chambers.

Nasiri et al. (2020) conducted a study on the effect of PRB position on Cr removal [19]. The placement of PRB in the middle has a higher Cr removal efficiency than the PRB position on the anode and cathode sides. However, PRB located on the anode and cathode side caused the removal in all parts of the soil to be more uniform than PRB located in the middle. A similar study was conducted by Xiao et al. (2020) in removing uranium on polluted soil [11]. The most optimum PRB placement is placed on three sides, namely at the anode, cathode, and middle. Different things are shown by Zanjani et al. (2012) research to remove Ni, where the most optimum PRB location is on the cathode side [29].

4. Conclusions

Bibliographic analysis was carried out using the PRISMA method and supported by VOSviewer software to map and tabulate information more concisely and efficiently. Based on the results of the elimination of scientific articles, 26 relevant scientific articles were obtained. Scientific articles on electrokinetic remediation with PRB on the heaviest metal contaminated soil in 2020 with 27% addition. China, Iran, and Australia are the countries that contributed the most to the research. Environmental Science and Pollutant Research is the journal with the most publications related to EK-PRB for heavy metal contaminated soil. The most studied elimination target is Cr. The author most actively collaborating on research related to EK-PRB in heavy metal polluted soil is D. Li. The mapping results using VOSviewer show three cluster groups, namely the electrokinetic remediation mechanism cluster, the use of electrokinetic remediation, and factors that affect electrokinetic remediation. It was found that in determining the type of PRB material and the position of PRB in the electrokinetic remediation process, it is necessary to consider the type of pollutant to be cleaned so that it can produce significant removal efficiency by minimizing the negative impact of physical and chemical reactions that occur during the electrokinetic remediation process.

Acknowledgments

This research is funded by SAPBN Undip 2021 under scheme of Program Penelitian Kolaborasi Indonesia (PPKI) Tahun 2021 Number 117-07/UN7.6.1/PP/2021

References

[1] Moghadam M J, Moayedi H, Sadeghi M M and Hajiannia A 2016 *Environ Geochem Health* **38** 1217-1227
[2] Wang Y, Li A and Cui C 2021 *Chemosphere* **265** 129071
[3] Camenzuli D, Freidman B L, Statham T M, Mumford K A and Gore D B 2013 *Polar Res* **32(1)** 1751-8369
[4] Saeedi M, Li L Y, and Moradi Gharehtapeh A 2013 *Int J Environ Res* **7(1)** 39-50
[5] Bruni A, Serra F G, Gallo V, Deregibus A, and Castroflorio T 2021 *Am J Orthod Dentofacial Orthop* **159 (4)** e343-e362
[6] Budihardjo M A, Ramadan B S, Putri S A, Wahyuningrum I F S and Muhammad F I 2021 *Sustainability* **13** 6562
[7] Wang Y, Huang L, Wang Z, Wang L, Han Y, Liu X and Ma T 2019 *Chem Eng J* **373** 131-139
[8] Cappai G, De Gioannis G, Muntoni A, Spiga D and Zijlstra J J P 2012 *Chemosphere* **86** 400-408
[9] Yu X, Muhammad F, Yan Y, Yu L, Li H, Huang X, Jiao B, Lu N and Li D 2019 *R. Soc Open Sci* **6** 182138
[10] Xu Y, Li J, Xia W, Sun Y, Qian G and Zhang J 2019 *Environ Sci Pollut Res* **26** 3392-3403
[11] Xiao J, Pang Z, Zhou S, Chu L, Rong L, Liu Y, Li J and Tian L 2020 *Sep Purif Technol* **244** 116667

[12] Yao W, Cai Z, Sun S, Romantschuk M, Sinkkonen A, Sun Y and Wang Q 2020 *J Soil Sediment* **20** 1526-1533
[13] Fu R B, Liu F, Zhang C B and Ma J 2013 *Environ Eng Sci* **30(1)** 17-22
[14] Zheng Y, Yan Y, Yu L, Li H, Jiao B, Shiau Y and Li D 2020 *Environ Sci Pollut Res* **27** 5572-5583
[15] Xu Y, Xu X, Hou H, Zhang J, Zhang D, Qian G 2016 *Environ Sci Pollut Res* **23** 6517-6523
[16] Xu Y, Xia W, Hou H, Zhang J and Qian G 2017 *Environ Sci Pollut Res* **24** 20479-20486
[17] Suzuki T, Kawai K, Moribe M and Niinae M 2014 *J Hazard Mater*
[18] Wang J, Hou L, Yao Z, Jiang Y, Xi B, Ni S and Zhang L 2021 *Chem Eng J* **406** 126822
[19] Nasiri A, Zanjani A J and Darban A K 2020 *Environ Pollut* **266** 115197
[20] Hu H 2013 *Adv Mater Res* **675** 188-191
[21] Xu J C, Ma Q, Chen C, Wu Q T and Long X X 2020 *Chemosphere* **259** 127441
[22] Zhou H, Xu J, Lv S, Liu Z and Liu W 2020 *Sep Purif Technol* **239** 116544
[23] Ding L, Lv W, Yao K, Li L, Wang M and Liu G 2017 *Environ Sci Pollut Res* **24** 3430-3436
[24] He C, Hu A, Wang F, Zhang P, Zhao Z, Zhao Y and Liu X 2021 *Chem Eng J* **407** 126923
[25] Ghobadi R, Altaee A, Zhou J L, McLean P and Ganbat N 2021 *J Hazard Mater* **402** 123891
[26] Ghobadi R, Altaee A, Zhou J L, Karbassiyazdi E and Ganbat N 2021 *Sci Total Environ.* **794** 148668
[27] Ghobadi R, Altaee A, Zhou J L, McLean P and Yadav S 2020 *Chemosphere* **252** 126607
[28] Ribeiro A, Mota A, Soares M, Castro C, Araujo J and Carvalho J 2018 *Key Eng Mater* **777** 256-261
[29] Zanjani A J, Saeedi M and Weng C H 2012 *Environmental* **5(2)** 28-35
[30] Zhou M, Zhu S, Yi Y and Zhang T 2016 *Clean Tech. Environ Policy* **18** 2691-2699
[31] Zhu S, Han D, Zhou M and Liu Y 2016 *Electrochim Acta* **198** 241-248
[32] Wen D, Fu R and Li Q 2021 *J. Hazard Mater* **401** 123345

IOP Conf. Series: Earth and Environmental Science **894** (2021) 012035 doi:10.1088/1755-1315/894/1/012035

Community based solid waste management to reduce open burning incidents: A case study of waste treatment facility in Gajahmungkur District, Semarang City

B S Ramadan[1], H S Huboyo[1], E Sutrisno[1], A M Sari[1], B P Samadikun[1], N Hardyanti[1], A Karmilia[1]

[1]Department of Environmental Engineering, Diponegoro University, Semarang Indonesia 50275

bimastyaji@live.undip.ac.id

Abstract. Ngudi Kamulyan is a domestic waste treatment facility which imply 3R (reduce, reuse, recycle) in Gajahmungkur District, Semarang City-Indonesia. This facility is already in status quo due to the lack of human resources and this Covid-19 pandemic. Therefore, this facility has chances to improve its services after the pandemic. The activation of this facility can reduce the number of mismanage waste and open burning incidents. Performance improvement planning is carried out based on the five aspects of waste management. This study was conducted using qualitative analysis and questionnaire survey to the neighbourhood leaders and facility officers. Community participation planning was done with activities focused on improving the 3R habit in the community. There is a need for continuous supervision from related parties, such as Gajahmungkur District Semarang City Government and the Environmental Agency of Semarang City. That supervision will be necessary to make the waste treatment facility and operational activities of 3R could run consistently.

1. Introduction

Solid waste management constituted one of the most environmental and human health problems facing the Semarang City government. Waste that cannot be processed and managed will be transported from the temporary storage area to the landfill and then disposed of safely to minimize the impact on the surrounding environmental footprint [1]. However, in reality, most of the waste that goes to landfills in Indonesia does not go through a sorting process first, so it ends up piling up openly in the landfill. Waste management treatment like this can shorten the life of landfills as the population increases, which is directly proportional to an increase in the volume of waste [2]. Therefore, waste treatment and management are needed that can help processing waste before it enters the landfill. Regional-scale solid waste treatment facility is already constituted in several areas in the Indonesian cities. This facility is hoped as a communal waste management which could promote 3R (reduce, reuse, and recycle) activities at district scale [3]. Community awareness is important to reduce the number of waste burning incidents.

The policy of managing urban waste or regional waste in Indonesia defines that waste management, outlined in 5 (five) subsystem components: operating engineering, institutional, regulatory, financing, and community participation aspects [2]. This paper focused on the planning of community participation for waste management in the waste treatment facility at regional-scale level and promote Ngudi Kamulyan waste treatment facility (WTF) as the case study area. Community participation is very supportive of a waste management program region. Community participation in waste management is a process where people, as consumers at once, waste service producers, and citizens affect the quality and smooth infrastructure available for them. Community participation is essential because participation is a valuable tool to obtain information about the local community's condition, needs, and attitudes. Participation will make people trust more in the development of the projects/programs. One alternative for regional-scale waste management that can be built to reduce the generation of waste transported to landfills is to manage waste in the form of sorting waste that still has use-value to be further processed or recycled into valuable products. WTF in Ngudi Kamulyan is one of the pilot waste temporary storage area 3R in Semarang City, which serves 300 families in Sampangan Village, Gajahmungkur District, with the amount of waste entering before being disposed of to landfill of 92.74 tons/year in 2017 [4].

In addition to reducing waste generation at the landfill and mismanagement of waste (such as open burning and backyard disposal), it is necessary to implement further waste management to increase the economic value of waste, such as compaction, purification, drying, and waste counting [5]. It is necessary to redesign the Ngudi Kamulyan WTF to improve performance. The final result of this recommendation is expected to be used as an academic reference in the development and optimization of waste treatment facility at similar regional-scale in Indonesia.

2. Methodology

This study was conducted starting with a literature study related to the general description of the Ngudi Kamulyan WTF then continued with surveys and direct observations in the field. A data and information collection survey were carried out to understand three things: regional development policies in the field, realities on the field, and the development of ideas, ideals, and aspirations from all parties interested in implementation in the field. The stages of data collection include; Institutional survey, field survey, interviews, and questionnaires. The neighbourhood leaders (4 person) and WTF officers (5 person) were interviewed to gain their insights about waste management practice. This type of research is qualitative research. Methode used to examine the object's condition that natural where the researcher is as key instruments, data collection techniques triangulated (combined), data analysis is inductive and the results qualitative research emphasizes meaning rather than generalization. This paper focuses on community participation in the construction and management of Ngudi Kamulyan WTF.

3. Results and discussion

3.1. Overview of the Ngudi Kamulyan WTF, Gajahmungkur District

Ngudi Kamulyan is one of the active WTF in the city of Semarang in carrying out its function as a place for collecting, sorting, and recycling activities on a regional scale. This WTF is located on Akasia street, Sampangan Village, Gajahmungkur District. WTF was established in 2008 in a Self-help community (KSM) with the management chairman. WTF has also promoted the establishment of a waste bank that actively accepts waste from residents at any time. The service area of Ngudi Kamulyan WTF currently only covers the area around the WTF, especially Sampangan Village. The waste management process at the WTF did not work effectively. It was temporarily stopped due to the ongoing Covid-19 pandemic and the lack of human resources that did not allow the Ngudi Kamulyan WTF to run correctly. Ngudi Kamulyan WTF has a total area of about 60 m^2. Facilities and infrastructure were available at the facility to support 3R activities include signage, waste transport vehicles, office space, sorting room, and composting room. The physical condition of the Ngudi Kamulyan WTF is still very decent. However, due to the ineffective operation of the facility, the incoming waste is not adequately organized and is scattered around the WTF.

Figure 1. The physical condition of Ngudi Kamulyan WTF.

3.2. Community participation in Ngudi Kamulyan WTF

Community participation in 3R-based waste management on the individual scale is lacking. Because the community accommodates their waste without separating dry, wet, and hazardous waste [6]. People who care about the environment join a self-help group community whose management consists of from neighborhood and citizen association, community leader of Ngudi Kamulyan and WTF officers is voluntary without getting a reward or salary. Ngudi Kamulyan WTF already has workers from residents. These workers carry out activities such as waste sorting, composting and also run a waste bank program. Community participation is quite good, with participation provided in the form of personnel, facilities, and operational costs, all from residents and the results of the waste bank program. Therefore, community participation in waste management based on the 3R scale excellent group. The participation of the Ngudi Kamulyan WTF community in the development and still lacking during this pandemic due to restrictions on public activities because the government must always involve the community in village development so that development will be right on target and according to community needs. For this reason, it is necessary to optimize community participation to increase the movement of activities at the Ngudi Kamulyan WTF to achieve common goals for the health of the environment and the Ngudi Kamulyan community.

3.3. Planning improvement of community participation aspects of Ngudi Kamulyan WTF

Community participation plays a significant role in the urban waste management system because the community is a person or group directly affected by or interested in a decision with legal responsibility [7]. In this paper, Ngudi Kamulyan WTF is planned for a service period of 10 years, with the calculation of waste generation increasing every year. The plan can extend its service time to more than 10 years if there is awareness from the community or stakeholders to manage and process the waste generated [8]. In order to realize the excellent participation of the Gajahmungkur district community in waste management and processing, it is necessary to have various activities that are the responsibility of each relevant stakeholder.

The district government should counsel and disseminate information regarding constructing a sub-district WTF in the Gajahmungkur district. The district should also construct waste management regulations from the Semarang city government to neighborhood association stakeholders. This activity aims to make the WTF built, and the applicable waste management regulations are known [9] and understood by every representative of the Gajahmungkur District community. The district could promote the appeal and applicable waste management regulations by installing infographic billboards and bulletin boards at strategic places in every neighborhood association in Gajahmungkur District. Continuous monitoring should also conducted to evaluate the implementation of waste management regulations that apply to each village. The district government should also affirm the reduction of single-use products such as plastic, and food packaging for all business actors in Gajahmungkur District. The district government Innovate a simple garbage vending machine in public facilities in Gajahmungkur District, which can record the value or points obtained by consumers from the waste deposit activities carried out. These points can be exchanged at the nearest waste bank in Gajahmungkur District. The points earned are well recapitulated with the QR code system. Internalization is also important to deliver of appeals and applicable waste management regulations into the school curriculum in the form of adiwiyata schools. This program is supervised by the Semarang City Education Office and the Semarang City Environment Service, and Ngudi Kamulyan KSM supports its implementation. Implementation of this program begins with determining a pilot school in each kelurahan, then continues to add the number of adiwiyata schools gradually until it reaches the 100% target.

Each sub-district level government is responsible for conducting education on waste management. The environmental care movement's primary focus is activities for managing organic, inorganic, and hazardous waste from sources with the 3R concept (Reduce, reuse, recycle) and application of sustainable lifestyle [10]. The sub-district level government should also support the establishment of a waste bank that can accept hazardous waste and composting facilities for organic waste [11]. In every village by empowering local communities as managers. Waste bank activities focus on producing recycled products that can become the characteristics of each village. The lowest level of the government should ensure that the appeal letter has been well understood and implemented in each neighborhood association in Gajahmungkur district and reports on the progress of the application appeal letter to Gajahmungkur district regularly every month. Implementation of reward and punishment system for each household should be conducted to the respective neighbourhood.

Local community can socialize the Ngudi Kamulyan WTF development plan to the people around WTF, such as Sampangan Village, and involving the opinions of the surrounding community in decision making. The community can empower the community around Ngudi Kamulyan WTF as operational employees of the facility. Formulation a community-based 3R waste management strategy is important, including hazardous waste management, with the Gajahmungkur District scale referring to the applicable regulations. The local community must ensure the implementation of the waste management strategy with the 3R concept in the community, especially the Sampangan Village community where the Ngudi Kamulyan WTF was established. Increasing the role is carried out in training and support for the local community, which is the responsibility of the Environmental Service and Gajahmungkur District Government. The community can enforce waste management regulations in the Neighborhood association, village, and a subdistrict that apply to the community of Gajahmungkur District. Scavengers and waste transporters should also be educated by the community who work in the Gajahmungkur District in terms of sorting waste by type.

4. Conclusion

Ngudi Kamulyan WTF in Gajahmungkur District could reduce the improper waste management in Semarang City. The role of the community in the management and construction of the Ngudi Kamulyan WTF was slightly hampered during this pandemic. Some activities have stopped due to limiting crowds and gatherings. Community participation in Ngudi Kamulyan is quite good by following the waste bank that has been formed, sorting waste, and composting. Nevertheless, it is still not optimal in developing the Ngudi Kamulyan WTF because there is still a lack of personnel working there. Planning for the aspect of community participation is carried out by maximizing the role of each relevant community representative in Gajahmungkur District, starting from the District, village, and neighborhood association, as well as KSM Ngudi Kamulyan itself. The activities carried out focused on increasing the 3R culture in the community and promoting the contents of circulars in the form of billboards in strategic places.

Acknowledgements

The authors would like to thanks to the Faculty of Engineering which support this research under strategic research grants number 195/UN7.5.3.2/HK/2021.

IOP Conf. Series: Earth and Environmental Science **894** (2021) 012035 doi:10.1088/1755-1315/894/1/012035

References

[1] Massreshaw A A 2017 *Int J Scient Eng Sci* **10(1)** 1-6
[2] Maharani A, Dewilda Y, Darnas Y, and Dewata I, 2019 *IOP Earth Environ Sci* **314** 012017
[3] Environmental Service Program, Medan, Bandung, Subang and Surabaya 2006
[4] Putra H P, Damanhuri E, and Sembiring E 2020 *J Mat Cycl Waste Manag* **22** 396–404
[5] Rathore and Sarmah 2019 *J Clean Prod* **211** 44-60
[6] Hou D, Al-Tavvaa A., Guthrie P, and Watanabe K 2012 *Environ Sci Tech* **46** 2494-2495.
[7] Lin T. Guo X H, Zhao Y, et al, 2010 *J Sustain Dev Ecol* **17(4)** 285-291
[8] Lishan X, Sha H, Zhilong Ye, Ouwen Z, Tao L 2020 *J Clean Prod* **278** 123569
[9] Shekdar AV 2009 *Waste Manag* **29**(14) 38-48
[10] Ağacsapan B, and Cabuk N S 2020 *Sustain Cities Society* **52** 101829
[11] Wang S, Shen W, Tang W, Wang Y, Duffield C F, Hui F K P 2019 *Renew Energy* **132** 326-334

Analysis of waste processing potential in Gunung Putri District, Bogor Regency

N U Fitri[1], D Indrawati[1*], Ratnaningsih[1]

[1]Department of Environmental Engineering, Faculty of Landscape Architecture and Environmental Technology, Trisakti University, Jakarta, Indonesia

dindrawati@trisakti.ac.id

Abstract. The increase in waste generation in Gunung Putri Subdistrict, Bogor Regency, causes an increased burden at the Landfill Site (TPA). One of the efforts to reduce the burden of Landfill Site (TPA) is to maximize the Waste Processing Site (TPS) by applying the circular economy concept. This study aims to analyze the generation, composition, and potential of waste processing to have economic value. The sampling method used is a combination of purposive sampling and stratified random sampling. The results showed that with a population of 304,672 people and a waste generation rate of 0.48 kg/person/day, it is estimated that the waste generation in Gunung Putri District is 147,156.58 kg/day. The waste material in Gunung Putri Subdistrict consists of 52.86% organic waste components and 47.14% non-organic waste. Organic waste is dominated by food waste by 50.38%. From the results of the characteristic test of the waste, it is known that the water content is 63.16%, the ash content is 29%, the C/N ratio is 7.03%, and the calorific value of the waste is 1.924 kcal/kg by considering the amount of generation, composition, and characteristics of waste, which are then used with standards or criteria for waste processing technology from various references. Organic waste in Gunung Putri Subdistrict has good potential to be processed using the Black Soldier Fly (BSF) and organic waste processing with Peuyeumisasi Method (Biodrying) and recycling for non-organic waste.

1. Introduction

Gunung Putri sub-district is one of the 40 sub-districts in the Bogor Regency area, divided into ten villages that experience an increase in population each year, causing an increase in the amount of waste generated. The problem of solid waste in Gunung Putri District is that it still uses the old pattern. Namely, waste collected from the source, transported and disposed of to the Landfill Site (TPA). Around 6,000 – 8,000 tons of waste produced in one day are transported to the TPA without managing and treating waste optimally. The increase in waste generation increases the burden on the landfill. One of the efforts to reduce the burden of the TPA is to maximize the Waste Processing Site (TPS) by applying the circular economy concept.

Waste processing explained that waste is a national problem, so that its processing needs to be carried out in a comprehensive and integrated manner from upstream to downstream to provide economic benefits, be healthy for the community, and be safe for the environment, and can change people's behaviour [1]. Black Soldier Fly (BSF) and Refused Derived Fuel (RDF) technology can be used for processing organic waste, and others. Meanwhile, non-organic waste can be recycled back into valuable goods and generate economic value.

IOP Conf. Series: Earth and Environmental Science **894** (2021) 012036 doi:10.1088/1755-1315/894/1/012036

2. Methodology

2.1. Location
Research conducted for sampling began in April 2021 in the Gunung Putri District, Bogor Regency.

2.2. Data collection
Collecting data, composition, characteristics, and sources of waste using the method of SNI 19-3964-1994 concerning Methods for Collection and Measurement of Samples of Generated and Urban Waste Materials and Stratified Random Sampling Methods. A sampling of data was carried out from residential and non-residential sources. The data needed are primary in the form of field observations and laboratory tests. To determine the number of waste samples originating from the calculation, the number of people is calculated based on SK SNI 19-3964-1994 [2] with the formula equation.

$$S = Cd \sqrt{Ps} \tag{1}$$

where:

S = Number of samples

Ps = Population (people)

Cd = Coefficient for medium/small town = 0,5

To calculate the number of samples taken from the comparison of the proportion of residential households consisting of high income, middle income, and low income.

The number of samples taken from the data on the total number of settlements in Gunung Putri District is 70 families, with a comparison:

- High income (60%) = 42 Households
- Middle income (30% = 21 Households
- Low income (10%) = 7 Households

2.3. Data processing and analysis
Data analysis and discussion were written descriptively. The calculation of waste generation and composition refers to SNI 19-3964-1994. Following are the stages of analysis carried out.

1. Calculation of waste generation and composition
2. The potential for waste processing is determined based on the amount of generation, composition, and characteristics of the waste in Gunung Putri District.

3. Result and discussion

3.1. Gunung Putri sub-district
Gunung Putri sub-district is one of the sub-districts located in the eastern part of Bogor Regency. Gunung Putri District is geographically located at coordinates 106'55 - 107'15 East Longitude and 6'15 - 6'35 South Latitude. Gunung Putri sub-district has 56 km2 or 5,600 hectares consisting of 10 (ten) villages, 1091 sub-districts, and 273 hamlets. In 2021 the population of Gunung Putri Subdistrict is 304,672 people.

3.2. Waste generation
The calculation results show that the average waste generation data per person per day in Gunung Putri District is 0.48 kg/person/day or 1.80 l/person/day. The calculation of waste generation data can be seen in Table 1.

Table 1. Weight and volume.

Source of Waste	(kg/person/day)	Produced Volume (litres/person/day)
Residential area	0.51	1.90
Shops	0.45	2.61
Restaurant	0.28	1.16

Source of Waste	(kg/person/day)	Produced Volume (litres/person/day)
Office	0.25	1.26
Market	0.25	0.53
Average	**0.48**	**1.8**

In Table 1, the data from the research on the value of waste shows that residential waste has a significant per capita generation. Judging from the residential area of 79% of the total area of Gunung Putri District, the load generated by residential waste is the largest.

3.3. Composition of waste

Based on direct measurements in the field, the data obtained from the composition are processed into the resulting composition presentation as follows:

Table 2. Composition of waste.

Composition of Waste	Residential	Non-Residential	Composition of Total
Organic	58.58%	42.18%	50.38%
Paper	10.33%	19.53%	14.93%
Wood	0.80%	4.17%	2.48%
Fabrics/Textiles	0.27%	0.49%	0.38%
Rubber/Leather	1.82%	0.00%	0.91%
Plastic	13.08%	20.65%	16.86%
Metals	0.09%	0.00%	0.04%
Glasses/Glass	4.84%	3.73%	4.29%
E-waste	1.74%	0.00%	0.87%
Others	8.46%	9.24%	8.85%

3.4. Proximate and ultimate waste analysis

Measurement of the characteristics produced in Gunung Putri District was carried out at the Environmental Laboratory, Trisakti University. When viewed from measuring the water content of waste in Gunung Putri District, organic waste has a water content of 63.16%. The results of the water content analysis of samples from Gunung Putri District can be seen in table 3.

Table 3. Moisture content.

Type	Composition	Moisture Content (%)
Organic	Food Waste	63.16%
	Twig Leaves	23.87%
Non-Organic	Paper	
	a. HVS Paper	2.87%
	b. Carton	6.46%
	c. Cardboard	4.51%
	Plastic	
	a. PP	14.05%
	b. PET Bottle	0.79%
	c. LDPE	4.58%
	d. PS	0.15%
	e. Others (7)	7.07%

The results of the ash content and volatile content analysis in samples from Gunung Putri District can be seen in table 4.

Table 4. Ash content and volatile content.

Type	Sample	Ash Content (%)	Volatile content (%)
Organic	Food Waste	29%	71%
	Twig Leaves	14%	86%
Non-Organic		Paper	
	a. HVS Paper	48%	52%
	b. Carton	7%	93%
	c. Cardboard	33%	67%
		Plastic	
	a. PP	80%	20%
	b. PET Bottle	98%	2%
	c. LDPE	8%	92%
	d. PS	17%	83%
	e. Others (7)	8%	92%

Testing the calorific value of organic waste food scraps and grass leaves in Gunung Putri District using a bomb calorimeter resulted in a calorific value of 1,924 kcal/kg. The results of C-organic content and total N content were used to calculate the C/N ratio so that the obtained value was 7.03.

3.5. Potential of waste processing

The following are the analysis results for the potential for waste processing in Gunung Putri District, which can be seen in table 5.

Table 5. Waste processing potential.

Composition	Percentage (%)	Recovery Factor	Volume (m^3/day)	Recovery/ (m^3/day)	BSF (m^3/day)	TOSS Peuyeumisasi (m^3/day)	The residue (m^3/day)
Organic Waste							
Food Waste	45.60%	100%	250.06		250.06		
Twig Leaves	4.78%	100%	26.21			26.21	
Non-Organik Waste							
Paper	14.93%	[a]85%	81.88	69.60			12.28
Wood	2.48%	100%	13.63			13.63	0.00
Fabrics/Textiles	0.38%	100%	2.09			2.09	0.00
Rubber/Leather	0.91%	100%	5.00			5.00	0.00
Plastic	16.86%	100%	92.48	92.48			0.00
Metals	0.04%	[a]90%	0.23	0.21			0.02
Glasses/Glass	4.29%	[a]70%	23.52	16.46			0.23
E-waste	0.87%	-	4.76				4.76
Others	8.85%	-	48.54				48.54
Total	100%		10.33	178.76	250.06	46.93	61.08

[a] Arie Herlambang, Djoko Heru Martono (2008)

The characteristics of the waste that were tested in this study were only the types of waste that could be used as compost, namely leaf waste and food waste. As for non-compostable waste, inorganic waste, and residual waste. Waste characteristics that need to be known are physical properties, chemical properties, namely initial analysis, final analysis, and energy content. Proximate analysis is divided into moisture content, ash content, and fluctuating content. The primary analysis is displaying carbon (C) and nitrogen (N) content to determine the C/N ratio. In comparison, the energy content is the calorific value of the waste. Tests are helpful to assist in developing methodologies for measuring and reporting the properties of waste materials and can be carried out with proper analysis and appropriate design, the overall characteristics of the waste required.

Based on the measurement results, the waste composition in Gunung Putri Regency consists of 58.58% organic waste and 42.18% non-organic waste. Organic waste is dominated by food waste by 45.60%. One way to overcome the accumulation of organic waste is to use household or organic waste that is not rotten such as vegetables or wilted fruits, as a growth medium for Black Soldier Fly (BSF). The cultivated BSF will be used as an alternative feed for livestock, such as poultry and fish. Provision of quality animal feed is one of the determining factors for the success of 2 industries and is the most significant component in these business activities, which is 50-70% [4]. BSF is a fly (Diptera) that belongs to the family Stratiomyidae. This fly can be found in tropical and subtropical areas (46° N - 42° South Latitude) [5].

The water content of the remaining food meets the optimum water content of BSF larvae feed ranging from 60-90%. Meanwhile, for food waste that will be used as organic fertilizer, based on laboratory tests for carbon and nitrogen levels, the C/N ratio of organic waste is 7.03. Based on the C/N of organic waste, food waste does not meet the value range, which is 30-35. So, it is necessary to add organic fertilizer containing carbon to increase the C/N ratio of the initial waste.

Using local waste processing technology (TOSS), the peuyeumization method (bio drying) is one of the renewable energy technologies. The waste that is utilized through physical conversion into pellets or briquettes can then be used as natural fuel to be converted into gas in a gasifier engine to produce syngas that can be used for household needs or conversion into electricity through a generator [7]. The waste is separated to get combustible waste (high NCV) such as biodegradable waste, etc., which is then fermented with the help of a bio-activator, then chopped and made into pellets or briquettes. Furthermore, pellets or briquettes are dried to increase their calorific value [7].

The maximum ash content for briquettes is a maximum of 8% [8]. Based on laboratory analysis, the ash content of twig leaf litter is 14%. It means that the total burned waste remains at 14%. The greater the ash content, the greater the number of drugs produced. The results of low ash content indicate that the waste has a high calorific value because low ash content results in less combustion so that combustion can run perfectly. So, with the addition of other non-organic materials such as wood, cloth and rubber will get a smaller ash content.

Recycling is waste that can be processed and reused because it has economic value. The recycling process uses waste materials to be returned to the same item or another form. Recycling or recycling of waste is carried out for paper, plastic, metal, and glass waste. Recycling waste is one of the methods provided that the waste used is waste that can be reused, has a high economic value, does not use paper or plastic coated with oil, for the non-organic waste it is carried out before being recycled, and the selection or grouping of waste according to the type of waste.

4. Conclusion

Based on the composition and characteristic testing in moisture content, ash, calorific value and C/N ratio, Gunung Putri Subdistrict waste can be processed using Black Soldier Fly (BSF) TOSS Peuyeumisasi Method (Biodrying), and Recycle or sold to the dealership. Organic waste for food can be processed using Black Soldier Fly (BSF) technology, organic waste for leaves and non-organic waste for twig leaves, wood, cloth and rubber can use TOSS processing technology. Meanwhile, plastic, paper, metal and glass waste can be processed in recycling (sold to dealers).

References

[1] The Republic of Indonesia 2008 *Law of the Republic of Indonesia Number 18 of 2008 concerning Waste Management. State Gazette of the Republic of Indonesia Year 2008, Number 69* (Jakarta: State Secretariat)

[2] National Standards Agency 1994 *SK SNI 19-3964-1994 Regarding Methods of Taking and Measurement*

[3] Central Bureau of Statistics Gunung Putri District 2021 *Gunung Putri District in Figures 2017* (Gunung Putri District)
[4] Wardhana and April D 2016 *Wartazoa* **26(2)**
[5] Harlystiarini 2017 *Utilization of Black Soldier Fly (Hermetia illucens) Larva Flour as a Protein Source to Replace Fish Meal in Laying Quail (Cortunix cortunix japonica) Ration* (Nutrition and Feed Sciences: Bogor Agricultural University)
[6] Diener S, Zurbrügg C, Gutiérrez FR, Nguyen DH, Morel A, Koottatep T, Tockner K 2011 Black army fly larvae for prospects and constraints of organic waste processing. In: Alamgir M, Bari QH, Rafizul IM, Islam SMT, Sarkar G, Howlader MK, editor. *Proceedings of WasteSafe 2011-2 nd International Conference on Solid Waste Management in Developing Countries*; 2011 13-15 February; Khulna, Bangladesh. Khulna (BD): Research Gate. 52-59.
[7] Iyus H 1, Mohamad H 2020 *Journal of Community Service*
[8] National Standardization Body 2004 *SNI 19-7030-2004 concerning Compost Specifications from Domestic Organic Waste* (Jakarta: Indonesian National Standards Agency)

IOP Conf. Series: Earth and Environmental Science **894** (2021) 012037 doi:10.1088/1755-1315/894/1/012037

Analysis of groundwater quality in Cimahi City of West Java Province

L O L Putri[1] and E Wardhani[2]

[1,2]Department of Environmental Engineering, Institut Teknologi Nasional (Itenas), Bandung, Indonesia

luvinaoktavia308@gmail.com

Abstract. Cimahi City is currently experiencing a clean water crisis, the cause of which is increasing population growth, so that water needs increase and land conversion, which results in reduced groundwater infiltration. The primary source of clean water in this city is groundwater. Population growth and residential development have an impact on pollution originating from domestic waste. These two things cause the quality and quantity of groundwater in Cimahi City to be increasingly critical. This study aims to analyze the quality of groundwater used by residents of Cimahi City. The data used is secondary data from the Environmental Service in 2019. Sampling was carried out at thirty points in Cimahi City spread over 14 urban villages, namely Citereup, Pasir Kaliki, Cibereum, Cibabat, Cipageran, Cimahi, Leuwigajah, Setiamanah, Karang Mekar, Cibeber, Melong, Cigugur, Padasuka, and Utama. The number of sampling is 1-2 samples/urban village. Based on the results of the analysis of parameters that do not exceed the required quality standards based on the Regulation of the Minister of Health of the Republic of Indonesia No. 32 of 2017 concerning the requirements and supervision of pristine water quality, namely: turbidity, Manganese, nitrate, detergent as MBAS, and total coliform. Based on the results of calculations with the water quality index, it is stated that the quality of groundwater in Cimahi City is in the normal-poor category.

1. Introduction

Cimahi City is a buffer zone for the capital city of West Java Province, namely Bandung City. The city is experiencing rapid population growth, which impacts increasing the need for land for settlements, the generation of domestic wastewater, garbage and others. The increase in population also causes the need for clean water to continue to grow. The primary source of clean water in Cimahi City is groundwater because local water companies have not served clean water for all residents in this city. It causes the exploitation of groundwater to increase, which impacts the occurrence of a decrease in groundwater reserves so that it reaches a critical level. A total of 1,174 households meet clean water needs using water from regional drinking water companies, with the highest user being Cimahi Tengah District. A total of 1,104 households in Cimahi City use other sources as their source of clean water. Residents of Cimahi City who use healthy water as a water source in South Cimahi District as many as 7,289 families, Central Cimahi Village as many as 970 families, and North Cimahi District as many as 927 families. In Cimahi City, 340 households use rainwater as a water source [1].

The generation of domestic wastewater resulting from daily human activities is one of the environmental problems in this area. Based on the characteristics of domestic wastewater, it is

IOP Conf. Series: Earth and Environmental Science **894** (2021) 012037 doi:10.1088/1755-1315/894/1/012037

categorized into greywater such as water used for washing in the kitchen, laundry, bathing and blackwater in the form of human waste. Domestic wastewater that is not managed correctly is a source of groundwater pollution [2]. This pollution is influenced by hydrogeological conditions, soil type, groundwater table depth, and the distance of the well from the local sanitation location [2]. The possibility of poor quality of healthy water can also be caused by sanitation that does not meet household standards, namely the absence of individual septic tanks or septic tanks that do not comply with regulations, thereby contaminating soil and groundwater [2].

This study aims to analyze groundwater quality in Cimahi City by calculating water quality and mapping the distribution of groundwater quality parameters that do not meet quality standards. The benefits of this research can provide an overview of groundwater pollution to assist local governments in making decisions to tackle groundwater pollution. This study uses quality standards based on the Regulation of the Minister of Health of the Republic of Indonesia No. 32 of 2017 concerning the requirements and supervision of clean water quality. Calculation of water quality using the water quality index (WQI) [3]. The WQI method has been widely used to analyze groundwater quality in several countries such as Iraq, India, and Palestine [3],[4],[5]).

2. Methodology

This study uses secondary data on the quality of resident well water from the Environmental Service of Cimahi City in 2021. The number of samples is 30 scattered in 14 villages, namely Citereup, Pasir Kaliki, Cibereum, Cibabat, Cipageran, Cimahi, Leuwigajah, Setiamanah, Karang Mekar, Cibeber, Melong, Cigugur, Padasuka, and Utama. The analyzed parameters consisted of 5 physical parameters: TDS, turbidity, colour, odour, and temperature. Eleven chemical parameters, namely: pH, iron (Fe), Fluoride (F), hardness, Manganese (Mn), Nitrate as N, Nitrite as N, Sulfate, Zinc (Zn), detergent as MBAS, and Organic Substances ($KMnO_4$) and one microbiological parameter, namely total coliform. The quality standard used refers to the Regulation of the Minister of Health of the Republic of Indonesia No. 32 of 2017 concerning the requirements and supervision of clean water quality are contained in Appendix I. Table 1 presents the locations of sampling points, and Figure 2 shows the distribution of sampling points.

Table 1. Location of groundwater quality sampling points in Cimahi city.

Coordinate 1 - 5	Village	Citeureup	Pasir Kaliki	Cibeureum	Cibabat	Cipageran
	East	107°32'53"	107°33'14"	107°33'42"	107°33'21"	107°32'14"
	South	06°51'34"	06°53'15"	06°54'40"	06°52'48"	06°51'17"
Coordinate 6 - 10	Village	Cipageran	Cibabat	Cimahi	Leuwigajah	Cibeureum
	East	107°32'27"	107°33'14"	107°32'27"	107°31'31"	107°33'52"
	South	06°51'16"	06°53'41"	06°52'13"	06°33'55"	06°42'21"
Coordinate 11 - 15	Village	Setiamanah	Citereup	Pasir Kaliki	Leuwigajah	Baros
	East	107°32'10"	107°32'57"	107°33'42"	107°31'31"	107°32'31"
	South	06°52'31"	06°51'50"	06°53'18"	06°33'55"	06°53'26"
Coordinate 16 - 20	Village	Karang Mekar	Cibeber	Karang Mekar	Cimahi	Melong
	East	107°32'33"	107°31'076"	107°32'37"	107°32'26"	107°33'30"
	South	06°52'37"	06°53'349"	06°52'32"	06°52'19"	06°55'17"
Coordinate 21 - 25	Village	Cigugur Tengah	Padasuka	Cigugur	Melong	Setiamanah
	East	107°33'16"	107°31'51"	107°33'16"	107°33'48"	107°32'12"
	South	06°53'20"	06°52'35"	06°53'33"	06°54'56"	06°52'26"
Coordinate 26- 30	Village	Baros	Utama	Cibeber	Utama	Padasuka
	East	107°32'20"	107°32'228"	107°31'150"	107°32'161"	107°31'38"
	South	06°53'24"	06°54'484"	06°53'351"	06°54'455"	06°52'41"

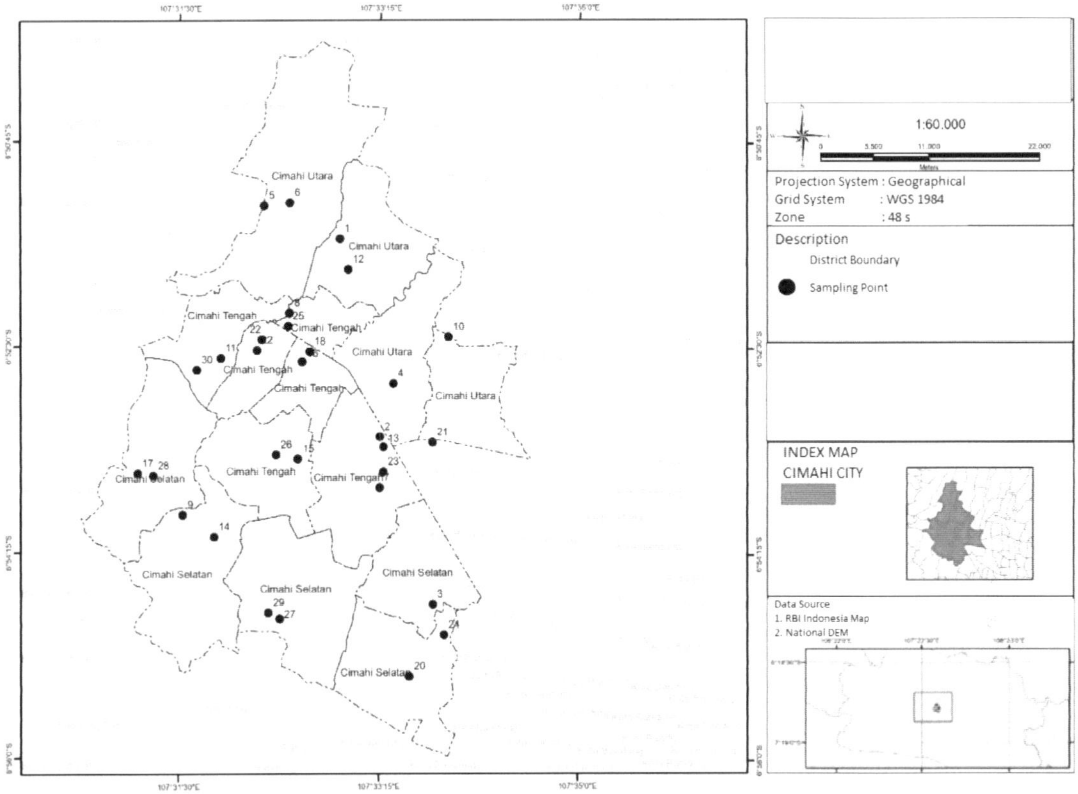

Figure 1. Location of Cimahi City groundwater quality sampling point.

WQI is a number that describes water quality by collecting the results of measuring water quality parameters [4],[5]. WQI is used to determine the level of water pollution based on nutrient parameters. The determination of the level of pollution based on WQI uses equation 1.

$$WQI = \frac{\sum_{i=1}^{n} C_1 P_1}{P_i} \quad (1)$$

Where: WQI = Water Quality Index, Ci = concentration of water quality parameters (i) the normalization results, Pi = the relative weighting of each parameter (i), and n = number of parameters. The determination of WQI is carried out by normalizing the data by transforming the value of the measurement results from n water quality parameters that have different units into values without units on a scale of 0-100. The weighting of water quality parameters is carried out to describe the importance of each parameter as an indicator of water quality. The weight of each parameter is given a value of 1 to 4 according to its level of importance for aquatic organisms. A value of 4 is given to the parameter that is considered the most important for aquatic organisms, and a value of 1 is given to the parameter that has the most negligible impact on the organism's life [6]. The resulting WQI values are classified into five classes, as presented in Table 2 [6].

Table 2. Criteria for waters based on the water quality index.

WQI Value	Water Quality
91-100	Excellent
71-90	Good
51-70	Medium
26-50	Bad
0-25	Very Bad

IOP Conf. Series: Earth and Environmental Science **894** (2021) 012037 doi:10.1088/1755-1315/894/1/012037

.3. Result and discussion

Based on the analysis results of 30 samples examined for groundwater quality, only 6 locations met the quality standards and were declared worthy of being a source of clean water. The rest there were 1 to 3 parameters that did not meet the quality standards. Locations that meet the quality standards are at sampling points 2, 9, 15, 20, 21, and 28. There are six sampling points where the Nitrate and Total Coliform parameters do not meet the quality standards, namely 1, 4, 11, 16, 21, and 26. 11 sampling points where the Total Coliform parameter does not meet the quality standard, namely 6, 7, 8, 13, 14, 19, 23, 24, 28, 29, 30. There are two sampling points where the Manganese and Total Coliform parameters do not meet the quality standard. Namely 2 and 22. The rest are sampling points of 3 parameters that do not meet the quality standards, namely Turbidity, Nitrate and Total Coliform. Sampling points of 5 parameters that do not meet the quality standards, namely manganese and detergent as MBAS. Sampling points of 10 parameters that do not meet the quality standards, namely turbidity, the sampling point is 17 parameters that do not meet the quality standards, namely Turbidity and Manganese, the sampling point is 18 parameters that do not meet the quality standards, namely pH and Total Coliform. The sampling point is 27 parameters that do not meet the quality standards, namely Turbidity and Manganese.

Based on the analysis of 21 sampling points, or 70% of the groundwater has been contaminated with E Coli, this is due to groundwater contamination by domestic waste originating from septic tanks. Until now, in Cimahi City, the domestic wastewater management system is still not good [1]. Based on the research results in Cimahi City, there are five rivers polluted by domestic waste, the parameters that do not meet the quality standards are the same as in groundwater [7],[8]. The study concluded that the water quality of the five rivers in this city is categorized as heavily polluted, and there is a high possibility of contamination from rivers to groundwater [7],[8]. It carries water and causing groundwater to be unfit for human consumption. Figure 2 presents a profile of the concentration of pH, turbidity, manganese, nitrate, and detergent as MBAS and the total amount of Coliform in Cimahi City groundwater.

Turbidity of water is caused by suspended solids, both inorganic and organic. The groundwater quality measurements in Cimahi City, sampling location point 3, 70.9 NTU, sampling point 10 is 49.4 NTU, and sampling location 17 is 34.8 NTU. It can be due to the presence of inorganic substances that usually come from weathered plants and animals. Industrial effluents can also be a source of turbidity. When compared with the parameter of the quality standard of the Minister of Health of the Republic of Indonesia, Number 32 of 2017 is 25 NTU. Of the three groundwater samples tested, all three exceed the applicable quality standards.

Based on the quality measurement results, there are 3 points where the Manganese concentration has exceeded the quality standard, namely the sampling location at point 5, the result is 5 mg/L, sampling point 12 is 0.593 mg/L, and at the sampling location point 17, the result is 0.745 mg/L. The high Manganese content in groundwater is predicted because of the decay of rocks containing Manganese [9]. The high content of this element in groundwater can reduce the aesthetics of the water because this element can settle in clothes, bathroom walls and other solid markers.

Groundwater whose nitrate parameter does not meet the quality standard is located at the sampling location point 1 the result is 14 mg/L, the sampling location point 3 is 16.2 mg/L, the sampling location point 4 is 12.6 mg/L, the point sampling location is 11 the result is 11.3 mg/L. Moreover, the location of sampling point 16 the result is 17.7 mg/L. The high level of nitrate in groundwater is caused by the amount of household waste input influenced by the density of residential houses; the age of a settlement. Where the older the age of a settlement, the greater the nitrate in the area. A high concentration of nitrate in groundwater is hazardous for human health who consume it.

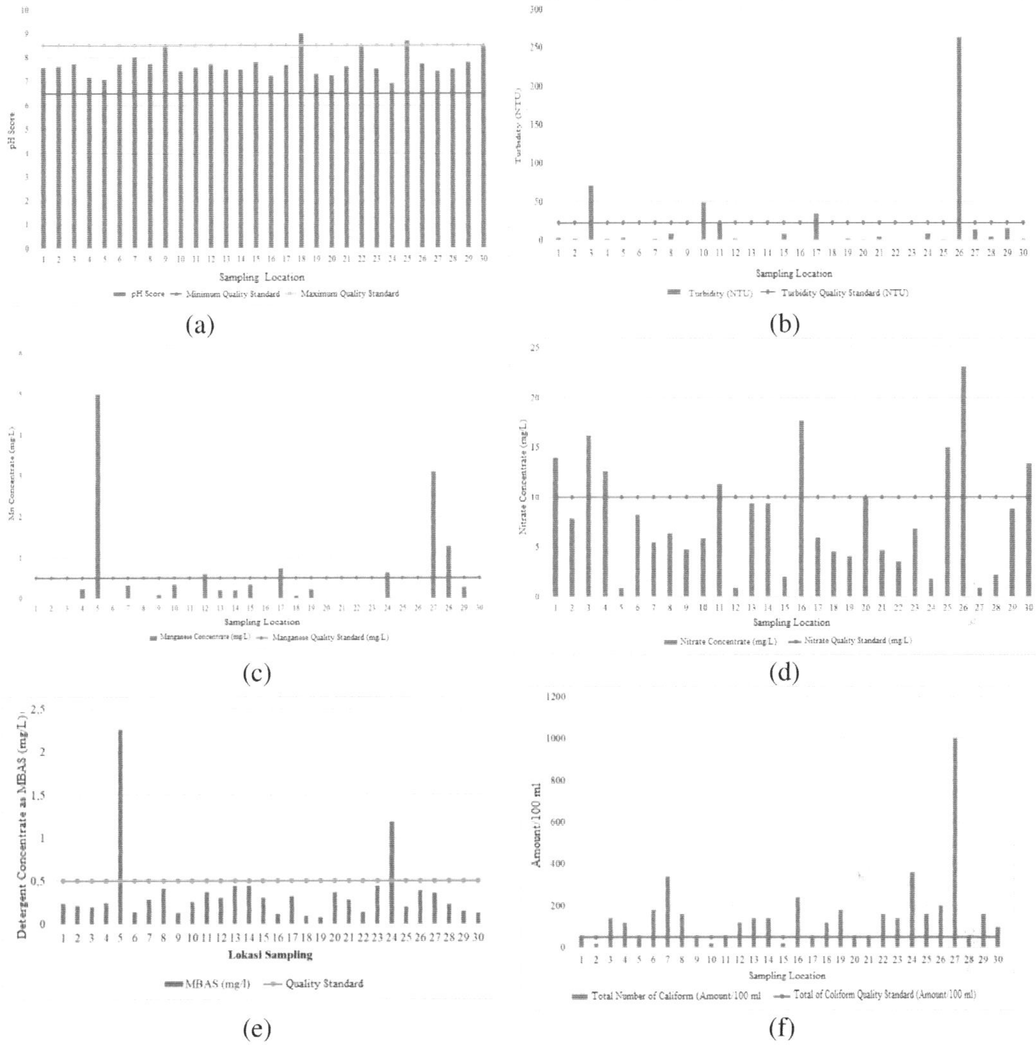

Figure 2. pH parameter profile (a), turbidity parameter profile (b), manganese parameter profile (c), Nitrate parameter profile (d), detergent as MBAS profile (e), total coliform parameter profile (f).

Based on Figure 2, there are 2 points where the detergent concentration as MBAS does not meet the quality standards, namely at sampling points 5 and 24. Detergent as MBAS found in groundwater is predicted to come from domestic wastewater contamination caused by leakage of septic tanks or due to poor construction. Not good enough to cause seepage. Steps to make understanding the position of polluted groundwater mapping is carried out as shown in Figure 3. Based on the figure, the red dots indicate parameters that do not meet the quality standards, while the black ones meet the quality standards. This mapping facilitates decision making to determine priority areas for handling groundwater pollution so that the efforts made are right on target.

Groundwater quality in Cimahi City is calculated using WQI. This method has been widely used in several locations and can provide an overview of groundwater pollution that occurs. Table 2 describes the results of the WQI calculations. Based on this method, it can be concluded that the groundwater quality in Cimahi City is categorized as usual to bad. Groundwater quality categorized as poor is found in three points, namely 3 in Cibeureum Village, 16 in Karang Mekar Village, and 26 in Baros Village.

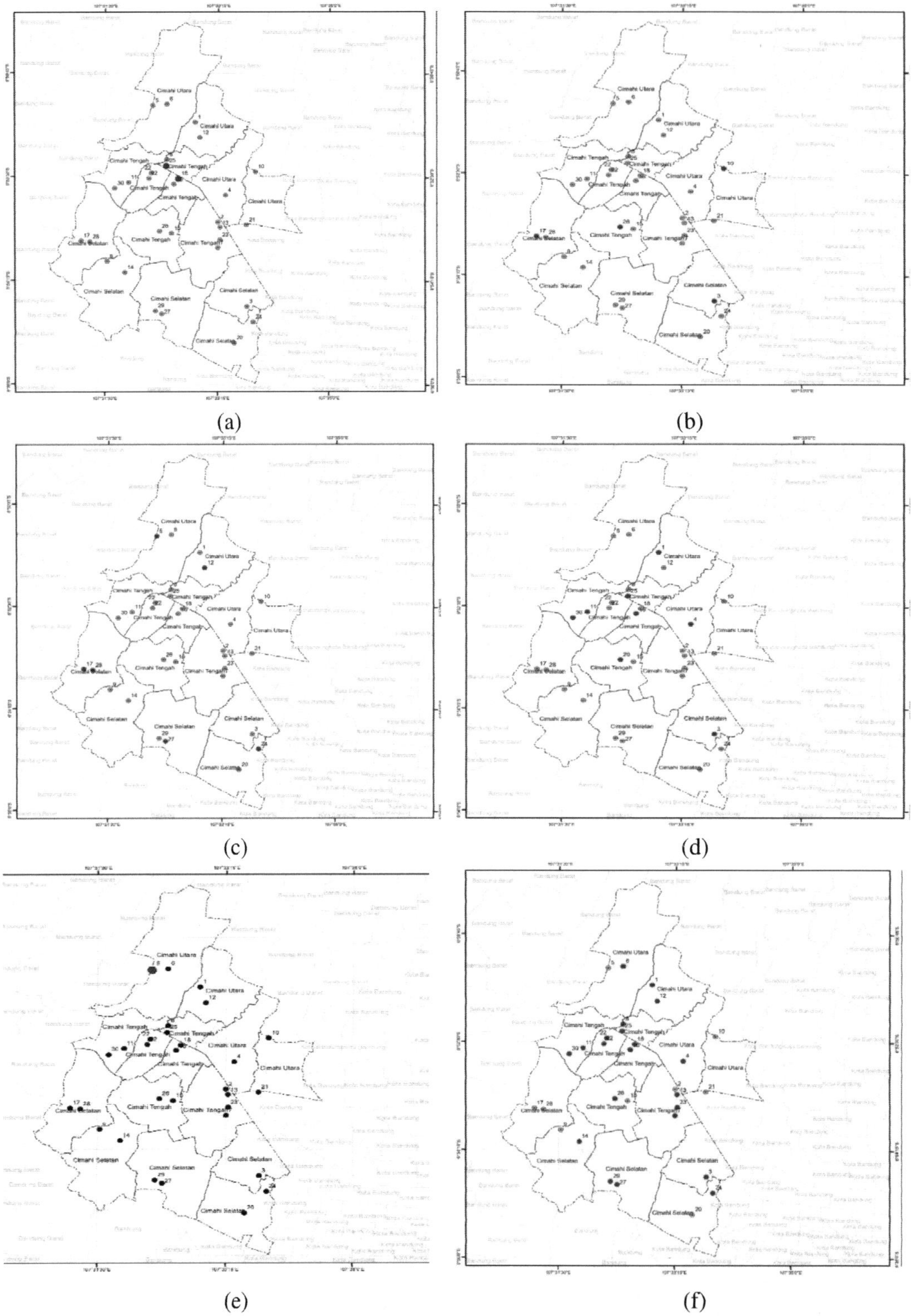

Figure 3. Distribution Map of pH parameter (a), turbidity parameter (b), manganese parameter (c), nitrate parameter (d), detergent as MBAS parameter (e), total Coliform parameter (f).

Based on Table 3, the WQI score ranges from 39-68. The lowest score is at point 26, and the highest score is at point 15. Based on the results of the WQI calculation, the sanitation system that must be immediately repaired is Cibeureum, Karang Mekar, and Baros Villages. Countermeasures are carried out by planning an excellent domestic wastewater management system using an offsite or on-site system that meets the applicable criteria. The selection of an appropriate domestic wastewater treatment system must refer to the Regulation of the Minister of Public Works and Public Housing of the Republic of Indonesia No. 04/PRT/M/2017 concerning the Implementation of a Domestic Wastewater Management System.

Table 3. Groundwater quality index in Cimahi City.

No	**Quality Index**						**WQI**	**Water**
Sample	**Temperature**	**pH**	**Turbidity**	**TDS**	**Nitrate**	**Total Coliform**	**Score**	**Quality**
1	17	88	90	60	45	60	58	Medium
2	16	88	93	80	58	63	65	Medium
3	16	88	29	47	42	41	45	Bad
4	16	88	93	73	48	42	57	Medium
5	14	88	90	52	97	55	65	Medium
6	16	88	99	49	56	38	55	Medium
7	16	84	93	55	65	32	54	Medium
8	14	88	78	68	60	40	56	Medium
9	16	84	99	58	70	55	62	Medium
10	16	88	40	58	65	63	57	Medium
11	16	88	59	42	49	50	51	Medium
12	16	88	93	50	97	42	62	Medium
13	17	88	99	54	53	41	56	Medium
14	17	88	99	54	53	41	56	Medium
15	17	88	80	66	95	20	68	Medium
16	16	88	96	32	40	36	50	Bad
17	17	88	50	55	60	55	55	Medium
18	16	49	96	60	70	42	53	Medium
19	16	88	93	60	70	38	58	Medium
20	16	88	96	45	53	55	58	Medium
21	17	88	86	75	70	55	63	Medium
22	12	84	96	47	90	40	59	Medium
23	17	88	96	71	60	41	59	Medium
24	14	55	78	51	96	32	52	Medium
25	17	84	96	57	43	40	54	Medium
26	16	88	5	49	34	37	39	Bad
27	16	88	69	62	97	22	55	Medium
28	16	88	88	58	95	55	66	Medium
29	16	88	67	51	56	40	52	Medium
30	16	84	93	61	46	44	55	Medium

4. Conclusion

Based on the study results, it was concluded that the groundwater quality of Cimahi City was polluted by domestic waste. It is based on the measurement results of Total Coli, Detergent as MBAS, turbidity and Nitrate parameters that do not meet quality standards at several sampling points. Based on the results of the WQI calculation, the water quality in this city belongs to the standard category at 26 points and 4 locations, including the wrong category. Locations that fall into the bad category are Cibeureum, Karang Mekar, and Baros, thirsty for immediate efforts to control groundwater pollution.

IOP Conf. Series: Earth and Environmental Science **894** (2021) 012037 doi:10.1088/1755-1315/894/1/012037

References

[1] Pemerintah Kota Cimahi 2020 *Dokumen Informasi Kinerja Pengelolaan Lingkungan Hidup Daerah (DIKPLHD) Kota Cimahi Tahun 2020* (Wali Kota Cimahi)
[2] Pemerintah Kota Cimahi Kota Cimahi 2018 *Rencana Pembangunan Jangka Menengah Daerah Kota Cimahi Tahun 2017-2022* (Cimahi)
[3] Moustafa E B, Prabin K, Marijke H and Florimond D S 2020 *Water* **12** 262
[4] Awachat A R, Salkar V D 2017 *Int J Eng Res Technol* ISSN 0974-3154 **10(1)**
[5] Hussein A A, Ayad K H, Asmaa S A 2020 *Iraq Periodicals of Engineering and Natural Sciences* **8(3)** August 2020 p1482-1490
[6] Anggraini Y and Wardhani E 2021 *J Serambi Engineering* **6(1)** 1478-1487
[7] Mochammad A R A and E Wardhani 2021 *J Serambi Engineering* **6(2)** p 1842-1853
[8] M Viqi R and E Wardhani 2021 *Serambi J Engineering* **6(2)** 1917-1925
[9] Wardhani E, Notodarmodjo S, Roosmini D 2018 *MATEC Web of Conf* **147(2018)** 08003

IOP Conf. Series: Earth and Environmental Science **894** (2021) 012038 doi:10.1088/1755-1315/894/1/012038

Estimating backyard waste burning emission: A case study of Tembalang Campus, Diponegoro University

E Sutrisno[1], B S Ramadan[1], H S Huboyo[1], N Ikhlas[3], A Karmilia[1]

[1]Department of Environmental Engineering, Diponegoro University, Semarang Indonesia 50275

[2]Department of Environmental Engineering, Institut Teknologi Pembangunan Surabaya, Surabaya Indonesia 60188

endrosutrisno@lecturer.undip.ac.id

Abstract. This study analyzes the distribution and amount of CO, CO_2, HC, and NOx gas emissions resulting from the open burning of backyard waste in Diponegoro University. The burned waste sample is taken from surrounding environment of Diponegoro University (Tembalang Campus) by taking 4 random samples (3 kg each). Emissions from CO, CO_2, HC, and NOx gases were obtained from the 24 minutes combustion test. Furthermore, the gaseous pollutant emitted is measured using a gas analyzer. The burned waste comprised 73.77% of organic waste, 17.45% of plastic; 4.33% of paper; and 4.45% of other waste. The emission test results show that the highest CO, CO_2, HC, and NOx emissions have occurred at 14, 20, 18, and 18 minutes, respectively. The combustion test reveals that an enormous amount of CO, CO_2, HC, and NOx gas is emitted during uncontrolled waste burning. Because backyard wastes burning produce significant gaseous pollutants, several efforts are needed to reduce this practice.

1. Introduction

Pollution due to emission outputs from human activities continuously causes environmental damage [1]. Problems related to environmental pollution are becoming more complex as time goes on. One of the problems that have high urgency in Indonesia is the waste sector. In Indonesia, waste management tends to be not entirely optimal at the landfill [2]. Even based on one of the data from similar research that has been done previously, the distribution of waste based on its management, Batu City, Indonesia still has a figure of 28.5% of waste burned openly. In 2021 the emissions generated from waste burning activities in Batu City have a quantity of CH_4 emissions of 0.032 Gg CH_4 or equivalent to 0.671 Gg CO_2-eq, N_2O emissions of 0.0007 Gg N_2O or 0.230 Gg CO_2-eq, and CO_2 emissions of 0.676 Gg CO_2, bringing the total emission to 1,575 Gg CO_2-eq. and there are still 60 tons/day of unmanaged waste [3].

Waste burning activities have the potential to produce CO_2, CH_4, and H_2O gas emissions. These gases are included as greenhouse gases that contribute significantly to climate change [4]. GHG emissions contribute significantly to the cause of global warming [5]. Environmental impacts that will occur include extreme weather and climate. Obradovich's (2018) and Ernyasih (2018) reveal that climate change and extreme weather are investigated further on human health. One of the direct impacts of extreme weather can cause acute respiratory infection (ARI). ARI cases from year to year continue to increase up to 11.2% [6,7]. Waste treatment by burning should be minimized or eliminated. Seeing more than 60 tons of unmanaged waste in Semarang City per day, which has excellent potential for this waste to end up burning in open spaces, triggering the danger of contributing greenhouse gas emissions from waste burning activities contributing to greenhouse gases that harm the environment, one of which is CH_4 gas which has a level of danger 21 times more dangerous than CO_2 [8]. This study will analyze the emissions from waste burning activities and plan emission reduction strategies in Diponegoro University as pilot study. This research were not estimating the number of waste piles that may be burned in Diponegoro University.

2. Methodology

The first method is waste sampling which started by visual survey through Diponegoro University, Tembalang campus. Then, around 3 kg of waste were taken in 4 random places in the campus. The determination of waste that will be burned was taken based on a short interview with the clean workers. The waste are mixed and then weighed according to its composition. The waste is put back together in an burning chamber and burned to determine the emission. Laboratory test studies were used to analyze the burned waste composition and potential emissions of nitrous oxide (NO_x), CO, carbon dioxide (CO_2), and hydrocarbons (HC). The collected waste are burned in the burning chamber within 24-30 minutes, or waited until the waste is completely burned. At the combustion time, the temperature is calculated using a thermogun and the concentration of oxygen and gas that comes out using a gas analyzer. After burning, 10 minutes of time are needed to release the high temperature and hot gaseous. After cooling, the fly ash and bottom ash are taken out until nothing is left in the incinerator. This burning test is repeated three (3) times to increase the accuracy of the data.

3. Results and discussion

Based on field observations, it is known that the condition of waste burned in Diponegoro University is still very mixed between wet and dry waste. The waste composition is in the form of plastic, twigs/leaves, cloth, paper, metal, rubber, and other compositions. The highest burned waste consisting of twigs/leaves/branches which has 73.77% from total waste that is burned in Diponegoro University, followed by plastic, paper and other waste about 17.45%; 4.33%; and 4.45%, respectively. The sample composition of the waste can be grouped based on the combustible and non-combustible fraction. The combustible waste fraction is in the form of samples with the composition of twigs/leaves, plastic, and cloth, for the non-combustible fraction in samples with other compositions, paper, and rubber. At the same time, the fraction of waste samples that cannot be burned is the metal waste. As it can be seen in Figure 1, around 300 ppm of CO and 12,000 ppm of CO_2 are released to the atmosphere during the combustion test. The NOx and HC also show that 120,000 ppm of NOx and 80 ppm of HC are emitted during uncontrolled waste burning which indicate a serious effect of uncontrolled waste burning. During the test, it is shown that the highest emission of CO, CO_2, HC, and NOx is occured after 12 minutes and slowly reduced after passing 20 minutes of burning. Since the waste as fuel is significantly reduced, the emission is also decreasing by time.

Regarding this result, the right strategy to reduce the emissions from the open burning of waste is essential. This strategy is needed to change the mindset and paradigm of the clean workers through an educational approach. The imposition of sanctions on people who burn waste openly and improvement of the waste management system through the provision of waste facilities is vital to reduce the waste burning on the campus [9]. Suppression of the amount of waste generated at the source through the simple 3R (Reuse, Reduce, Recycle) program involving clean workers to do sortation and composting may be another effective strategy to reduce waste burning. This waste reduction activity is expected to reduce emissions from the open burning process because the amount of waste can be reduced through proper backyard waste management [10, 11].

Permadi and Oanh reported that open burning significantly contributes to toxic emissions in two big cities (Jakarta and Bandung) in Indonesia compared to total anthropogenic emissions. However, the estimation may have high uncertainty due to the lack of activity inventory data for open waste burning. In Indonesia, household open burning is predicted to have a Pfrac ranging from 0.04-0.20. This number needs to be thoroughly explored [12]. Further, tree pruning and backyard waste dominated the composition of waste burned. According to Alves et al., there are many possibilities for the emission of gaseous and particulate compounds from combustion. Mixed waste of different branches and leaves can emit 1,499 g/kg of CO_2, 147 g/kg of CO, 1.34 g/kg of PM, and 0.25 g/kg of organic carbon [13]. Therefore, the existence of plastic waste in burned waste piles is concerning. Burning of the waste can result in the emission of a small amount of dioxins and brominated flame retardants, which are toxic and harmful to human health [14]. Open burning of plastic waste was also found to increase the concentration of dioxin-like compounds in the Surabaya and Palembang landfill sites [15]. Overall, waste burning was found to directly contribute to air pollution via many forms of harmful compounds.

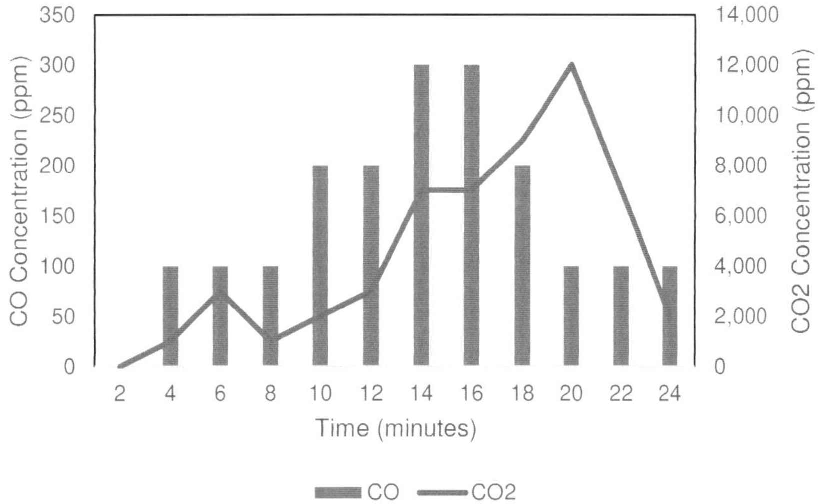

Figure 1. CO and CO_2 concentration of backyard waste burning.

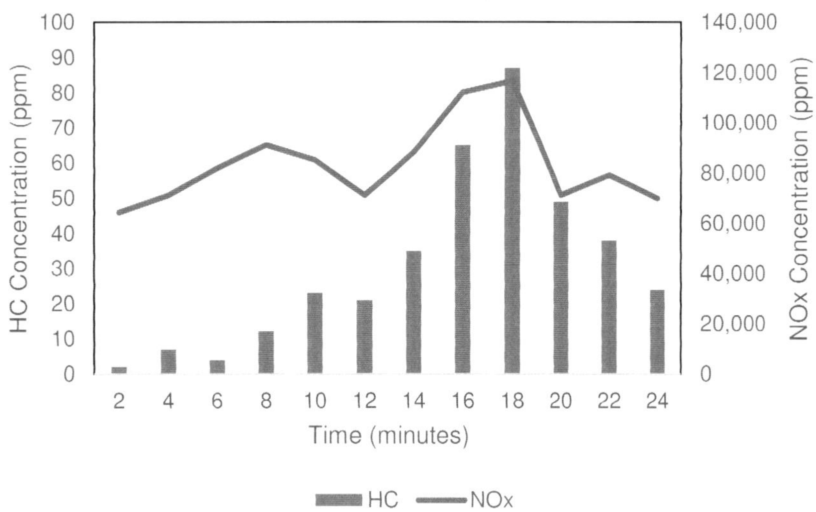

Figure 2. HC and NO_x concentration of backyard waste burning.

4. Conclusion

Research on the estimation emission of backyard waste burning in Diponegoro University was conducted by taking 4 random samples in Diponegoro University. After field observations, the composition of the waste burned is backyard waste which taking up to 70% of total waste that is burned and followed by plastic and paper waste. The burning simulation test reveal that there is significant emission from backyard waste burning. The highest gaseous emission is NOx gas and followed by CO_2, CO, and HC indicating incompleted combustion. Future research should estimate the number of piles that are burned in Diponegoro University and also Semarang City.

Acknowledgements
This research was financially supported by Faculty of Engineering, Universitas Diponegoro under Strategic Research program number 195/UN7.5.3.2/HK/2021.

References

[1] Budihardjo M A, Hadiwidodo M, Wardhana I W, Tuasykal M R N A, Samadikun B P, and Ramadan B S 2021 *IOP Conf Ser: Earth Environ Sci* **802** 012004
[2] Budihardjo M A and Wahyuningrum I F S 2018 *E3S Web Conf* **159** 01029
[3] Lestari J A 2017 *ITS Repository TESIS–RE142541*
[4] Krecl P, de Lima C H, Dal Bosco T C, Targino A C, Hashimoto E M, and Okawa G Y 2021 *Sci Total Environ* **765** 142736
[5] Pan L, Lin T, Xiao L, Zhao Y, and Cui S 2010 *Int J Sustain Dev World Ecol* **17**(4) 342–349
[6] Ernyasih E, Fajrini F, and Latifah N 2018 *Jurnal Ilmu Kesehatan Masyarakat* **7**(3) 167-173
[7] Obradovich N, Tingley D, and Rahwan I 2018 *Proceedings of the National Academy of Sciences of the United States of America (PNAS)*
[8] Syafrudin, Budihardjo M A, Yuliastuti N, and Ramadan B S 2021 *Evergreen* **8**(1) 23–35
[9] Tangwanichagapong S, Nitivattananon V, Mohanty B, and Visvanathan C 2017 *Int J Sustain High Educ* **18** (2) 203-217
[10] Budihardjo M A, Sutrisno E, Purwono, Fatimah M, *E3S Web Conf* **159** 01033
[11] Budihardjo M A, Hadiwidodo M, Huboyo H S, Aulia F R 2018 *E3S Web Conf* **73** 07017
[12] Permadi D A and Oanh D A K 2013 *Atmos Environ* **78** 250–258
[13] Alves C A, Vicente E D, Evtyugina M, Vicente A, and Pio C 2019 *Atmos Res* **226** 110–121
[14] Ni H, Lu S, Mo T, and Zeng H 2016 *Environ Pollut* **214** 70–76
[15] Ferronato N and Torretta V 2019 *Int J Environ Res Public Health* **16**(6)

Assessment of domestic waste management in Demak Regency, Indonesia

Syafrudin[1,2], B S Ramadan[1,2], W D Nugraha[1], G Samudro[1], R Ardiana[1]

[1]Environmental Sustainability Research Group, Department of Environmental Engineering, Diponegoro University, Semarang Indonesia 50275
[2]Department of Environmental Engineering, Diponegoro University, Semarang Indonesia 50275

syafrudin@lecturer.undip.ac.id

Abstract. Increasing the amount of waste generation is a problem in every city. This research aims to know the waste transportation and management system to analyze its compliance with the applicable national and local regulations in Indonesia. Demak Regency produced 208 kg/day of waste which this number represent 35.18% of waste generated in Demak Regency from domestic and non domestic sector. The waste is transported and not sorted at sources, as the same with many other cities in Indonesia. Therefore, a proper waste management system is needed for Demak Regency that includes five aspects, which include technical, institutional, financial, regulation, and community aspects. The Environmental Agency of Demak Regency is the institution in charge of waste management in Demak Regency. The analysis shows that improvements are needed to achieve appropriate waste management system in Demak Regency. Therefore, efforts should be made to improve waste management in the Regency, for example, by improving technical performance, human resources, and promoting community-based waste management in the studied area.

1. Introduction

Population growth has triggered increasing waste generation, as this condition produce a serious problem in many cities, including Demak Regency [1,2]. In 2019, Demak Regency had a population of 1,158,772 people which are divided into 14 districts [3]. Therefore, appropirate waste management system must be developed for reducing more complex problems that can impact the health quality, public welfare and pollute the environment [4]. The technical issues of waste management system are divided into several concern including: storage, collection, transfer, transportation, processing, and final processing techniques [5]. The service areas of waste transportation at regency-scale is determined according to the function parameters and regional values, population density, existing service areas, environmental conditions; income level of the population; and topography [6]. The level of service is based on the number of residents served, the area, and the waste amount lifted to the landfill. Waste transportation carries waste from the waste collection site or directly from the waste source to the landfill site. The pattern of waste transportation can be made by direct individual method (door to door) and communal collection from the waste collection site [7].

Urban waste in Demak Regency comes from domestic and non-domestic waste (shopping centers, markets, and schools). This condition managed should spur the improvement of the quality of waste management services in the Demak Regency. However, there are still many obstacles in implementing proper waste management system, including the lack of a waste collection site to serve all of the areas in the Demak Regency. Therefore, the waste community based solid waste management under waste bank helps the Environmental Agency of Demak Regency to minimize waste thrown into the landfill. Based on these backgrounds, this paper aims to assess the waste transportation and management system to be analyzed per the applicable regulations in Indonesia.

2. Methodology

This research was conducted by observing the existing condition of the Demak Regency solid waste management system. At the same time, the interview was done by asking the responsible officers about the waste management system in Demak Regency. Meanwhile, supporting literatures was obtained through analyzing the documents of the waste management division, Environmental Agency, Demak Regency.

3. Results and discussion

Demak Regency is divided into 14 districts where only a few districts are served by waste transportation system. The total area of Demak Regency is 897.43 km^2, while the area served by the waste collection site of the Sayung Area to the Kalikondang landfill is 17 km^2. The volume of Demak Regency's domestic waste is predicted as much as 208 m^3/day, while the volume of waste transported to the Kalikondang landfill is 73.173 m^3/day. Based on the transported volume, the service level of the Environmental Agency is only 35.18%. Based on these data, the level of waste service in Demak Regency is not fulfilled the national regulation which targeting 75% of waste service level.

According to State Stastical Beureau of Demak Regency data, the population of Demak Regency is 1,158,772 people (in 2019) so that the waste generation rate per capita can be calculated as follows:

Waste generation rate = (kg/day)/person
= (208×1000kg/day)/1,158,772 people
= 0.2 kg/person/day

The waste generation rate per capita in Demak Regency is below the assumption applied, which is 3.25 l/person/day. The solid waste generated depends on various factors, such as economic activity, habits, population, and level of commercial or industrial activity. Many cities in Indonesia posing an enormous amount of solid waste, even reaching more than double the population growth rate. This phenomenon is significantly associated with rapid urbanization. The waste problem in regency is different comparing to metropolitan cities in Indonesia (14). The waste generation is also lower than in metropolitan cities [1,7].

3.1 Household waste collection

Waste collection is carried out independently in each neighborhood where different capacities is found in each places [2]. This situation is evidenced by the waste generation at the Demak waste collection site as much as 24 m^3/day which equipped with 3 containers. Besides, at the SMAN 2 Demak waste collection site, waste collected is found as much as 6 m^3/day with 1 container. A hired-cleaning person who paid by the residents / community bring waste from the household to the nearest waste collection site using garbage carts. Sometimes, residents are also immediately dispose the waste to the nearest waste collection site without being sorted. In some conditions, especially where the household are near the landfill site, people using three-wheeled fleet and carts to bring and dispose the waste directly to the landfill site.

Figure 1 Three-wheeled fleet.

3.2 Transfer and transport of waste

The Environmental Agency of Demak Regency carries out waste service regarding the technical operations of transporting and managing waste from the waste collection site to the landfill. Meanwhile, waste management from the source to the waste transfer station is carried out by the neighborhood or the community itself. Communities collectively transfer their waste to the waste collection site by hiring cleaning person who usually use three-wheeled fleet or carts to transport their waste [2]. Arm-roll and dump trucks are used to transport the waste from waste collection sites to the landfill. Therefore, leaves and other waste from street cleaning are done by the Environmental Agency and transported from the waste collection site to the landfill [9]. The Environmental Agency has provided 25 waste collection sites spread across Demak Regency as communal containers. Each waste collection site consist of 1 to 3 containers. Therefore, there are 34 containers with a capacity of 8 m^3.

The total capacity of existing containers in the Demak Regency is 272 m^3. Thus, the need for additional containers can be calculated.

Waste collection container requirement

= (Number of production × number of population × number of days)/1000

= (0.2 L / person / day × 1,158,772 person × 1 day)/1000

= 238 m^3

Based on the calculations, 25 waste collection sites available in Demak Regency have only 272 m^3 of the total capacity. This situation means that the existing waste collection sites cannot accomodate the total waste generated in Demak Regency. Therefore, it is necessary to add 5 more containers with a capacity of 8 m^3. The pattern of waste transportation in the Demak Regency is a direct transportation system. Waste is collected at the nearest waste collection system and then large vehicles such as arm-roll trucks and dump trucks, will transport the waste directly to the Kalikondang landfill. Waste collection is carried out every Monday to Saturday. Waste collection site which has 2 containers occasionally wait the container is full before it is transported to the landfill. They are waiting the waste container full to reduce the cost of waste transportation. The figures 2 and 3 depict the transportation system used in Demak Regency.

There are 8 units arm-roll trucks (ART) available in Demak Regency which has good and proper condition. The dump trucks (DT) consists of 5 units which are also counted in good and proper condition. Containers or garbage bins that are ready to be transported all have lids, but some containers do not have covers, so they are required to cover them with nets for containers that are still open. In the operation of ART, the driver is assisted by 1 person to carry out the operation. While in the operation of the DT, the driver is assisted by 5 workers to carry out the operation. Each driver and crew is responsible for the operational vehicle for transporting the waste he carries.

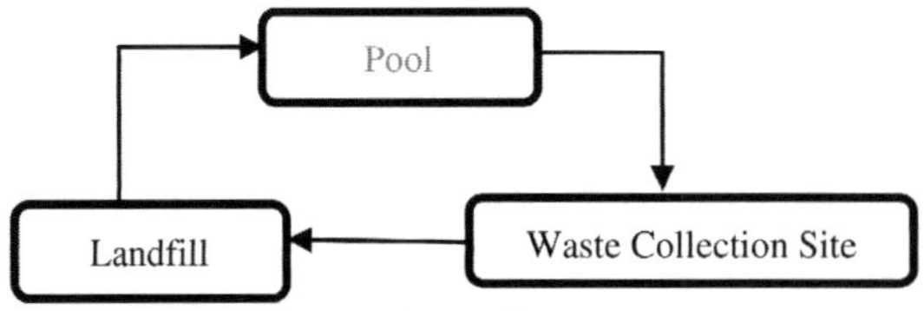

Figure 2. Arm-roll truck transport system

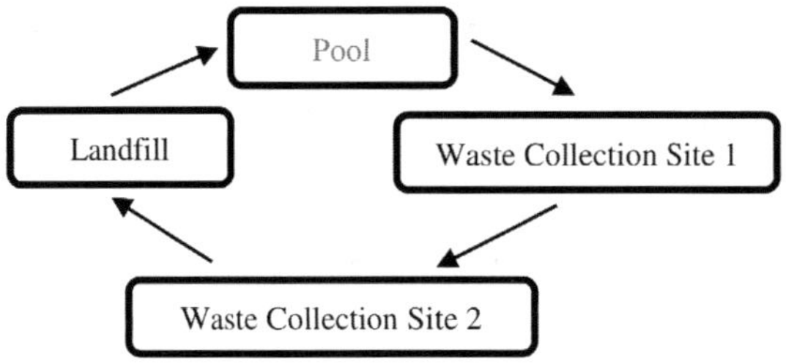

Figure 3. Dump truck transport system.

3.3 Waste transportation route

The waste transportation does not have any technical issues regarding to the route of the truck from waste collection site to the landfill. The road on each route is enough for each truck to pass through. This situation happened in all waste transport station in Demak Regency.

3.4 Retribution

The waste management system in Demak Regency funded by the cleaning services retribution. Another source of funding is coming from regional expenditure budget (*anggaran pendapat belanja daerah* – APBD). Retribution rates used in waste management and sanitation services in Demak Regency are based on Demak Regency Regional Regulation Number 4 of 2012 concerning Public Service Retribution. Based on the Demak Regency Regional Regulation, the households must pay the retribution of Rp. 500/month to 1500/month. This retribution are used as waste transportation from the waste collection site to landfill. Most residents do not want to pay dues because they do not put their waste to the waste collection site and they think that they have enough backyard to dispose their waste. Besides, people who are live in residential area should also pay for about Rp. 20,000/month for waste collection from their household to the household collection site.

3.5 Kalikondang and Candisari landfill

Waste management at the Kalikondang landfill is currently still operated using open dumping system, where the waste is only piled up and compacted without any daily cover. Kalikondang landfill has been operated from 07.30 – 14.00 WIB. While roads and gates to Kalikondang landfill are good, but the offices for staff are inadequate. Many gates around the waste pile are broken and gas pipes are no longer functioning. There is no weighbridge so the waste generation cannot be recorded. Waste is only an estimation of the total truck capacity. The leachate processing installation at the landfill has not operated for a long time, where the leachate is only collected without any further treatment. When it rains heavily, the leachate will overflow into the surrounding plantations or rice fields.

Kalikondang landfill will be closed in early 2021. Therefore, Demak Regency government has prepared a new landfill in Berahan Village, Wedung District, with a sanitary landfill system. The construction of this landfill has not been completed, so the temporary disposal is located in the Candisari landfill. Candisari landfill is also operated using open dumping system. Waste disposal in this landfill is not recorded, where the incoming trucks immediately disposes without recording. The infrastructure in this landfill has not been fulfilled where all the fences in landfills have not been made. There are no staff offices, leachate treatment installations, gas pipes, workshops. Landfill leachate often flows into the rice fields, causing the rice field owners to suffer losses because the leachate contaminates the rice fields.

3.6 Waste bank and waste treatment site facility

Demak Regency has 91 waste banks actively manage inorganic waste in their environment. However, several waste banks are not running well due to inappropriate management system. Demak Regency also have 9 waste treatment site, but none of the 9 temporary waste transfer stations is active. Based on the results of a field survey at the Bango waste treatment site, which is located in Bango Village, Demak District, this treatment site only operated for 1 month in 2018. This situation was due to a lack of human resources and inconvenient waste management system. Besides that, the surrounding community had low self-awareness where people immediately put their waste into the collection system. The community does not do any sorting first, so the officers have to sort it out before processing it. Organic waste in this treatment site is converted into compost, while inorganic plastic waste is collected and collectors take the waste out from the facility.

4. Conclusion

Environmental agency of Demak Regency has been managed 25 waste collection site where the service level is reach 35.18% of the total waste generated in Demak Regency which is below the targets of a minimum waste service level of 75%. Moreover, some waste trucks have a shortage of the fleets and containers due to the large amount of waste that does not meet the availability of minimum containers. The waste containers and trucks should be added. Because the service level is low, there are still many illegal dumping founds on the roadside. The operational of Kalikondang landfill is not following the local and national government regulation. However, waste management through waste bank practices in the Demak Regency area has been going well. Those waste banks is potentially used as community based waste management which can reduce the number of waste in Demak Regency.

Acknowledgements

Faculty of Engineering, Universitas Diponegoro grants this study under *Penelitian Strategis Fakultas Teknik Universitas Diponegoro* number 195/UN7.5.3.2/HK/2021.

References

[1] Arikan E, Simsit-Kalender Z T, and Vayvay Ö 2017 *J Clean Prod* **142 (1)** 403–412.
[2] Wibisono H, Firdausi F, and Kusuma E M 2020 *IOP Conf Ser: Earth Env Sci* **447** 012050
[3] State Statistical Beureau 2020 Demak District in number *State Stastical Beureau Central Java*
[4] Yang N, Zhang H, Shao L, Shao L M, Fan L, and He P J 2013 *J. Environ Manag* **129** 510–521
[5] Sudibyo H, Majid A I, Pradana Y S, Budhijanto W, Deendarlianto and Budiman A 2017 *Energy Procedia* **105** 263–269
[6] National Standards Agency. 2002. SNI 19-2452-2002 concerning *Technical Procedures for Urban Waste Operations*
[7] Pereira T D S and Fernandino G 2019 *Ocean & Coastal Management* **179**
[8] Regulation of the Minister of Public Works of the Republic of Indonesia Number 21/PRT/M/2006 concerning *National Policies and Strategies for the Development of Solid Waste Management Systems*
[9] Suryadarma D, Poesoro A, Akhmadi, Budiyati S, Rosfadhila M, and Suryahadi A 2010 *Food Policy* **35** 79–86

The kinetics of Fe^{2+} heavy metal adsorption by microalgae *Desmodesmus* sp. beads

D A Widyaningrum[1], A Rinanti[2], R Hadisoebroto[3*]

[1,2,3]Environmental Engineering Study Program, Faculty of Landscape Architecture and Environmental Technology, Universitas Trisakti, Jakarta, Indonesia

rositayanti@trisakti.ac.id

Abstract. Industrial waste that contains high concentration of iron heavy metal (Fe^{2+}) needs to be reduced into a safer limit for water ecosystem using an environmental-friendly, sustainable and low cost technology. This literature study was aimed to gather information on Fe^{2+} metal ion removal with biosorption method by mobilized beads-shaped *Desmodesmus* sp. green microalgae. Biosorbent beads are made of *Desmodesmus* sp. green microalgae powder binded with Na-alginate polymer. *Desmodesmus* sp. microalgae were cultivated in an *Erlenmeyer* flask with wastewater contains Fe^{2+} and *Beijerinck* growth media controlled in batch system. Analysis was conducted to study the influence of pH, size, beads, beads concentration, and Na-alginate concentration towards Fe^{2+} heavy metal removal efficiency on biosorption process. Research results revealed that Fe^{2+} heavy metal was effectively adsorbed by 1 mm beads up until 46% at maximum capacity on pH level of 6. These beads contain 8% gr/L of microalgae powder and 9% gr/L of Na-alginate. The involved function groups in Fe^{2+} sorption and the beads biosorbent morphology changes were analyzed by using FTIR and SEM. This biosorption research was following by analysis of Freundlich isotherm and reaction kinetics order two with linear approach. This was to shows the beads possess adsorption kinetics with optimum Fe^{2+} absorption capacity. *Desmodesmus* sp. beads are highly potential to be used as biosorbent since the heavy metal removal process was selectively conducted on wide pH level and temperature ranges, fast adsorption kinetics, and low operational cost. The smaller the beads size the wider the surface area, which leads to higher removal efficiency rate.

1. Introduction

The toxic and persistent heavy metal nature and its tendency to be accumulated in the environment has become the main cause of declined ecosystem quality due to heavy metal pollution. To reduce heavy metal ion concentration, further processing stage should be implemented before removing it to the environment. Currently, there are three popular waste water processing technology namely physical, chemical and biological waste water processing. Physic-chemical methods that commonly used to remove Fe^{2+} from the environment are adsorption process, reverse osmosis, oxidation process, membrane removal, ion changes, and chemical precipitation [1, 2, 3]. The advantage of this method is that the dose of the absorbent can be arranged to effectively and efficiently remove the metal compound at high/low pollutant concentration. However, this method has weaknesses due to its high energy consumption, high cost, low metal removal efficiency, not environmentally friendly, and can produce toxic side product [3, 4].

Another method that can be utilized to reduce polluting heavy metal compound is biotechnology method through biosorption, which is a method to remove heavy metal compounds by living or dead microorganism assistance through ion change process, complex formation development, and heavy metal absorbance. The best microorganism to play this role is microalgae, because its cellular wall contains active function groups that play important role in metal ion binding process [5]. Factors that influence biosorption process are acidity level (pH), waste concentration, temperature, biomass concentration in the solution, and contact time [3, 7, 8]. On a passive manner, heavy metal compound will be absorbed by biosorbent through physic-chemical binding on the surface of cellular walls, but in an actively manner it will occur through cellular metabolism [9].

Microalgae as autotroph microorganism has the potential to be utilized as biological material (biosorbent) because of its capacity to absorb non-organic compounds in biosorption process, which makes heavy metal solution that contains high level of affinity can be easily bound by it [3, 4, 10, 12]. Complex formulation of cation heavy metal will interact with anion active group at cellular cell [12], which place microalgae as the most potential compound to absorb and accumulate heavy metal in waste water [13, 14]. Researcher [4] discovered that Ca-alginate mobilized microalgae are effectively potential to absorb Cu^{2+} heavy metal ion from a solution in a batch system.

2. Methodology

2.1. Desmodesmus sp. biomass and growth media preparations

Secondary data in this research refer to literature [3]. *Desmodesmus* sp. green microalgae were isolated from Parkview Lake freshwater in Johannesburg, South Africa. 1 liter of Beijerinck growth media that contains stock 1 solution (1.5 gr of NH_4NO_3, 0.2 gr of KH_2PO_4, 0.2 gr of $MgSO_4.7H_2O$, and 0.2 gr of $CaCl_2.2H_2O$), stock 2 solution (9.07 gr KH_2PO_4), stock 3 solution (11.61 gr of KH_2PO_4), and micronutrients (10 gr of H_3BO_3, 5 gr of $MnCl_2.5H_2O$, 50 gr of EDTA, 1.5 gr of $CuSO_4.5H_2O$, 22 gr of $ZnSO_4$. H_2O, 1.5 gr of $CoCl_2.6H_2O$, 5 gr of $FeSO_4.7H_2O$, and 1 gr of $(NH_4).6Mo.7O_{24}.4H_2O$). Incubated *Desmodesmus* sp. in an Erlenmeyer flask with a batch system containing modified *Beijerinck* medium, streamed with CO_2 on a 50 mL/minutes rate, placed on room temperature of 25±2°C, completed with artificial lighting from TL lamp. Mixing process was conducted by utilizing Boeco, MS 300 (German) magnetic stirrer. *Desmodesmus* sp. cultivation was observed each day for 25 days by utilizing UV-VIS 4802 spectrophotometer with 680 nm wavelength.

2.2. Immobilized biosorbent with Na-alginate polymer

Desmodesmus sp. biomass was harvested on the exponential phase at the 18th day. 50 mL of *Desmodesmus* sp. biomass was separated from the solution through centrifugation on 4,000 rpm speed for 30 minutes. Growth media was poured and biomass was filtered to remove water excess. To create beads or immobilized biosorbent, 3% of Na-alginate powder was dissolved into aquades and stirred by using magnetic stirrer for one hour. After that, filtered 1.5 gr of biomass powder was mixed into 50 mL of Na-alginate, while stirring it on a magnetic stirrer to accelerate mixing process. A mixture of biomass and Na-alginate was inserted into a syringe and added with a drop technique into 0.2M 500 mL $CaCl_2$ solution, homogenized on magnetic stirrer to avoid beads pasting. The beads formed in the CaCl2 solution were soaked for 24 hours so that the formation of the beads took place completely. The beads were picked up with a filter, and rinsed for 3 times to remove remaining $CaCl_2$. The beads were stored in a duran schott glass bottle and covered by wet cotton on the lid to create humid environment.

2.3. Fe^{2+} heavy metal ion biosorption

Biosorption process of Fe^{2+} using beads was conducted in batch system on pH variations of 4, 5, 6; contact time of 15, 60, and 120; size of beads at 1, 2, 3, and 4 mm. Five grams of *Desmodesmus* sp. Beads were poured into 50 mL of Fe^{2+} solution with initial concentration of 120 mg/L on optimum level of pH. All research stages were conducted in 1L Erlenmeyer flask at room temperature. The Fe^{2+} concentration analysis prior and post sorption was measured on model PG-990 (Leicestershire, UK)

Atomic Absorption Spectrophotometry (AAS). Function groups analysis on biosorbent beads was conducting by utilizing Fourier Transform Infrared (FTIR, Bruker, Tensor 27, Germany) spectroscopy. Biosorbent beads surface morphology analysis was measured on a Scanning Electron Microscope (SEM, Hitachi S-4160).

2.4. Biosorption modeling

The maximum Fe^{2+} absorption by biosorbent beads was determined by adjusting adsorption data with Langmuir isotherm and Freundlich isotherm. Langmuir isotherm (equation 1) utilization is meant to produce linear regression coefficient value (R^2) by plotting Ce/qe value on Ce value [8] and implemented it on research data as shown below:

$$\frac{Ce}{qe} = \frac{1}{qm \,.\, L_L} + \frac{Ce}{K_L} \tag{1}$$

Ce: metal ion concentration at balanced state (mg/L); qe: number of metal ion absorb at balanced state (mg/g); qm: maximum absorption capacity (mg/g); K_L: Langmuir empirical constant

Freundlich isotherm (equation 2) [8] applied on research data is as below:

$$q_e = K_f \, C_e^{1/2} \tag{2}$$

qe : absorbed metal ion on biomass surface at balanced state (mg/g); Ce: metal ion concentration at balanced state (mg/L); Kf: Freundlich adsorption constant

The next stage is to determine Freundlich isotherm constant based on a graphic plotting between qe and Ce, and logarithmized it into:

$$\log q_e = \log K_f + \frac{1}{n} Ce \tag{3}$$

Difference in absorption time was implemented to determine the correct required time by biomass to bind metal ion [9]. Adsorption kinetics refer to reaction order one (equation 4) and reaction order two (equation 5) [9, 10, 11].

$$\ln (qe - qt) = \ln qe - K_1 \,.\, t \tag{4}$$

$$\frac{t}{qt} = \frac{1}{K_2 \,.\, q_e^2} + \frac{t}{qe} \tag{5}$$

qe: number of absorbed metal ion at balanced state (mg/g); qt: number of absorbed ion during t time (mg/g); K_1, K_2: reaction order one and two equation reaction constant rate; t : time (minutes).

3. Result and discussion

Microalgae growth pattern cultivated in a photobioreactor batch is influenced by nutrition, type of microalgae, initial cell population in culture media, and other environmental factors [6]. *Desmodesmus* sp. microalgae cultivation was conducted on 1L Erlenmeyer flask by temperature setting, CO_2 gas stream with air flow of 50 mL/minute, and artifical lighting made from TL lamp and magnetic stirrer utilization as mixing tool (Figure 1). Microalgae growth pattern wa started from adaptation phase happened on day 1 (Figure 2). During this phase, microalgae are still adapting to their new environment which lead to insignificant cell divisions. Exponential phase occured on the 10th to the 20th day, marked by rapid cell division, on a constant metabolism activity and balanced growth between microalgae cell increase and nutrition supply. Biomass permanent was conducted with centrifugation method conducted on the 18th day where significant biomass increase happened between exponential phase ranges. Fe^{2+} metal biosorption process is influenced by pH variable, contact time, and size of biosorbent beads (Figure 3 and Figure 4).

Figure 1. *Desmodesmus* sp. Microalgae Cultivation on 1L Erlenmeyer Flask [3].

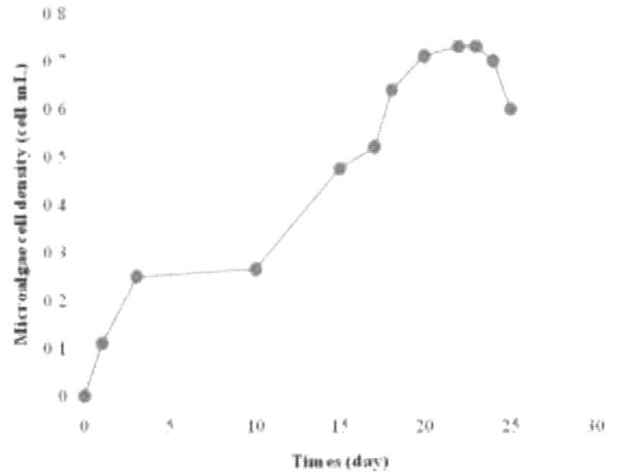

Figure 2. *Desmodesmus* sp. microalgae growth pattern on modified Beijerinck medium [3].

Figure 3 shows that on pH level of 2 and 6 Fe^{2+} removal efficiencies on each level reached 43% and 46%, before finally declined at pH level of 7 into 38%. On lower pH level, there can be a rivalry between hydrogen (H^-) and metal ion to attach themselves on biosorbent active side, due to function group protonation and positive charged biomass surface. However, as pH level increases, proton concentration will decrease and deprotonated that would lead to a more negative biomass [5]. The increased pH level happened along with metal ion binding on biosorbent surface [5, 15].

Figure 4 shows that Fe^{2+} metal ion absorption reached 48% by 1 mm beads in just 30 minutes, followed by 2 mm beads with 34% of Fe^{2+} metal absorption in 120 minutes. Researchers [14. 16. 17] explained that the smaller the size of the biosorbent, the larger the surface contact between biosorbent and heavy metal ion will be. Besides that, the surface area is also proportional to the number of pores contained on a biosorbent unit, which create higher level of heavy metal ion removal [15].

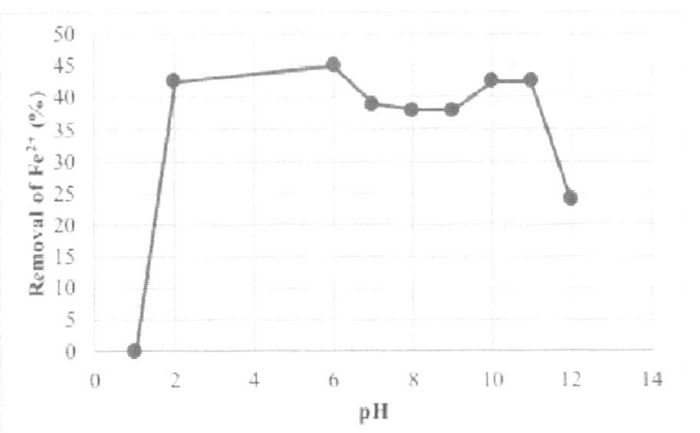

Figure 3. Fe^{2+} removal under variation of pH levels on biosorbent beads [3].

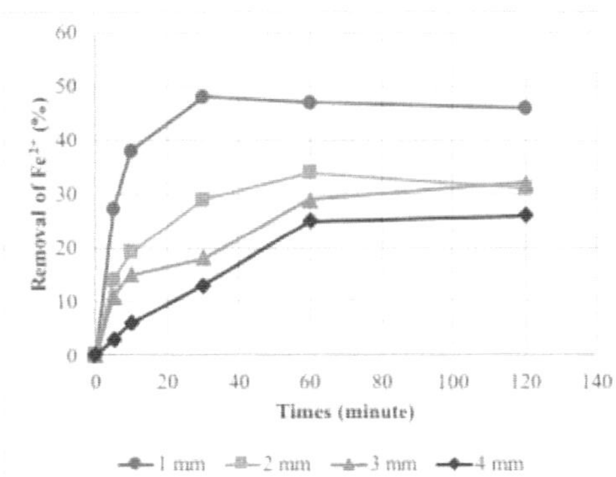

Figure 4. Fe^{2+} removal with variation of size of biosorbent beads [3].

FTIR analysis shows that the involved active function groups in adsorbing Fe^{2+} metal ion are carboxylate acid, hydroxyl, sulphonate acid, and alcohol. Biosorbent morphology structure chage after biosorption process were analyzed on SEM, which marked by the more visible pores and amorphous structure [14, 17]. Biosorption process has rapid reversible nature between heavy metal ion binding on cellular wall, both on living and dead cells. The basic principle of biosorption process with microalgae utilization is by substituting function group with heavy metal, so one of compounds contained in function group can be replaced by heavy metal ion [12, 17].

Based on the secondary data, we can create biosorption models based on Langmuir and Freundlich adsorption isotherm modelling and adsorption kinetics (reaction order one and two).

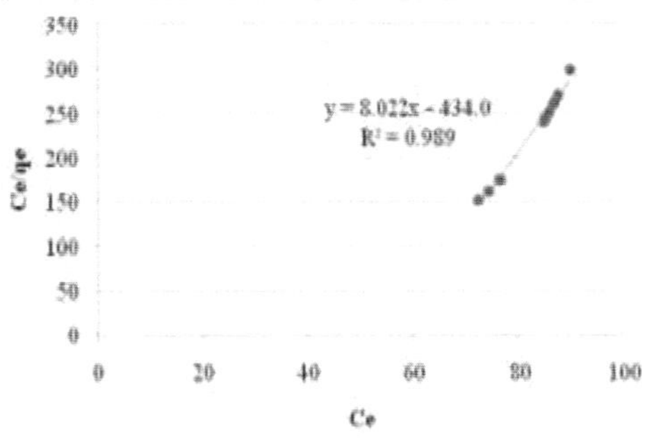

Figure 5. Langmuir isotherm curve.

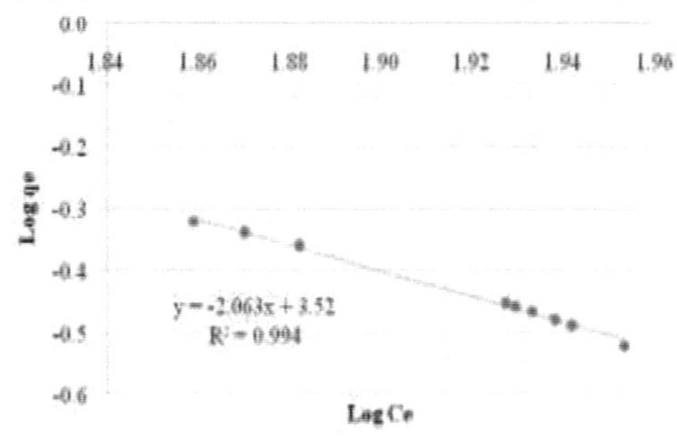

Figure 6. Freundlich isotherm curve.

Langmuir isotherm curve produce an equation of y = 8.0229x – 434.05, with linear coefficient value (R^2) of 0.9891 (Figure 5). Based on the equation, we were able to obtain the adsorption capacity value (qm) at 0.1246 and Langmuir empirical constant value (K_L) at -0.0185. On the other hand, Figure 6 shows Freundlich Isotherm curve with an equation of y = -2.0631x + 3.52, and linear regression coefficient value (R^2) of 0.9947. Based on the equation, we were able to obtain Freundlich constant factor value (n) at -0.4847 and Freundlich (Kf) Constant value of 3,311.31. Based on the findings, Fe^{2+} metal biosorption process by mobilized *Desmodesmus* sp. followed Freundlich Isotherm model, because the obtained R^2 value is closer to 1. This means that Freundlich Isotherm model is able to decently interpret the data, because the sorption mechanism occurred physically with multilayer formation and heterogeneous. Biosorption occurs in an effective way because microalgae possess active function groups supported by broad surface area, which enable Fe^{2+} metal ion to be optimally adsorbed [16, 17]. Adsorption kinetics is defined as heavy metal removal rate caused by heavy metal contact with microalgae biosorbent. This research was conducted with reaction order one and two which were plotted on a curve to produce linear regression coefficient value (R^2) (Table 1).

Table 1. Adsorption kinetics recapitulation.

Order	pH	R^2	qe (mg/g)	Reaction Rate Constant
One	4	0.0101	0.1912	0.0250 /minutes
	5	0.2157	0.4850	0.0130 / minutes
	6	0.0064	0.2246	0.0018 / minutes
Two	4	0.9486	0.1926	0.0861 /g.mg/ minutes
	5	0.9580	0.2720	0.1294 /g.mg/ minutes
	6	0.9880	0.4547	0.7924 /g.mg/ minutes

Based on Table 1, we can see the linear regression coefficient value (R^2) comparison results. Fe^{2+} metal concentration was decently adsorbed on reaction order two at pH level of 6, 120 minutes of contact time, with qe value of 0.4547 mg/g and reaction rate constant value of 0.7924/g.mg/minutes. This finding shows that adsorption kinetics reaction order two can interpret decent Fe^{2+} biosorption kinetics rate by biosorbent beads.

The implementation of biosorption process on pilot scale to process 75L of waste that contain Fe^{2+} metal by *Desmodesmus* sp., beads is shown on Table 2. The calculation is also completed with reaction design on Figure 7.

Table 2. Biosorption implementation recapitulation on pilot scale.

Description	Value	Units
Fe^{2+} Volume	75	L
Biosorbent beads required	7,500	gram
Biomass powder required	121	gram
Microalgae harvest required	605	L
Reactor volume required	1,500	L
Reactor height	2	m
Reactor diameter	1	m
Reactor design volume	1,571	L
Detention time	69.88	minutes

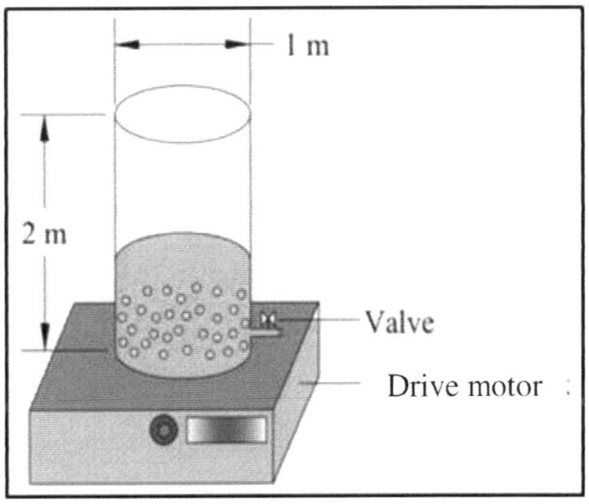

Figure 7. Reactor design.

Based on calculation results by utilizing adsorption capacity obtained from Freundlich isotherm value and reaction order two adsorption kinetics on pH level 6 condition, we can calculate the required biosorbent beads to process 75 L of $Fe2^{+}$ metal ion with initial metal concentration of 50 mg/L into final concentration of 7 mg/L, which is at 7,500 grams for 69.88 minutes.

4. Conclusion

Na-alginate mobilized *Desmodesmus* sp. microalgae were able to remove Fe^{2+} heavy metal by utilizing 1mm biosorbent beads in controlled environmental condition, with initial metal concentration at 120 mg/L, on 60 minutes of contact time. Biosorption process was marked by heavy metal binding into microalgae function group located on cellular wall surface. Based on the findings, biosorption process of Fe^{2+} metal by mobilized *Desmodesmus* sp. followed Freundlich Isotherm model and adsorption kinetics reaction order two can interpret decent Fe^{2+} biosorption kinetics rate by biosorbent beads. The implementation of biosorption process on pilot scale to process 75 L of waste that contains Fe^{2+} heavy metal design to be conducted in a 1,500 L reactor, requires 7,500 gram of biosorbent beads that for 69.88 minutes to remove similar level of Fe^{2+} heavy metal content at the same removal efficiency level.

References

[1] Siwi W P, Rinanti A, Silalahi M D S, Hadisoebroto R, Fachrul M F 2018 *IOP Conf Ser: Earth Environ Sci* **106(1)**

[2] Rinanti A, Jonathan D, Silalahi M D S, Fachrul M F, Hadisoebroto R 2018 *IOP Conf Ser: Earth and Environ Sci* **106(1)**
[3] Diale P P 2015 *The Biosorption of Heavy Metals by Immobilized Green Microalgae Desmodesmus sp. Disertasi* (Johannesburg: Faculty of Engineering and the Built Environment, University of the Witwatersrand)
[4] Rinanti A, Jonathan D, Silalahi M D S, Fachrul M F, Hadisoebroto R 2018 *IOP Conf Ser: Earth and Environ Sci* **106(1)**
[5] Shanab S, Essa A, Shalaby E 2012 *Plant Signal Behav* **7(3)** 392-399
[6] Ahalya N, Ramachandra T V and Kanamadi R D 2010 *Indian J Biotechnol* **7** 159-169
[7] Ahmad A, A H Bhat and Azizul B 2019 *Environ Technol* **40(14)** 1793-1809
[8] Tchobanoglous G, Franklin L, Stensel, H D 2003 *Wastewater Engineering Treatment and Reuse (Fourth Edition) Metcalf and Eddy* (New York: Mc Graw-Hill)
[9] Manzoor Q, R Nadeem, M Iqbal, R Saeed, T M Ansari 2013 *Bioresour Technol* **132** 446-452 (Elsevier)
[10] Xu J, Z Chao, Y Zhang, Z Lou, Z Xu, X Wang 2018 *Chemosphere* **195** 351-364 (Elsevier)
[11] Moreira V R, Y A R Leborn, S J Freire, L V S Santos, F Palladino, R S Jacob 2019 *Microchem J* **145** 119-129 (Elsevier)
[12] Rinanti A, Fachrul M F, Hadisoebroto R, Silalahi M 2017 *Int J GEOMATE* **13(36)** 6-10
[13] Goher, Mohamed E, et al 2016 *J Elementol* **21(3)** 703-714
[14] Ahmad A, Bhat A S, Buang A 2018 *J Clean Prod* **171** 1361-1375
[15] Saleh T A, Tuzen M, Sarı A 2019 *J Environ Chem Eng* **7(6)** 103436
[16] Siregar S H, Wijaya K, Kunarti E S, Syoufian A, Suyanta *IOP Conf Ser: Mater Sci Eng* **509** 012136
[17] Rinanti A, Fachrul M F, Hadisoebroto R, Silalahi M D S 2018 *Int J of GEOMATE* **15(52)** 45 - 52

Heavy metal removal and acid mine drainage neutralization with bioremediation approach

M S M Sihotang[1], A Rinanti[2*], M F Fachrul[3]

[1,2,3]Environmental Engineering Department, Faculty of Landscape Architecture and Environmental Technology, Universitas Trisakti, Jakarta, Indonesia

astririnanti@trisakti.ac.id

Abstract. Mining Industry can improve the national economic situation; however, it also can damage the environment, mainly because of its wastewater that contains heavy metal and acidic solid compounds. When exposed to free air, sulfide minerals can be naturally oxidized and create acid mine drainage (AMD), a highly acidic waste that can mobilize heavy metals towards the environment. This literature study will discuss practical and sustainable biological processing to remove AAT. Sulfate Reducing Bacteria (SRB) were isolated from AMD polluted soil and grown inside an AMD-containing batch reactor. The environmental conditions (temperature, AMD concentration, SRB concentration, and contact time) were controlled during this research. The implementation of pH sampling was conducted every day, and the heavy metal final result was measured with an Inductive Coupled Plasma Optical Spectrophotometry or ICP-OES. SRB produced Hbiogenic$_2$S that reacts with heavy metal and creates metal sulfide sediment. The remediation process by SRB will create biogenic alkalinity as an SRB side product that plays a significant role in neutralizing acidic water. Remediation is also influenced by organic compounds such as animal waste, rice, hay, or coconut husks. In this research, SRB plays a significant role as biosorbent that utilizes organic compounds as electron sources. The iron removal efficiency in AMD reached 96% and occurred on a contact time of 144 hours. To reach similar efficiency removal on a pilot scale, we planned AMD bioremediation on a tube-shaped reactor with 7.3m^3 with 3.5 m height and 0.88 of each reactor radiuses. This bioremediation study has provided an alternative solution for environmental management quality due to AAT pollution in water and groundwater around mining areas.

1. Introduction

The mining industry is one of the essential activities to improve the economic situation in a country. However, on the other hand, it can also disrupt the environmental balance, mainly because of its waste rich with heavy metal contents and its highly acidic nature known as acid mine drainage (AMD). Rain that pours down over mining areas will distribute heavy metals and other compounds and create reactivity in soil, which leads to erosion and leaching. Leaching would cause land fertility problems due to lowered pH levels in soil [1, 2].

Factors that influence AAT formation are acidity, alkalinity, morphology, pyrite formation, paleoenvironments, water and oxygen, and bacteria [3]. Minerals such as pyrite, chalcopyrite, pyrrhotite, and arsenopyrite can be naturally oxidized when exposed to air and create acidic waste that dissolves and mobilize toxic heavy metal compounds into the environment [4]. Based on that, mining activity that produces AMD rich in sulfate will lower pH levels due to high metal compounds [5], [6], [7], [8]. AMD

IOP Conf. Series: Earth and Environmental Science **894** (2021) 012041 doi:10.1088/1755-1315/894/1/012041

processing can be conducted with the chemical method through soil neutralization by utilizing chemical compounds such as lime ($CaCO_3$), hydrated lime ($Ca(OH)_2$), soda-ash (Na_2CO_3), or caustic soda (NaOH). Chemical processing can be accompanied by physical processes such as coagulation, flocculation, and sedimentation [5], [8]. AAT neutralization by utilizing alkali compounds such as lime, limestone, chemical compound, and industrial waste is easy, simple, and requires a relatively short time to remediate AMD pollutants in the environment. However, physical and chemical AMD processing has the potential to create a large amount of secondary mud that requires further processing [7], [9], [10].

Biological processing of AMD pollutants in the environment is a promising alternative. It has several advantages because it comes with a relatively low cost and is environmentally friendly [6], [8], [10]. The bioremediation method is considered environmentally friendly because it utilizes microbes naturally living in soil and water and does not cause any threat to other living things, including humans. Passive processing alternative by utilizing indigenous microbes metabolism on polluted location is generally conducted in situ or around it. BPS consortium is highly effective in optimizing the bioremediation process in AMD polluted environments [5], [10].

Control parameter that supports the growth rate of bacteria community is dissolved oxygen, temperature, pH level [12], Hydraulic Retention Time (HRT), and carbon sources [13], [14]. In AMD bioremediation process utilizes organic compounds as electron donors to support the BPS growth rate [13], [15]. Inside the reactor, environmental enrichment is implemented to accommodate bacteria needs to grow decently. If necessary, exogenous bacteria that are synergistic with indigenous bacteria can improve the insecticide degradation rate. This process is known as bioaugmentation [13], [14], [15].

2. Methodology

2.1. Acid Mine Drainage (AMD) preparation and Sulphate Reducing Bacteria (SRB) Cultivation

This research refers to [16] and [17] as primary literature sources. AMD was obtained from Sangão River, in Santa Carina State Carbon Basin (28°45'38.7"S 49°25'58.1"W). The samples were gathered on unsterilized polypropylene bottles with no headspace and stored at 4°C temperature. This research utilizes (three) carbon sources: rice husk, coconut husk, and pig farm wastewater treatment sludge. Organic carbon sources were washed on running water, dried, and cut into preferred sizes. Carbon sources were dried and ground into fine powder as organic substrate. 1 gram of organic substrate was mixed with 10 mL of nitrate acid (HNO_3) and heated at 120-140°C temperature.

2.2. Experiment

The reactor filled with a mixture of three carbon sources, as shown in Table 1, reached 3L and occurred at a constant temperature of 30°C.

Table 1. Ratio of every organic compound in the reactor.

Reactor	**Volume (L; % *v/v*)**			
Number	**PWTS**	**RH**	**CHC**	**Total Volume (L)**
1	1.0; 33.3	1.0; 33.3	1.0; 33.3	3
2	1.8; 60	0.6; 20	0.6; 20	3
3	0.6; 20	1.8; 60	0.6; 20	3
4	0.6; 20	0.6; 20	1.8; 60	3

PWTS = Pig Farm Wastewater Treatment Sludge; RH = Rice Husk; CHC = Coconut Husk Chips.

2.3. Sample Analysis

Heavy metal, pH, and sulfate concentration in the sample were gradually analyzed. The acidity rate was measured on a pH meter—bacteria growth analysis was conducted with the Total Plate Count (TPC) method by utilizing the spread plate technique. Sulfate removal as a BPS performance indicator is

IOP Conf. Series: Earth and Environmental Science **894** (2021) 012041 doi:10.1088/1755-1315/894/1/012041

measured on a spectrophotometer turbidimetry. Heavy metal compounds (Fe and Mn) were measured on a Perkin Elmer Inductive Coupled Plasma (ICP) optic spectrophotometry or ICP-OES.

3. Result and discussion

The influence of the carbon source factor on AMD bioremediation, as explained, is shown in Figure 1 [17]. Figure 1. shows significant pH improvement that almost reaches neutral with an average value of 7.31 from the initial pH of 4.16. The best result occurred in reactor 2 (Table 1), on observed 30°C temperature for 200 days. Carbon sources composition on reactor 3 shows the lowest pH level at only 6.82. Figure 1. b shows the highest Fe metal removal occurred on reactor 4 (Table 1) that reached a removal efficiency of 95% than any other reactors (Figure 1. c). The Lowest Mn removal occurred on reactors 1, 3, 4. Reactor 2 and 3 show the highest and the lowest removal rates. Reactor 2 shows the best Fe and Mn removal efficiency at more than 95% up to 99%.

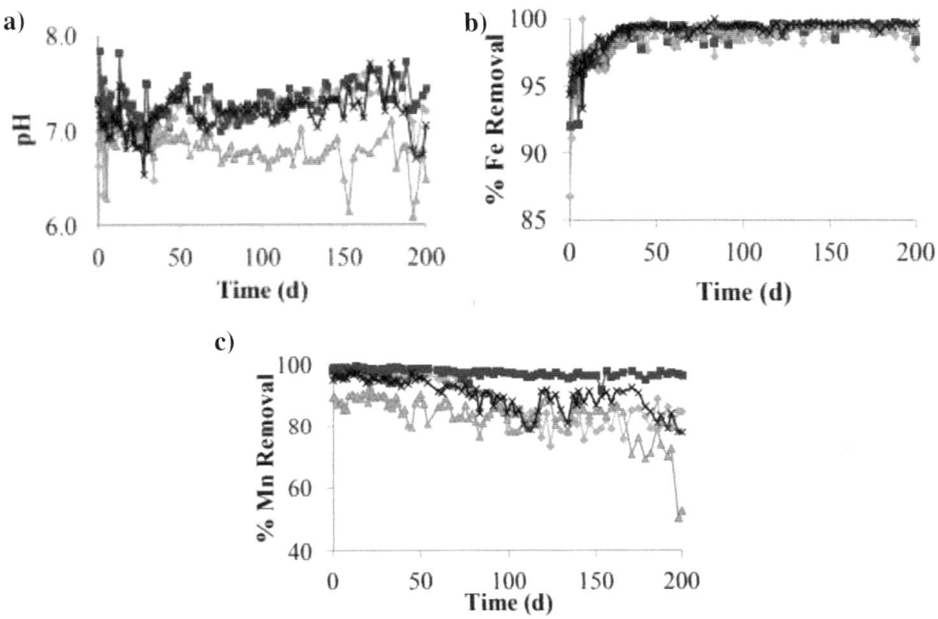

Figure 1. The influence of carbon source towards AMD bioremediation [17].

◆= Reactor 1, ■= Reactor 2, ▲= Reactor 3, ✖ = Reactor 4

According to [19], the biostimulation process with the addition of carbon sources such as sheep waste, compost, and hay through batch experiment was able to remove 98% of Fe heavy metal. Similar results were also shown in research [17] that removed 99% Fe heavy metal. In 2020, research by [19] on local beach sand as inoculum shows the highest level of bioremediation with pH level increased from 2.85 to 6.80 and Fe_2^+ removal efficiency higher than 82% by utilizing Sulphate Reducer Bio Reactors (SRBRs) on 30°C temperature in an anaerobic condition to stimulate BPS growth rate.

Compost can accumulate heavy metals and contains dissolved organic compounds as an important carbon source for BPS growth [17]. Organic husks such as coconut, rice, and hay, contain a lower rate of dissolved organic carbon than compost. Because of that, compost utilization as an organic carbon source produced a higher removal efficiency rate than organic husks made from plants. Organic carbon improves reduction process, removes Fe^{3+} and increase pH level up until it reaches neutral [1], [13], [15].

Figure 2 explains that BPS possesses a specific growth rate (μ) at around 0.014 to 0.083-hour jam^{-1}, with specific substrate utilization (q) rate of from 0.226 to 4.697 $hour^1$. The relationships between μ and q can be seen in Figure 3. The relationships between q and μ values result in a total growth value (Y_T) of 0.016 and Kd of 0.0123 on Fe metal of 64.46.

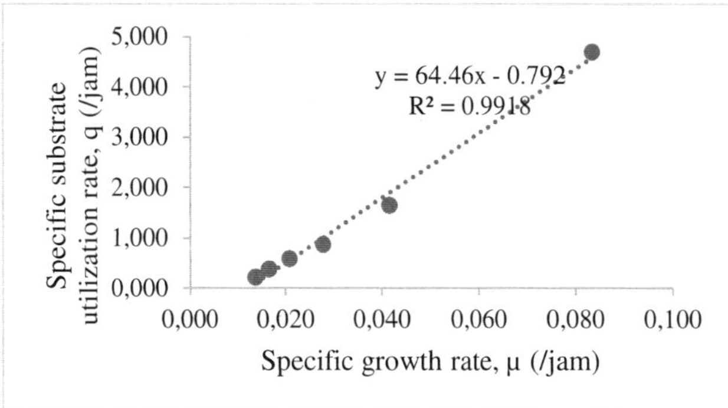

Figure 2. The relationships between specific growth rate (μ) and specific substrate utilization rate (q).

Table 2. Bacteria-specific growth rate (μ) and specific substrate utilization rate (q) measurement results.

Time (hour)	Removal %	Fe (mg/L)	Amount of Biomass (gr/L)	q (/hour)	μ (/hour)
0	0	100	0	0	0
24	62	38	1.1	4.697	0.083
48	75	25	1.9	1.645	0.042
72	88	12	2.8	0.873	0.028
96	90	10	3.2	0.586	0.021
120	94	6	4.1	0.382	0.017
144	96	4	5.9	0.226	0.014

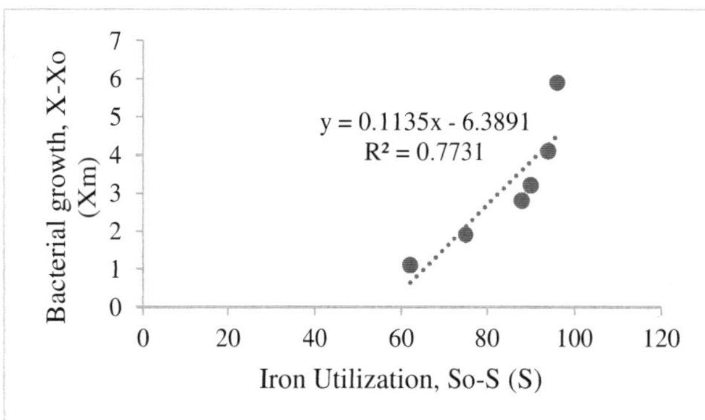

Figure 3. The relationships between substrate limitation and bacteria growth rate.

Table 2 shows that the highest iron metal removal occurred on 144 hours of contact time with a removal percentage of 96% with an initial dissolved iron amount of 100 mg/L into just 4 mg/L, by optimum effective bacteria growth at 5.9 gr/L. The lowest removal efficiency at 62% occurred on 24 hours of contact time, with an initial dissolved iron concentration of 100 mg/L into 38 mg/L. This condition occurred because 24 hours of contact time is not considered as optimum contact time. After all, the effective bacteria lag phase generally occurs on the first 24 hours of contact time [13], [15].

Figure 3 shows the relationship between bacteria growth numbers in limited culture (X_m) and specific substrate utilization (S). Relationship graphic between X_m and S will create a linear line with Yobs slope of 0.1135 hour^{-1}. The maximum substrate utilization rate value (q max) is at 4.697 with a maximum substrate utilization rate value (q max/2) of 2.348 (Figure 4), which leads to reaction order determination in this research. Based on the graphic between μ and substrate, we obtained Ks value of 25.05 (Figure 5). The kinetic measurement of dissolved iron concentration removal in AMD can be seen in Table 3.

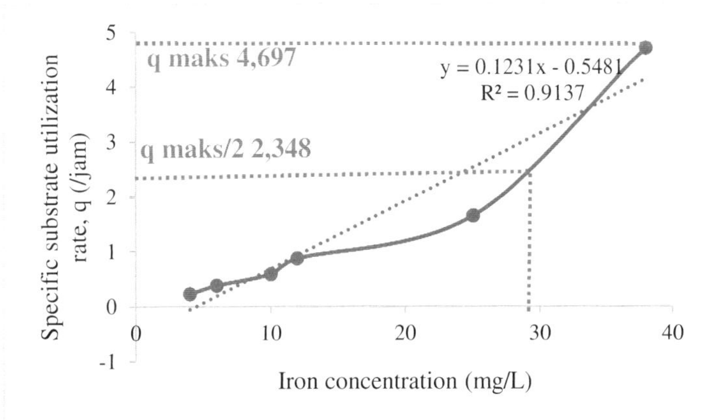

Figure 4. The relationships between substrate utilization rate and iron concentration.

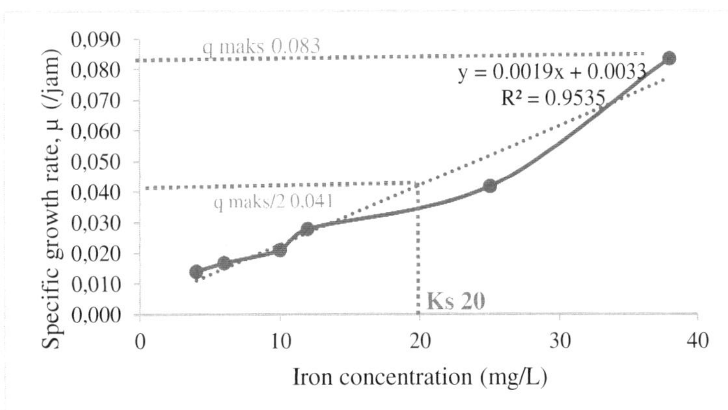

Figure 5. The relationships between specific growth rate (μ) and substrate.

Table 3. Kinetic calculation recapitulation of dissolved iron concentration removal in AAT.

Parameters	Calculation Results
Y_T (hour^{-1})	0.016
Kd (hour^{-1})	0.0123

Yobs ($hour^{-1}$)		0.1135	
Ks		25.05	
Reaction Order Equations			
	0	1	2
Slope (a)	0.0694	0.0052	0.0006
Intercept (b)	72.792	1.2805	0.0205
Regression (R^2)	0.646	0.8746	0.9931

Table 4. Pilot-scale implementation calculation.

Parameter Design	Units	Values
Reactor volume	7.3	m^3
Reactor height	3	m
Base diameter	1.76	m
Base radius	0.88	m
Reactor height	3.5	m

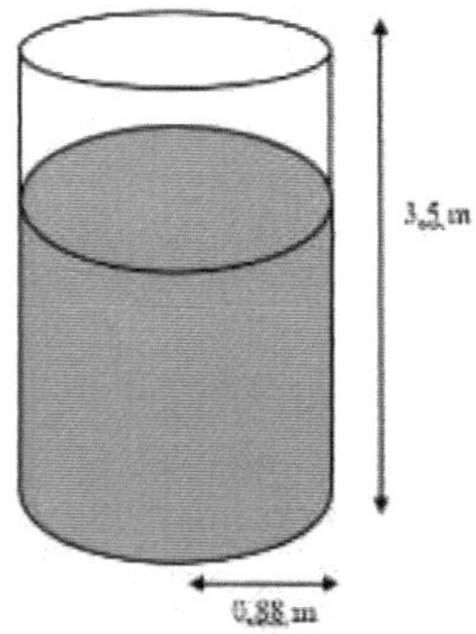

Figure 6. Tube-shaped pilot-scale reactor design.

Fe metal removal in AMD utilized two orders. The regression value closest to 1 was shown by an R^2 value of 0.9931 with the Y= 0.0006 x – 0.0205. AMD processing implementation calculation on the pilot scale can be seen in Table 4, and the reactor design can be seen in Figure 6.

4. Conclusion

Indigenous Desulfovibrio, Desulfotomaculum, and *Pseudomonas* bacteria in AMD are SRB that we can neutralize pH and remove heavy metals. BPS can survive on highly acidic conditions, with a temperature range of 25 to 35°C. The organic compound as electron source or electron donors such as animal waste, rice, hay, or coconut husks can assist BPS to reduce sulfate by neutralizing AMD pH levels up to 6-7. Fe metal kinetic removal on AMD follows reaction order 2 with the equation of Y= 0.0006 x – 0.0205 that resulted in an R^2 value of 0.9931. On pilot-scale implementation, a tube-shaped reactor unit with a volume of 7.3 m^3 with 3.5 m of height and 0.88 m radius can remove up to 96% of iron metal.

References

[1] Villegas-Plazas M, Janeth S, Howard J 2019 *Journal of Environmental Management* **251** 109581
[2] Hille R v, Neehal M, Tynan M, Nikhil N, Geoff M, Susan H, Ritva M 2016 *Mining Meets Water – Conflicts and Solutions.*
[3] Acharya B S and Gehendra K 2020 *Journal of Hydrology* 125061
[4] Bwapwa J K 2018 *Environmental Management and Sustainable Development* **7(1)**
[5] Acharya B S and Gehendra K 2020 *Journal of Hydrology* 125061

[6] Sheoran A S, V Sheoran R P, Choudhary 2010 *Minerals Engineering* **23(14)** 1073–1100
[7] Sinharoy A, Kannan P, Piet N L Lens 2020 *Current Pollution Reports*
[8] Sánchez-Andrea I, Jose L S, Martijn F M, Bijmans, Alfons J M, Stams 2014 *Journal of Hazardous Materials* **269** 98–109
[9] Yildiza M, Tulay Y, Cemile S A, Adem Y, Anna H K, Deniz U 2019 *Minerals Engineering* **133** 52–59
[10] Hussain A, Ali H, Arshad J, Javed I Q 2016 *3 Biotech* **6(2)**
[11] Ayangbenro A S, Oluwaseyi S O and Olubukola O B 2018 *Frontiers in Microbiology* **9**
[12] Laroche E, Corinne C, Lidia F-R, Angélique D, Vincent T, Odile B, Fabienne B-B, Catherine J and Marina H 2018 *Front. Microbiol.* **9** 3169
[13] Putri A N, R Ratnaningsih, A Rinanti *IOP Conference Series: Materials Science and Engineering* **1098 (5)** 052072
[14] Zhang M and Haixia W 2014 *Minerals Engineering* **69** 81–90
[15] Imaniar K R, T Tazkiaturrizki, A Rinanti *IOP Conference Series: Materials Science and Engineering* **1098 (5)** 052071
[16] Alvarado L C R 2016 *Optimization of the electron donor supply to sulphate reducing bioreactors treating inorganic wastewater* (Degree of Doctor in Environmental Technology. Erasmus Joint Doctorate Programme in Environmental Technology for Contaminated Solids, Soils and Sediments (ETeCoS3))
[17] Kijjanapanich P 2013 *Sulfate reduction for remediation of gypsiferous soils and solid wastes. Earth Sciences* (Université Paris-Est)
[18] Ruehl M D and Sage R H 2020 *Minerals Engineering* **157** 106554
[19] Oporto C, Baya G and Vandecasteele C 2019 *Environmental Technology* 1–33

IOP Conf. Series: Earth and Environmental Science **894** (2021) 012042 doi:10.1088/1755-1315/894/1/012042

Endosulfan insecticide removal planning with bioaugmentation-landfarming bioremediation method

A Pipit[1], Ratnaningsih[2], A Rinanti[3*]

[1,2,3]Environmental Engineering Study Program, Faculty of Landscape Architecture and Environmental Technology, Universitas Trisakti, Jakarta, Indonesia

astririnanti@trisakti.ac.id

Abstract. Endosulfan is a toxic organochlorine insecticide and is persistent in the environment. Endosulfan residue can be accumulated underground and lower soil quality, pollute water sources, and create bioaugmentation. This research aims to gather required information and study the potential of bacteria consortium consists of *Bordetella* sp., *Bordetella petrii,* and *Achromobactery xylosoxidans* to remediate endosulfan polluted soil. Bioremediation on laboratory scale conducted in a soil reactor, the pH level of 7, 20% humidity, and adjusted temperature to field temperature. Endosulfan was added into a reactor with a concentration of 2mg/g. The bacteria consortium utilized endosulfan as a nutrient source to decently grow up until this research was finished on the 30th day. Maximum removal occurred on upper layer soil with 99% of alpha-endosulfan and beta-endosulfan removal rates. Pilot-scale removal can be implemented with landfarming bioremediation. Two (2) processing beds were prepared with 15m of length, 7.5m of width, and 0.5m of height. This method was able to remove 99% of endosulfan in just 457.75 hours. This research can be implemented to remediate endosulfan polluted soil through the bioremediation method by utilizing bacteria consortium.

1. Introduction

The increasing population growth has driven the agricultural sector to be able to ensure food supply fulfillment. The success of the agricultural sector requires various supporting factors such as the availability of fertilizer, farming tools, and chemical materials such as insecticide. The insecticide is widely used as a pest-controlling agent to prevent crop damages [1, 2]. The organochlorine insecticide is an environmental contaminant because it is persistent and cannot be naturally degraded [3]. The insecticide is categorized as a manufactured xenobiotic compound that can pollute the environment and is difficult to degrade, accumulating and bringing negative impacts to the environment [4].

Endosulfan is a toxic insecticide but is globally used as a pest-controlling agent to increase agricultural productivity. The chemical composition of endosulfan is $C_9H_6Cl_6O_3S$ [5]. Technically, endosulfan is synthesized of two stereoisomers which are 70% alpha-endosulfan and 30% beta-endosulfan [6]. Chronic toxicity can be found in neural and behavioral disorders (neurotoxin) [7]. Environmental pollution due to excessive insecticide utilization has drawn a large number of attention [8]. Excessive insecticide utilization in agriculture has caused contaminations in the land, groundwater, and surface water [9]. Direct insecticide application through spraying activity can lead to direct exposure in the land. As a result, endosulfan biodegradation by bacteria, chemical hydrolysis, photolysis, and vaporization will occur, but endosulfan can spread through surface water flow [10]. Insecticide residue left on soil can be dangerous for the environment because it will affect land fertility and be highly toxic for living things [11, 12]. Insecticide can enter the food chain and will reach humans through biomagnification along the food chain [13].

There are several applicable technologies to solve insecticide exposure in the environment, one of which is through a physic-chemical method such as coconut shell active charcoal that can lower underground endosulfan residue concentration up to 80.36%-95% [14]. Besides that, researcher [15] has proven that photocatalytic degradation can remove up to 95% of endosulfan. However, this method has weaknesses with its high cost and inability to degrade the recalcitrance compound properly, so it needs additional treatment [16].

Besides the physic-chemical method, the insecticide pollution problem can be solved with the biotechnology approach through bioremediation. Bioremediation is one of the technologies that can be used to overcome environmental pollution through indigenous bacteria utilization. One way to conduct bioremediation is through bioaugmentation, which is implemented by adding bacteria into a polluted environment to degrade insecticide [17]. The advantages of bioremediation are that it comes at a low cost than any other technology. It can transform insecticide into CO_2, H_2O, and biomass. A more straightforward explanation can perfectly and permanently remove pollution and be safe for the environment [18, 19]. Both external and internal factors influence the effectiveness of bioremediation. External factors include physical, chemical, and environmental aspects such as temperature, pH, humidity, soil type, oxygen availability, and chemical structure. Internal factors consist of bacteria activity utilized in the bioremediation process. Bacteria will produce a suitable enzyme to degrade insecticide compounds as substrate [21]. Examples of bacteria able to degrade endosulfan in soil are *Bordetella* sp. [22], *Bordetella petrii* [23], and *Achromobacter xylosoxidans* [24, 25]. Several pieces of research on endosulfan biodegradation in liquid media only focused on biodegradation in culture media. There are only a few studies that research on how endosulfan degradation in various layers of soil. This research aims to analyze endosulfan insecticide remediation through the enzymatic way by bacteria activity on various soil layers in a controlled environment can be a basis to conduct endosulfan bioremediation on a pilot scale.

2. Methodology

Endosulfan removal data on soil layers act as secondary data. The primary literature utilized in this research refers to [26]. We will calculate the implementation of research results on a pilot scale through landfarming bioremediation based on the secondary data.

2.1. Soil media preparation

Endosulfan insecticide contaminated soil was gathered from some locations, namely insecticide producer locations in Ambatur, farming land in Thiruvallur District in Tamilnadu, and insecticide contaminated land in Kasargod District in Kerala. These soil samples were taken from different levels of depth. Soil samples were inserted into a reactor with a 15 x 10 x 40 cm dimension and then divided into upper layer soil, middle layer soil, and lower layer soil.

2.2. Endosulfan removal on soil media

2 mg/g of endosulfan concentration and bacteria consortium consisting of *Bordetella* sp., *Bordetella petrii,* and *Achromobacter xylosoxidans* were put into contact with soil. Biodegradation was observed on the pH level of 7, 20% humidity, and adjusted temperature with field temperature for 30 days and five days of sample taking intervals.

2.3. Endosulfan removal efficiency

This method is used to measure degraded insecticide levels. The removal calculation is as follows:

$$\text{Endosulfan Insecticide Removal (\%)} = \frac{Co-Ca}{Co}\, x\ 100 \qquad (1)$$

Co represents initial insecticide concentration; Ca represents final insecticide concentration

3. Result and discussion

3.1. Endosulfan bioremediation on soil layers

Endosulfan bioremediation on various soil layers with a concentration level of 2 mg/g was conducted on the pH level of 7, 20% of humidity, and adjusted temperature with field temperature. Bioremediation was conducted for 30 days [26]. Figure 1 shows endosulfan removal graphics on various layers of soil.

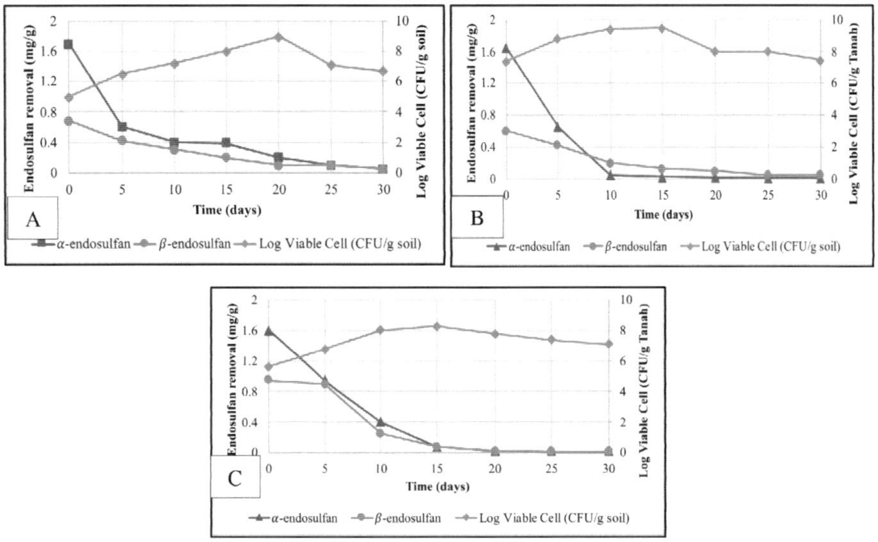

Figure 1. Endosulfan removal graphics on various layers of soil [26] (A) upper layer soil, (B) middle layer soil, (C) lower layer soil.

On the 30th day, up to 99% of alpha-endosulfan and beta-endosulfan removals occurred on lower layer soil. Endosulfan removal increased on deeper soil layer was caused by low organic carbon availability. On a lower layer of soil, endosulfan provides more organic carbon than those available in the soil. This situation drives the bacteria consortium to utilize endosulfan as its nutrient source and accelerate degradation [26]. Endosulfan sulfate and endosulfan diol are the most common metabolites in endosulfan biodegradation process [27]. Endosulfan diol can be mineralized into a more degradable compound and will eventually become CO_2 [28].

3.2. Bacteria consortium growth rate in soil that contains endosulfan

The specific substrate (q) utilization rate on the lower soil layer is at 0.0003 to 0.0009 hour^{-1}. Meanwhile, the specific growth (μ) rate is at 0.00030 to 0.00154 hour^{-1}. Based on Figure 2, the relationships between the values of q and μ will produce a total growth rate value (Y_T) of 2.70 hour^{-1} with a bacteria death constant value (Kd) of 0.000541 hour^{-1}. The maximum substrate utilization rate calculation and the relationship between specific growth rate and alpha-endosulfan substrate concentration are shown in Figure 3. The maximum substrate utilization rate value (qmax) is at 0.00154, and with $(\frac{qmaks}{2})$ value of 0.00077, the saturated constant value (Ks) is at 0.02 hour^{-1}.

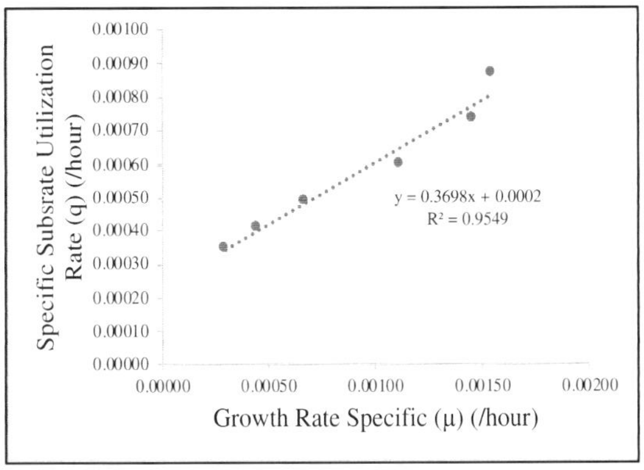

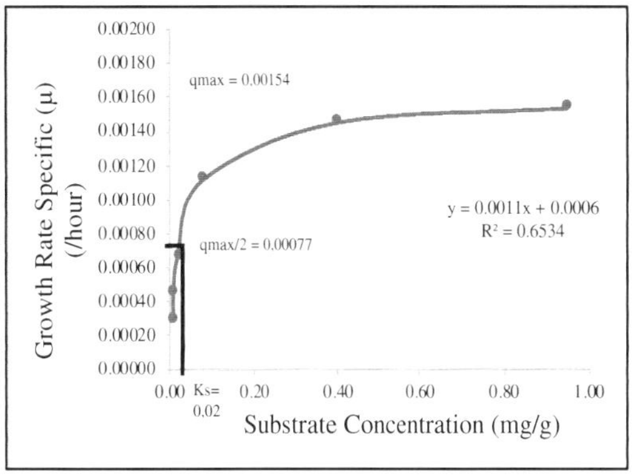

Figure 2. The relationships of specific growth rate and specific substrate utilization rate.

Figure 3. The relationships of specific growth rate (μ) and endosulfan substrate concentration.

Y_{obs} value of 0.7067 $hour^{-1}$ was obtained from the bacteria growth relationship graphic by utilizing endosulfan, as shown in Figure 4. Reaction order determination was conducted by creating a growth rate graphic. The reaction order selected based on the closest R^2 value is order reaction 1, as shown in Figure 5. Based on that, the bacteria consortium growth rate on the lower soil layer that contains alpha-endosulfan occurred on order 1 with an R^2 value of 0.9578 and a growth rate constant of 0.0013 $hour^{-1}$.

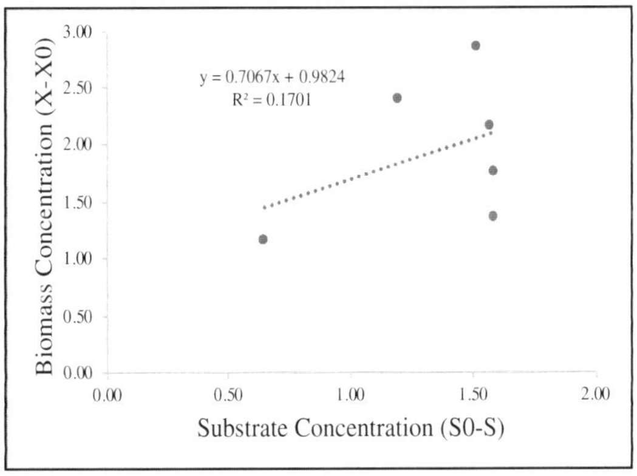

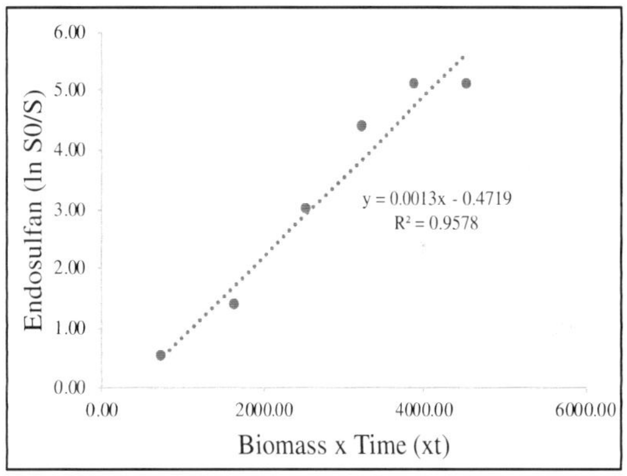

Figure 4. The relationships of bacteria growth and endosulfan utilization

Figure 5. The growth rate in order 1

3.3. Pilot Scale Endosulfan Removal Simulation

The research on this literature study is an initial stage of endosulfan bioremediation by bacteria consortium conducted on the laboratory scale, which means that implementation on a larger scale by utilizing landfarming bioremediation should be available. Based on the literature study results, the bacteria consortium is proven to remove 99% of 2 mg/g endosulfan in 30 days or 720 hours. Based on these results, if the initial endosulfan concentration is at 2 mg/g pollutes 100 m^3 of clay with a density level of 1700 kg/m^3, we can calculate that the weight of the soil is at 170,000 kg. Contaminated soil processing is planned to be conducted in 2 processing beds.

The bed is planned to be rectangle-shaped with a length-to-width ratio of 2: 1. One bed unitis expected to cover 50% of 100 m^3 soil volume or 50 m^3. With assumption that the soil height is at 0.5m, and the volume is at 0.5 m, it means that 50 m^3 = length x width x height = 2L x L x 0.5 m = $2L^2$ x 0.5 m. Based on the calculation, we obtained a result of L^2 = 50, which means that L = 7.071 m rounded into 7.5 m, the length is at P = 2 x 7.5 m = 15 m. The volume becomes: P x L x T = 15 m x 7.5 m x 0.5 m = 56.25 m^3. Processing bed design can be seen in Figure 6, with the following calculations:

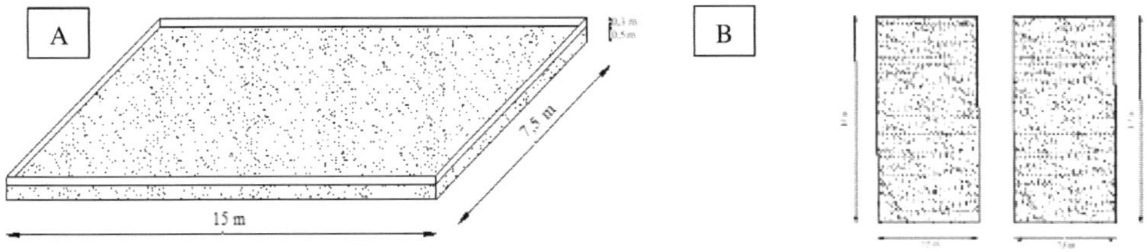

Figure 6. Processing bed design. (A) front view (B) top view.

On the landfarming method with a volume of 56.25 m^3, we can process 95,625 kg of endosulfan polluted soil. In this design, on processing bed unit will be processing 85,000 kg of soil. Based on that, two beds should process 170,000 kg of soil so all polluted soil amounts can be processed.

This literature study informs that 7650 grams of polluted soil with 2mg/g endosulfan can be degraded by 1500 ml of bacteria consortium in 30 days of 720 hours with a maximum level of removal at 99%. If this result is considered at pilot scale, and similar condition should occur, soil polluted by 2 mg/g of endosulfan concentration with soil weight of 85,000 kg in one bed, endosulfan degrading bacteria consortium required should be at:

$$\frac{\text{amount of soil (gr)(research)}}{\text{amount of bacteria consortium (ml)(research)}} = \frac{\text{amount of soil (gr)}}{\text{amount of bacteria consortium (ml)}} \qquad (2)$$

$$\frac{7650 \text{ gr}}{1500 \text{ ml}} = \frac{85{,}000 \text{ x } 10^3 \text{ gr}}{\text{x}}$$

Based on the calculation above, the required bacteria consortium is at 16,666.667 L/bed for each bed. If the two beds, bacteria consortium requirement is at 33,333.333 L. Based on this calculation result, the formulation of order 1 reaction rates Y = 0,0013 x – 0,4719 and growth curve formula on exponential phase is at Y = 0,0071 x + 6,08. To acknowledge the time (t) factor on order 1 in removing endosulfan, we can utilize the following calculation:

$$t = \frac{\ln\frac{So}{S}+0.4719}{0.0013} = 769.23 \text{ x } \ln\frac{So}{S} + 368$$

The maximum detention time required by bacteria consortium to degrade endosulfan is as follows:

$$T = \frac{769.23 \text{ x } \ln\frac{1.6}{0.01} + 368}{0.0071\, t + 6.08}$$

$$0.0071\, t^2 + 6.08\, t - 4271.84 = 0$$

$$t = \frac{-b \pm \sqrt{b^2-4ac}}{2a} = \frac{-6.08+\sqrt{6.08^2-4 \text{ x } 0.0071 \text{ x } (-4271.84)}}{2 \text{ x } 0.0071} = 457.75 \text{ hours} = 20 \text{ days}$$

Bacteria consortium requires 457,75 hours or 20 days to degrade 99% of endosulfan.

4. Conclusion

Based on this research, it is proven that *Bordetella* sp., *Bordetella petrii*, and *Achromobacter xylosoxidans* bacteria consortium can survive on 2 mg/g endosulfan polluted soil by utilizing the pollutant as a nutrient source. On a laboratory scale, bacteria consortium requires 30 days to reach up to

99% removal efficiency. The laboratory-scale research result is recommended to be utilized as a basis to conduct pilot-scale endosulfan removal. To obtain exact endosulfan removal rates using two processing beds, we required 33,333 L of bacteria consortium and 20 days of detention time.

References

[1] Anggreini C D, Tazkiaturrizki T, Rinanti A 2019 *J Phys Conf Ser* **1402(3)** 033004
[2] Huete-Soto A, Castillo-Gonzalez H, Masis-Mora M, Chin-Pampillo J S and Rodriguez-Rodriguez C E 2017 *J Hazard Mater* **321** 1 – 8
[3] Palanimanickam A and Sepperumal U 2017 *J Pure Appl Microbiol* **11** 221 – 227
[4] Akhdiya A, Wartono, Sulaeman E and Samudra I M 2018 *J AgroBiogen* **14** 37 – 46
[5] Kaur A 2012 *Studies on biodegradation of endosulfan by bacterial species from agriculturan soils* (India: Departemen of Microbiology Punjab Agricultural University)
[6] Kafilzadeh F, Ebrahimnezhad M and Tahery Y 2015 *Osong Public Health Res Perspect* **6(1)** 39 – 46
[7] Alen Y, Adriyani F, Suharti N, Nakajima S and Djamaan A 2016 *Der Pharmacia Lettre* **8(8)** 137 - 141
[8] Kumar M and Philip L 2017 *Microbe-Induced Degradation of Pesticides* ESE (Springer, Cham) p 59-81
[9] Sunaryo T, Widyatmoko H and Rinanti A 2018 *MATEC Web of Conferences* **197(13012)**
[10] Cycon M, Zmijowska A, Wojcik M and Seget Z P 2013 *J Environ Manag* **117** 7 – 16
[11] Ortiz-Hernandez M L, Sanchez-Salinas E, Dantan-Gonzales E and Castrejon-Godinez 2013 *Biodegradation - Life of Science* **10** 1 - 39
[12] Parte S G, Mohekar A D and Kharat A S 2017 *Afr J Microbiol* **11(24)** 992 – 1012
[13] Barba S, Villasenōr J, Rodrigo M A and Cañizares P 2018 *J Appl Electrochem* **48** 579-588
[14] Eunike K, Rinanti A, Ratnaningsih R 2018 *MATEC Web of Conferences* **197** 13013
[15] Mudhoo A 2018 *Chem Eng J* pp 1-67
[16] Ibrahim K E A and Solpan D 2018 *J Environ Chem Eng* **7(1)** p 102820
[17] Vernans A K R, Iswanto B, Rinanti A 2019 *J Phys Conf Ser* **1402(2)** 022102
[18] Megharaj M, Venkateswarlu R and Naidu 2014 *Bioremediation* pp 197 – 298
[19] Sunaryo T, Widyatmoko H, Rinanti A 2018 *MATEC Web of Conferences* **197** 13012
[20] Morillo E, Madrid F, Moreno A L and Villaverde J 2020 *Int J Pharm* p 119943
[21] Valentina C, Hadisoebroto R, Rinanti A 2018 *MATEC Web of Conferences* **197** 13016
[22] Goswami S and Singh D K 2009 *Biodegradation* **20(2)** 199-207
[23] Kong L et al 2018 *Ecotoxicol Environ Saf* **160** 75-83
[24] Li W et al 2009 *J Hazard Mater* **167(1-3)** 209-216
[25] Singh N S and Singh D K 2011 *Biodegradation* **22(5)** 845-857
[26] Odukkathil G 2014 *Bioremediation of endosulfan and other pesticide residues in agricultural soil* (India: Faculty of Science and Humanities Anna University)
[27] Giri K and Rai J P N 2012 *Int J Environ Stud* **69(5)** 729-742
[28] Sarat N and Barathi S 2013 *Int J ChemTech Res* **5(1)** 6-14

IOP Conf. Series: Earth and Environmental Science **894** (2021) 012043 doi:10.1088/1755-1315/894/1/012043

Desiccation cracks behaviour of leachate in bentonite - zeolite composite liner

I W Wardhana[1], M A Budihardjo[1,2], T Istirokhatun[1], N Ikhlas[3]

[1]Department of Environmental Engineering, Diponegoro University, Semarang, Indonesia 50275
[2]Environmental Sustainability Research Group, Department of Environmental Engineering, Diponegoro University, Semarang, Indonesia 50275
[3]Department of Environmental Engineering, Institut Teknologi Pembangunan Surabaya, Surabaya, Indonesia 60186

irawanwisnuwardhana@lecturer.undip.ac.id

Abstract. This study tested desiccation crack potential and migration of metals through pure bentonite and bentonite-zeolite composites to determine the best composition of the composite liner. Pure bentonite (B) and bentonite-zeolite composites of 2% (B2), 5% (B5), and 8% (B8) were used as controlled variables. The results showed that the addition of zeolite could not minimize the desiccation behavior in the liner. The value of crack intensity factor (CIF) of each sample B, B2, B5, and B8 was 3.44%, 3.51%, 3.58%, and 3.64%, respectively, indicating a moisture content of 29.95%, 34.54%, 30.88%, and 28.21%, respectively.

1. Introduction

The percolation of rainwater and a small amount of anaerobic waste in landfills causes the formation of leachate, which has a high concentration of metal, organic materials, minerals, and microorganisms. Landfill leachate can cause environmental contamination through infiltration/migration mechanisms of the soil and groundwater. Preventing leachate migration to the environment is the focus of many engineers [1,2]. A landfill liner system prevents air, soil, and water contamination caused by leachate [3]. This layer consists of materials with low permeability that can avoid infiltration, gas leakage, and movement [4]. Many innovations regarding this liner ensure a reduction of the advective and diffusive transport of contaminants to the environment. The landfill liner acts as a barrier designed to retain the migration of the contaminants through several types of processes, including physical and chemical adsorption, ion exchange, precipitation, mechanical filtration, and biological degradation. Per leachate retention, landfill liners must have less than 1×10^{-7} m/s of permeability value [5].

Natural clays (often bentonite), geomembrane, and mixtures of both (Geosynthetic Clay Liner/GCL) are often used as liners because their low hydraulic conductivity is less than 10^{-9} m/s [6]. GCL performs well compared to compacted clay liner (CCL), especially during the installation and hydraulic conductivity. However, GCL is limited by the possibility of leakage during improper installation, shear strength after the hydration phase, and the potential for loss of bentonite during the placement process, especially bentonite powder [7]. The use of CCL is preferable in developing countries because the installation costs are cheaper. However, clay material is very susceptible to desiccation crack, so finding the most effective and appropriate combination barriers [8]. Bentonite composite-zeolite/bentonite layer-embedded zeolite (BEZ) is recommended as a landfill liner because of its ability to absorb and prevent metals migration. BEZ also has lower hydraulic conductivity and volumetric shrinkage values than other materials [9].

Desiccation crack phenomena are essential parameters that must be understood whenever using clay material as a landfill liner. Rayhani et al. tested cracks in natural and artificial clay. They found that cracks and hydraulic conductivity of clays depend on the physic-mechanical properties of soil, especially the plasticity and swelling capacity. Cracks can cause an increase in the hydraulic conductivity value [10]. Desiccation cracks occur in three stages: the initial phase, the primary phase, and the steady phase. At the initial phase, cracks begin to appear gradually as the water content decreases. At the primary phase, the water content reaches critical values, and the cracks develop rapidly. At the steady phase, the water content approaches the soil shrinkage limit. The cracks develop slowly and near to a fixed condition [11]. The effectiveness of bentonite embedded zeolites (BEZ) compared to other materials, like sand [12] and polyacrylamide showed that BEZ has potential for use as a landfill liner. This study used the behavior of desiccation cracks to determine the effectiveness of BEZ material as landfill leachate retainers. Measurement of cracking was done by looking at crack behavior sequentially, which was not accommodated in previous studies by Kaya and Durukan. Bentonite and zeolite composite materials are expected to be a solution to minimize the desiccation cracks potential of the barrier layer.

2. Methodology

The nature of bentonite clay and zeolite used in this study is the same as in another study published elsewhere [5]. The study was conducted by varying the composition of zeolites, including pure bentonite (B) and bentonite-zeolite composites with zeolite content of 2% (B2), 5% (B5), and 8% (B8). A 1:1 ratio of bentonite to zeolite powder produced the bentonite embedded zeolite. The mixed powder was then weighed, to calculate the water content percentage. Each composite's ability to absorb water was tested for 26 days using pure water, natural, and artificial leachate. The water absorbance test was conducted to see the response of each composite to crack and water loss.

The development of the crack was recorded and documented daily. On the last day, kerosene was injected between the cracks, and the volume of kerosene that seeped was calculated. The documentation results were made black and white to be analyzed using MATLAB so that the percentage and width of the cracked area of the artificial liner were known. Temperature and air humidity were also calculated to determine the relationship between the desiccation crack and these two parameters. The area of composite cracks was processed with the help of MATLAB software; after air-drying for 26 days and oven-drying for 24 hours, cracks occurred in the composites. MATLAB can calculate the area by identifying colors, making it easier for researchers to calculate the cracked area. In this experiment, the cracked sample area was marked in black, and the uncracked sample area was marked in white. The crack intensity factor (CIF) is used as a parameter to evaluate cracks of bentonite dryness. CIF is defined as follows:

$$CIF = Ac / At \times 100\% \quad (1)$$

While Ac is the drying crack area and At is the total surface area.

3. Results and discussion

Cracks start from the surface to the lowest layer of the soil. At higher temperatures and relatively low humidity, drying–cracks begin faster in the surface layer. Only a small amount of water in the lower layers can be transported to the surface layer before cracks occur because of the generally low permeability of bentonite. This situation will cause a higher water content. If the clay stress decreases due to temperature rise and moisture decrease, cracks occur at the absorption of the higher water content. This result shows that a reduction of water content is also related to an increase in temperature [13].

Cracks in the liner layer are a significant concern in the performance of a landfill liner system. The liner layer consists of materials with low permeability to minimize infiltration and gas displacement. Liner layers are susceptible to cracks during the change between the dry and the rainy seasons over time.

The presence of cracks is a severe problem for long-term landfill performance [4]. Cracks usually occur in different widths that are problematic to analyze in an abbreviated time frame. Therefore, the crack intensity factor (CIF) is used to facilitate the measurement of cracks. CIF is calculated based on the crack surface area (Ac) ratio to the total dry land surface area (At). The measurement of the crack surface area is based on the width and length of the crack using a ruler or wire [14].

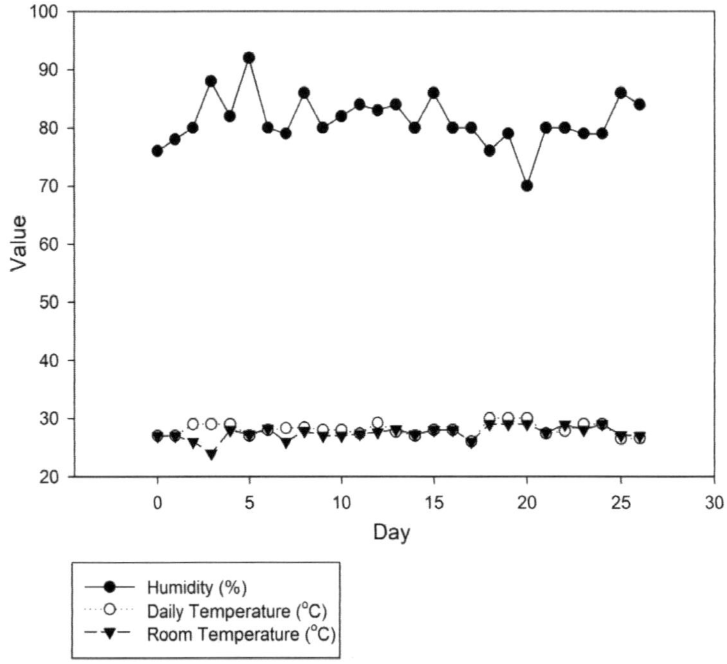

Figure 1. Humidity, daily temperature, and room temperature.

According to Nahlawi and Kodikara (2006), the bentonite used in this study has a higher water content than zeolite. Bentonite belongs to the montmorillonite group, which has an affinity for moisture. Therefore, bentonite has more water content than zeolite. The addition of zeolite in bentonite resulted in a significant increase in CIF. In this study, it was found that the rise in cracks was influenced by the addition of zeolite so that pure bentonite (B) was more suitable for use as a landfill liner compared to mixing zeolite (B2, B5, B8) because bentonite tends to have more water content and does not cause crack formation overall. Regarding the landfill liner, bentonite can be used as an alternative geomembrane replacement liner because bentonite has a lower CIF value and can expand again after going through the drying stage [15].

Temperature and humidity are factors influencing crack conditions in the sample. Air humidity is one element affecting the condition of the sample. The temperature and humidity of the air are very closely related because if the air humidity changes, and the temperature will also change. Air humidity is inversely proportional to air temperature. The higher the air temperature, the lower the air humidity. The elevated temperatures cause precipitation (condensation) of water molecules contained in the air. Thus the water content in the air decreases. The higher the air temperature, the lower the humidity, meaning humidity is inversely proportional to temperature; therefore, when the temperature rises, the humidity will decrease and vice versa. In the measurement process, heat transfer generally occurs from a measured place. What is read on the measuring device is the temperature after equality occurs. The results showed an increase in air temperature along with a decrease in humidity and vice versa. This situation happens because when elevated temperatures occur, condensation of water molecules in the air condenses, decreasing air water content. The average daily temperature ranges from 26°C to 30°C, while the average room temperature ranges from 24°C to 30.5°C. Increasing temperatures can increase the potential for crack formation because the level of soil moisture loss is high [14].

Desiccation crack testing includes water content and crack surface area testing. Testing of the water content showed that water content fluctuated, with the highest levels in the B2 sample and the lowest in the B8 sample. Increasing bentonite content in the zeolite bentonite mixture increases optimum water content, while the dry unit weight decreases [12]. If the bentonite content in the mixture increases, the optimum moisture content should also increase due to the bentonite's water-holding capability. Zeolite has considerable water content in bentonite-zeolite mixtures. This result can be seen from the increase in zeolite water content little by little until it reaches the optimum water content and then begins to decline rapidly. The limit of saturated levels of water for zeolite is 28%. The decrease in the water content of zeolite in a bentonite-zeolite mixture is relatively high. It is related to the capillarity of zeolite and chemical and electrical bonding with bentonite.

The crack intensity factor (CIF) results showed an increase in the percentage of crack intensity factors and the percentage of the zeolite mixture. The lowest crack intensity factor is in sample B, which is pure bentonite. The highest crack intensity factor is in the B8 sample. Based on this research, the sample suitable for landfill liner usage was pure bentonite because the cracks occurred less than when mixing zeolites, as the nature of zeolite being able to release H_2O faster results in composite materials drying faster, thus that more cracks arise. Consequently, leachate seeps into and contaminates the soil, so that the bentonite sample is chosen for this consideration.

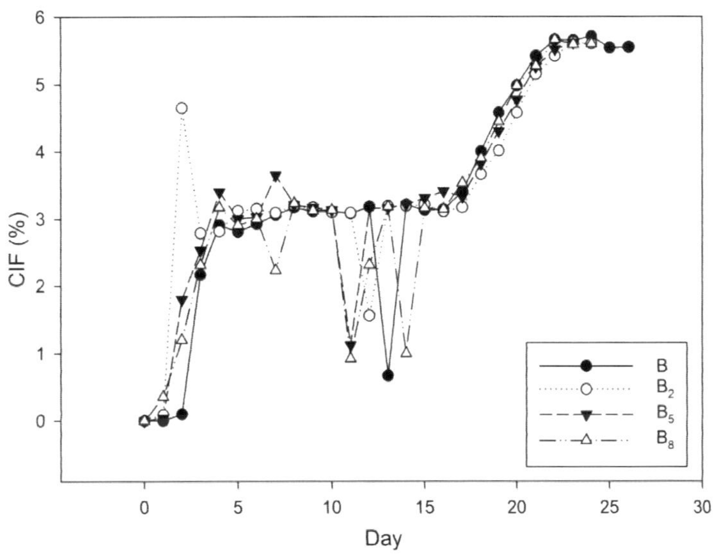

Figure 2. Crack intensity factors per day.

The percentage of CIF per day showed fluctuations in the four samples, but, overall, the percentage of CIF is highest in sample B and lowest in sample B2. Based on Tang et al.'s research, the water content during cracks ranged from 38% to 48%, and an increase in crack intensity factor (CIF) occurred with decreasing water content and pore volume. This result is consistent with the research data; the water content decreases along with the increase in CIF. Based on the research, the smallest CIF was produced in pure bentonite samples of 3.44% [16].

4. Conclusion

Composite-pure bentonite and bentonite-zeolite showed exciting results, indicating that the potential for drying cracks and migrating leachate concentrations was reversed between the two types of composites. Pure bentonite has a better performance keeping the liner layer from cracking while adding zeolite increases cracking. Bentonite embedded zeolite (BEZ) has the potential to be material for landfill liners, although new materials must be added to reduce the possibility for cracks in this mixture.

IOP Conf. Series: Earth and Environmental Science **894** (2021) 012043 doi:10.1088/1755-1315/894/1/012043

Acknowledgements
Faculty of Engineering, Universitas Diponegoro funded this research under research program *Penelitian Strategis Fakultas Teknik Universitas Diponegoro* number 195/UN7.5.3.2/HK/2021.

References

[1] Falamaki A, Eskandari M, Homaee M, and Gerashi M 2018 *KSCE Journal of Civil Engineering* **22** 3852–3859
[2] He J, Wang Y, Li Y and Ruan X C 2015 *Water Science and Engineering* **8**(2) 151-157
[3] Mishra, A K and Ravindra V 2015 *International Journal of Geosynthetics and Ground Engineering* **1**(2) 17
[4] Li J H, Li L, Chen R, and Li D Q 2016 *Eng Geol* **206** 33-41
[5] Budihardjo M A, Syafrudin S, Priyambada I B, Ramadan B S 2021 *J Ecol Eng* **22**(7) 132-141
[6] Deka A dan Sekharan S 2017 *Waste Manag Res* **35**(1) 40-46
[7] Budihardjo M A, Chegenizadeh A, and Nikraz H 2012 *International Conference on Civil and Architectural applications (ICCAA'2012) December 18-19, 2012. Phuket, Thailand*
[8] Wagner J F 2013. Clay Liners and Waste Disposal in *Developments in Clay Science*
[9] Varank G 2017 *Clean - Soil, Air, Water* **45**(1)
[10] Rayhani M H T, Yanful E K, and Fakher A 2008 *Eng Geol* **97**(1-2) 25-31
[11] Li J H and Limin M Z 2011 *Eng Geol* **123**(4) 347-358
[12] Kaya A and Durukan S 2004 *Appl Clay Sci*, **25** 83-91
[13] Budihardjo M A, Hadiwidodo M, Wardhana I W, Tuasykal M R N A, Samadikun B P, and Ramadan B S *IOP Conf Ser: Earth Environ Sci* **802** 012004
[14] Safari E, Ghazizade M J, Abduli M A, and Gatmiri B 2014 *Waste Manage* **34**(8) 1408-1415
[15] Nahlawi H and Kodikara J K 2006 *Geotech and Geol Eng* **24** 1641-1664
[16] Tang C, Cui Y, Shi B, Tang A, and Liu C 2011 *Geoderma* **166**(1) 111-118

IOP Conf. Series: Earth and Environmental Science **894** (2021) 012044 doi:10.1088/1755-1315/894/1/012044

An overview of the waste management system in Tembalang District, Semarang City

I W Wardhana[1], M A Budihardjo[1,2], T Istirokhatun[1], N Ikhlas[3], I Fadhilah[1]

[1]Department of Environmental Engineering, Diponegoro University, Semarang, Indonesia 50275
[2]Environmental Sustainability Research Group, Department of Environmental Engineering, Diponegoro University, Semarang, Indonesia 50275
[3]Department of Environmental Engineering, Institut Teknologi Pembangunan Surabaya, Surabaya, Indonesia 60186

irawanwisnuwardhana@lecturer.undip.ac.id

Abstract. It is necessary to observe and evaluate the performance of the authorized institutions, in particular, related to the sub-unit of the Environmental Agency, which manages the technical aspects (UPTD) and the five aspects of waste management to achieve better service to the people of Tembalang District. The UPTD is responsible for technical operations, including transporting domestic waste using arm roll and dump trucks, managing the cost budget plan, and supervising waste collection at waste collection site. Tembalang District has a minimum waste generation of 156 m^3/day with a generation rate of 0.814 l/person/day and a maximum of 216 m^3/day with a generation rate of 1.217 l/person/day. The implementation of the five management aspects in Tembalang District still needs to be evaluated from the operational, technical, and community participation aspects. This situation is related to the segregation and sorting of waste that has not been carried out, independent processing that has not gone well, and activities at the landfill that are not suitable. The community participation, public awareness, and concern for waste problems are still low where the independent waste management through waste treatment facility partnerships and waste banks are possible to do.

1. Introduction

Waste is the residue of human daily activities and/or natural processes in solid form [1]. Meanwhile, waste management is a systematic, comprehensive and sustainable activity that includes waste reduction and handling [2]. Waste that is allowed to accumulate and rot on the street or around the house will cause unpleasant odors, interfere with environmental aesthetics, and become a nest for disease vectors so that it can reduce the level of quality of health and public welfare [2]. Proper waste management needs to be done so that waste does not become a vector of disease and does not cause disturbance to the environment [3]. The management of urban waste in Indonesia is carried out under the supervision and responsibility of the Environmental Agency [4].

Tembalang District became the study area based on being the sub-district with the largest population based on the BPS in 2019, which has 206,271 people in the area. The Semarang City Environmental Agency assigns the operational, technical responsibility for waste management to the UPTD (Regional Technical Implementation Unit) for Hygiene in implementing solid waste management. Waste

IOP Conf. Series: Earth and Environmental Science **894** (2021) 012044 doi:10.1088/1755-1315/894/1/012044

management is carried out based on five aspects: legal, institutional, technological, financial, and community participation [5]. Based on these matters, the purpose of this work was to make an overview of the management activities, primarily technical operations under the responsibility of the UPTD, and review the suitability of the five aspects of overall waste management in Tembalang District with the established standards.

2. Methodology

This work is carried out for 30 calendar days from July 16, 2020, to August 14, 2020. The data are obtained from direct observations of waste management activities in Tembalang District and interviews with UPTD officers. Meanwhile, some data was obtained through literature review documents of the Regional Cleanliness and Waste Management UPTD Region III and the Semarang City Environmental Agency.

3. Results and discussion

3.1. Overview of waste management in the district

The services provided by the Semarang City Environmental Agency related to operational technicalities are assigned to the Regional Technical Implementation Unit (*Unit Pelayanan Teknis Daerah* - UPTD), with the service area divided into four sub-districts. The UPTD carries out technical services in Tembalang District for Cleanliness and Waste Management Region III. The UPTD Region III has the authority related to operational technology with the following details to fulfill its duties.

a. Waste transportation at waste collection site using arm roll trucks
   UPTD Region III has 6 arm roll trucks used to transport waste from waste collection site to Landfill.
b. Waste transportation at waste collection site using dump trucks
   To assist the operation of arm roll trucks, especially in cleaning up the waste scattered at the waste collection site, UPTD Region III has 1 fleet of dump trucks. The field coordinator will report waste collection site that require more priority to be cleaned to the dump officer.
c. Waste transportation using non- waste collection site dump trucks
   Non- waste collection site dump trucks are used for transporting waste sources on the road. However, because the UPTD Region III only has 1 dump truck for Tembalang District, non- waste collection site dump truck operations in the Tembalang District Area cooperate with third parties.
d. RKA proposal procedure
   The RKA is a planning and budgeting document related to the need for costs used for spending on programs and activities by the UPTD.
e. Street sweeping
   Clean forces carry out road sweeping from the Rapid Action Force (*Pasgat*) from the Semarang City Environment Service (protocol road) and clean forces from third parties in collaboration with the Semarang City Environment Service (non-protocol). A third party carries out the road sweeping in Tembalang District because it is a non-protocol road.
f. Monitoring of waste collection from waste sources to waste collection site
   Sources of domestic and non-domestic waste in Tembalang District are accommodated in the nearest waste collection site before being disposed of to the landfill. Tembalang sub-district has 25 waste collection site, including non-service (self-help/private) waste collection site, where Rowosari Sub-district does not yet have waste collection site.

3.2. Overview of the five aspects of waste management in Tembalang District

3.2.1. Legal aspect. Waste management in Semarang City, especially in Tembalang District, is carried out legally. Waste management is carried out based on applicable laws, starting from Law no. 18 of 2008, the regulation of SNI for solid waste, Permen PU No. 3 of 2013, until the Semarang City Regional

Regulation No. 6 of 2012 and Perwalkot Semarang No. 37 of 2015 as a guide to Regional Regulation No. 6 of 2012.

3.2.2. Institutional aspect. The institution authorized to handle and manage waste in Semarang is the Environmental Agency, which was established on the legal basis of Semarang Mayor Regulation No. 72 of 2016 concerning Position, Organizational Structure, Duties, and Functions, and Work Procedures of the Semarang City Environmental Agency.

Then to carry out tasks in the technical field, the Semarang City Environmental Agency is assisted by the UPTD for Cleanliness and Waste Management Region III, formed through the Semarang Mayor Regulation No. 77 of 2018.

3.2.3. Technological aspect. Tembalang District waste generation can be calculated through a calculation approach based on the size of the container at the waste collection site, which is a 6 m^3 container. Tembalang sub-district has 25 waste collection site and 36 containers, where the daily rhythm is different. From the analysis results, the minimum waste generation in Tembalang District is 156 m^3/day, and the maximum generation is 216 m^3. Then, the rate of waste generation can be calculated by dividing the waste generation per day by the total population, with population data that is known to be 191,672 people (has been reduced by Rowosari residents who are not served), and the minimum generation rate is 0.814 l/person/day, the maximum generation rate is 0.814 l/person/day 1,127 l/person/day.

Tembalang sub-district uses an individual housing pattern for the domestic sector, small offices such as government offices, and the Sendangmulyo Market. Meanwhile, the communal pattern is used in 3 other market sectors and hospitals. The suitability of the housing pattern to the existing provisions, namely SNI 19-2454-2002, the individual housing pattern of Tembalang District in terms of materials is appropriate. However, some have not been closed, and most have not been separated. Separated bins are only used in government offices, so the pattern of accommodation is not appropriate based on the provisions.

Figure 1. Example of Tembalang Subdistrict waste container.

The pattern of waste collection used in Tembalang District is an indirect individual pattern and an indirect communal pattern. In the domestic sector, waste is first collected by a collector with 1 – 3 days, transferred to a waste collection site, and then transported to the landfill. Meanwhile, in the market/hospital sector, waste from each source is first transferred to communal containers (done every day) and then transported to the landfill. The suitability of the Tembalang District waste collection pattern to the established standards is appropriate from the aspect of collection periodization but does not meet the waste separation aspect.

Before being transported to the landfill, Tembalang District uses a waste collection site per existing provisions. There are no waste collection site in Tembalang District, which have more than 200 m^2 of area. The existing condition of waste transportation in Tembalang District is quite different from the written waste collection site transportation schedule from the UPTD. This result is because of the many adjustments of the mobilization level and the truck driver's ability. Transportation that occurs in Tembalang District is carried out 3 to 5 times/day. The container has not been equipped with a leachate protective layer but has been closed using a net or the container cover.

Waste processing in Tembalang District relies on central processing at the Jatibarang landfill. This condition is because the independent processing has not been active in Tembalang District. Jatibarang landfill is a Semarang City landfill with waste processing technology in the form of utilizing methane gas into energy that can produce up to 300 kWh and waste reduction with BSF (Black Soldier Fly) technology with the ability to reduce waste to 300 kg/day. Jatibarang landfill is operated using Sanitary Landfill method with the final disposal of waste still mixed. Waste that has been disposed of leveled and stockpiled is covered with a tarp every day. It is also covered with soil every 5-7 days or when the waste has reached a height of 50 cm. Jatibarang Landfill has various facilities, including basic, environmental protection, facilities, and supporting facilities.

3.2.4. Financial aspect. As support in managing Tembalang District, there is a financial need by the management agency (Environmental Agency) through retribution from the community. The retribution is included the levy imposed on transportation waste from source to waste collection site and from waste collection site to landfill. The amount of retribution for transporting waste from the source to the waste collection site is determined by each region's respective neighborhood areas. A survey analysis was carried out on several residents of Tembalang District, and the results of transportation levies from sources to waste collection site varied, ranging from Rp. 10,000.00/month–Rp. 50,000.00/month. Meanwhile, transportation retribution from waste collection site to landfill is determined based on Perwalkot Semarang No. 18 of 2018 concerning Changes in Retribution Rates for Waste/Cleaning Services and Retribution for the Provision and Draining of Toilets. The collection of levies for transportation of waste collection site to landfill at Tembalang Sub-district houses on average has class V roads and class IV/III roads, which are Rp. 3,000.00 and Rp. 9.000,00.

3.2.5. Community aspect. The role of the community that is expected by the Environmental Agency in waste management in Tembalang District includes activities of sorting waste from the source, actively participating in paying waste management fees, actively participating in proposals/forums on the condition of the nearest waste collection site, being able to conduct intelligence on waste management towards waste management. Surrounding environment, and actively participate in waste management through waste treatment facility. From the results of the review and analysis of the participation of the Tembalang District community in waste management, the Table 1 illustrates the field conditions that occurred.

Tembalang District has a partnership in 2 waste treatment facility and 8 waste banks with existing conditions. Both waste treatment facility in Tembalang District are inactive, and 8 waste banks in Tembalang District are still active. Based on interviews, observations, and analysis, the Sri Rejeki waste bank is a sectoral waste bank at the neighbourhood level located on Jalan Baskoro RT 3 RW 7, Tembalang Village. Sri Rejeki Waste Bank was established in 2015. Activities in the Sri Rejeki waste bank are still in the form of buying and selling and administrative activities, including:

- Submission of inorganic waste (submitted at neighbourhood meeting)
- Weighing and recording
- Sales to collectors
- Utilization of money as a savings and loan cooperative
- Distribution of money and cooperative savings to customers every holiday

The facilities owned by the Sri Rejeki waste bank include land rental, scales, stationery, and waste storage inventory such as tables and cabinets.

Tabel 1. The form of community participation in Tembalang District in waste management.

Community Involvement	Existing Condition
Sorting waste from source	Not done yet, the waste is still mixed
Participate in payment of waste management retribution	The people are orderly paying the retribution
Actively participate in waste collection site aspiration forum	Actively complains about the condition of the TPS, but there is no solution, and participation in maintaining the cleanliness of the TPS is still lacking.
Participate actively in the intelligence of waste management	It has not been done. There are still many people who are not aware and care about the waste problem
Participate in the partnership of waste treatment facility and the waste bank	There are 2 waste treatment facilities and 8 waste banks in Tembalang District, but the waste treatment facilities are not active

3.3. Future research direction

It seems that waste management system in Tembalang District are well-managed by the UPTD. However, Rowosari sub-district has not been covered by the waste management system. In this case, community-based solid waste management should be prepared to ensure there is no waste left behind and unmanaged. In Indonesia, *Rukun Warga* (RW)/Neighbourhood Association (NA) is part of the hierarchy of the smallest community organization system to currently incentivize creative and innovative solid waste management activities through a savings system [6]. Therefore, waste bank activities should be carried out at a minimum, on the NA scale, to ensure that the distance between the waste bank and its customers is not prohibitive [7]. Formation and management of central and regional waste bank can help control the selling price of each solid waste piece to control price fluctuations. Increasing the number of waste bank will increase the chances of creating more zero waste communities in the city [8]. More waste banks in the area are needed to reduce the burden on landfills and reduce national GHG emissions. Waste banks management quality also needs to be improved through training, coaching, and guidance from the government and industry [9].

4. Conclusion

Based on the results of the analysis that has been done, several conclusions can be drawn. The UPTD Region III is responsible for the operational, technical implementation of waste management in Tembalang District, including transportation of land /other waste using arm roll or dump, management of financial budget drafts, road sweeping, and monitoring of waste collection at waste collection site. Tembalang District has a minimum waste generation of 156 m^3/day and a maximum generation of 216 m^3/day, with a minimum generation rate of 0.814 l/person/day and a maximum generation rate of 1.217 l/person/day. Of the five aspects, Tembalang District has fulfilled the legal, institutional, and retribution aspects well. Meanwhile, several activities such as segregation in the container, collection, and transportation of waste, promotion of independent processing, and several activities in the landfill that are not by regulations need to be evaluated to improve waste management in terms of operational, technical aspects. Community awareness and mobilization, especially activities through waste treatment facilities partnerships and waste banks, need to be carried out so that community participation can increase waste management in Tembalang District. Suggestions that the author can give regarding waste management in Tembalang District as a consideration for the Semarang Mandiri City Environment Service are: sorting waste starting from the source/individual houses to reduce the mixing of waste, provide a separate container, provide input to the Environment Agency to start collecting, moving, and

IOP Conf. Series: Earth and Environmental Science **894** (2021) 012044 doi:10.1088/1755-1315/894/1/012044

transporting based on waste material, invite the community to develop existing solid waste partnerships jointly.

Acknowledgements
The authors acknowledge Faculty of Engineering Universitas Diponegoro who funded this research program under *Penelitian Strategis Fakultas Teknik Universitas Diponegoro* number 195/UN7.5.3.2/HK/2021.

References
[1] Morone P, Koutinas A. Gathergood N, Arshadi M, Matharu A 2019 *J Clean Prod* **221** 10-16
[2] Samadikun B P, Sinttia D A B, Rezagama A, Sumiyati S, Huboyo H S, Ramadan B S, Hadiwidodo M, Nabila F 2021 *IOP Conf Ser: Earth Environ Sci* **623** 012077
[3] Abul S 2010 *Journal of Sustainable Development in Africa* **12(7)** 64-78
[4] Budihardjo M A, Wahyuningrum I F S 2018 *Matec Web of Conf* **159** 01029
[5] Abdul M, Syafrudin 2018 *Matec Web of Conf* **73** 07012
[6] Towolioe S, Permana A S, Aziz N A, Chin S H, & Pampanga D G 2016 *PLANNING MALAYSIA: Journal of the Malaysian Institute of Planners* **IV** 31–44
[7] Budihardjo M A, Wahyuningrum IFS, Muhammad F I, Pardede R 2019 *IOP Conf Ser: Earth Environ Sci* **337** 012028
[8] Zaman A U 2017 *Recycling* **2**(1) 1–15
[9] Permana A S, Towolioe S, Aziz N A, Ho C S 2015 *Habitat Int* **49** 197–205

IOP Conf. Series: Earth and Environmental Science **894** (2021) 012045 doi:10.1088/1755-1315/894/1/012045

Evaluation of flood risk as a step in determining the "Acceptable Risk" criteria. Case study: Dengkeng River, Klaten, Central Java

A Sarminingsih[1*], M Hadiwidodo[2]

[1,2]Environmental Engineering Department, Diponegoro University

aniksarminingsih@lecturer.undip.ac.id

Abstract. The planning of a flood control system in Indonesia is based on the planning criteria issued by the Ministry of Public Works. Flood control planning is based on flood discharge with a specific return period depending on the order of the river and the number of protected populations. Flood events in areas where the flood control system has been planned continue to occur almost every year, meaning that the probability of being exceeded is not as planned. This study is intended to evaluate the criteria for the magnitude of the designed flood discharge in flood control planning that considers the acceptable risk. Potential risks are evaluated against system reliability. The probability of failure of the flood control system occurs if the resistance is smaller than the load expressed as a performance function. By knowing the performance function associated with the level of flood risk, then the flood discharge can be selected with the appropriate return period according to the acceptable risk.

1. Introduction

A flood is a disaster that occurs in most of the floodplain areas. Several factors causing flooding include reduced river capacity, increased flood discharge due to changes in land use, damage to flood control buildings, and others. This event can be reduced if we understand that there are many uncertainties in hydrological analysis, which aims to determine the amount of flood discharge.

In general, there are two flood protection efforts, namely structural and non-structural approaches. Structural measures are divided into extensive and intensive measures. The extensive structural measures include reshaping the land surface, soil conservation, flow delay, and infiltration increase. The intensive measures consist of four categories: levees and dikes, water storage, increase of channel flow capacity, and floodplain polders and platforms. In planning for flood control buildings, it is necessary to have a planned flood discharge with a specific return period. The criteria for the amount of flood discharge planned for each country are generally different depending on their financial or economic capabilities. The magnitude of this flood discharge will ultimately affect the magnitude of the risk that may occur.

Some of the challenges of flood management are the increasing number of human activities, climate change, and policy changes, so to minimize losses due to flooding, it is necessary to carry out risk management because completely safe conditions are impossible to achieve. Considering the high level of losses caused by the flood disaster, strategic steps are needed to reduce the risk due to the disaster. One of them is doing flood risk mapping [1].

In determining the level of risk of an area from flooding and considering the factors of uncertainty that occur, it is also necessary to evaluate the magnitude of the planned discharge against the risks that may befall an area. Evaluation of the discharge plan is intended so that the risk level that occurs in an area is at least tolerable or even an acceptable level.

The study's purpose is to evaluate the determination of the criteria for the magnitude of the designed flood discharge in flood control planning that considers the acceptable risk. For the case study in this research, the Dengkeng river is used in Klaten Regency, Central Java Province.

2. Literature review

2.1. Flood risk

Risk is used in various contexts in a natural process perspective and social, economy, security, and environment in different definitions, understanding, and implementation. As a society, managing the uncertainty related to various aspects of the risk process is needed. Various approaches have been developed to define, analyze, and control the risk. Risk assessment is generally divided into three methods: qualitative method, semi-quantitative method, and quantitative method [2].

In this paper, risk assessment refers to the quantitative method because the assessment based on this method is about the concept and measurable. Risk assessment is generally affected by two main components, which are hazard vulnerability in which both can be defined as:

a. Flood hazard is defined as the probability of a flood's situation that can destroy the inside section. In a specific period, vulnerability means to be the consequences of the flood [3].
b. Vulnerability means loss level from a specific element that has risk or a group of elements that are produced from natural phenomenon with a certain magnitude and expressed in scale from 0 (no loss) to 1 (full loss) [4].

A complex interaction from partial factors creates a difficulty to calculate the risk level or potential damage impact correctly so that in practical application, vulnerability is represented by damage or loss curve, widely used in the insurance sector [5].

Potential or consequences of damage caused by flood can be stated as follows [6]:

$$D = V \times E \text{ temporary } V = f(H) \quad (1)$$

Where,

D is potential damage, V vulnerability; E is property value, and H is the flood depth.

2.1.1. Flood characteristics analysis. Analysis of flood characteristics was carried out to determine flood losses. The characteristic flood parameters reviewed were the inundation area, inundation depth, water velocity, and velocity x depth of inundation. Flood analysis was carried out with the help of HEC RAS 5.0.3 software with a 2-dimensional model.

2.1.2. Flood hazard analysis. Flood hazard is defined as the probability of a situation in which the flood can destroy the inside section, and in a specific time, while vulnerability means to be the consequences of the flood [7]. The risk analysis was assessed based on flood inundation modeling at the research site. Risk analysis is assisted by using a Geographic Information System to make it easier to read the information provided. In conducting a risk analysis, the reviewed things are the level of vulnerability and danger.

Based on the conceptual model of the risk process, in evaluating risk needs to consider several components [8]:

- Nature and probability of hazard (p)
- The degree of exposure of the receptors (number of people and property) to the hazard (e)
- The susceptibility of the receptor to the hazard(s)
- Receptor value (v)

So the risk can be written as follows:

Risk = (Probabilits) x (Consequence)

The level of flood hazard for each region should have different categories depending on the topography and characteristics of the existing floods in the area. Therefore the limit of the level of human hazard for Indonesia needs to be added with speed and multiplication criteria. Speed and depth [8].

Table 1. The limiting conditions of Flood Hazards for humans [8].

Parameters			Classification
D (m)	**V (m/s)**	**D x V (m2/s)**	
< 0,9	< 2,4	< 0,75	Low
0,9 < D < 1,2	2,4 < V < 3	0,75 < DV < 1,1	Moderate
> 1,2	> 3	> 1,1	Significant

2.1.3. Flood vulnerability analysis. Flood risk analysis involves flooding, flood hazard, and characteristics of elements at risk: socially or living systems, the built environment, and the natural environment. The extent of flood damage depends not only on the flood characteristics but also on the area's vulnerability to inundation.

3. Result and discussion

3.1. Evaluation of designed flood discharge

Evaluation of planned flood discharge is based on analysis of social and economic losses to determine risk criteria. In her study, Sarminingsih [8] created a risk level matrix that can intervene in the plan return period for the design so that the risk is minimally tolerable.

3.1.1. analysis of designed flood. The designed flood discharge was analyzed based on the maximum daily discharge recorded in the Paseban AWLR. The magnitude of the planned flood discharge for various return times is as shown in Figure 1.

3.1.2. Channel capacity analysis. Bank capacity analysis is carried out to determine the capacity of the river to hold water. The bankful capacity analysis was carried out on a cross-section of the Dengkeng River downstream, namely at the initial location where flooding began. Based on the bankful capacity analysis results on the cross-section of the starting point of the flood downstream of the Dengkeng River, it shows that the discharge capacity of the Dengkeng River is lower than the flow rate in each analyzed return period. The return period of bankful capacity in the initial cross-section of the flood downstream of the Dengkeng River is analyzed based on the relationship between the magnitude of the discharge and the return period so that the return period of bankful capacity is 1.4 years. An overview of the channel capacity evaluation is shown in Figure 2.

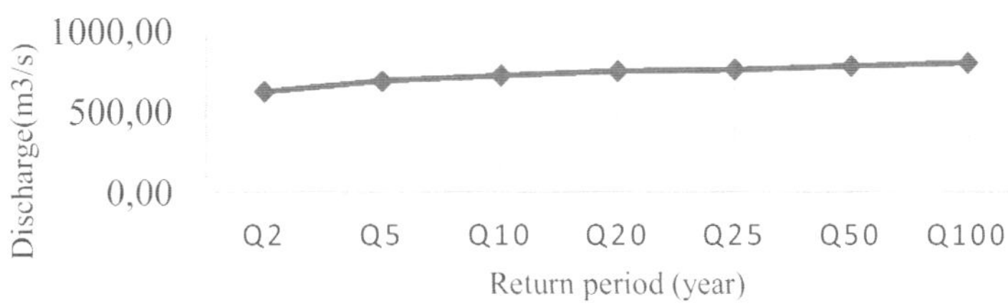

Figure 1. Flood discharge in many return periods of Dengkeng River.

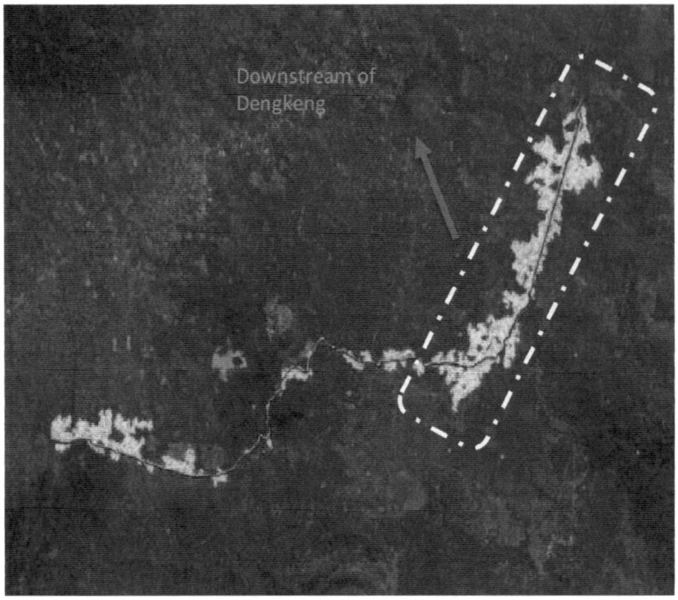

Figure 2. Evaluation location of bankfull capacity.

3.1.3. Inundation evaluation of flood plain. Based on the results of inundation analysis on the floodplain, the discharge of the 2-year return period has caused flooding in residential areas and cultivated areas. The number of the inundated area is shown in Figure 3.

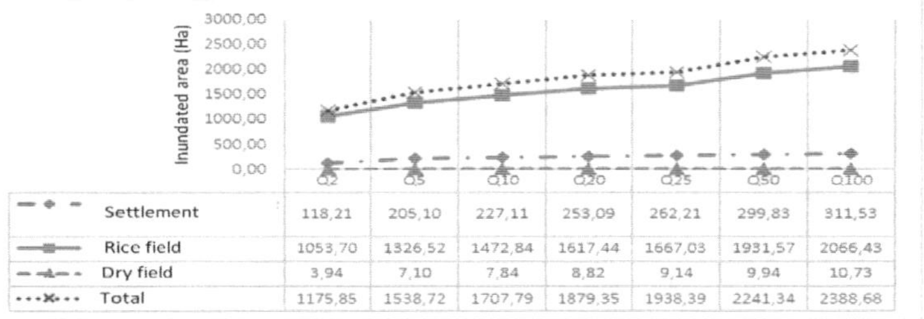

Figure 3. Flood inundation area on flood risk elements.

3.2. Hazard evaluation for humans

The classification of flood hazards for humans is based on the multiplication of flow velocity and inundation depth [8].

Table 2. Classification of flood hazards for adult humans.

Parameters			**Classification**
D (m)	**V (m/s)**	**D x V (m2/s)**	
< 0,9	< 2,4	< 0,75	Low)
0,9 < D < 1,2	2,4 < V < 3	0,75 < DV < 1,1	Moderate
> 1,2	> 3	> 1,1	Significant

Flood classification by considering depth and speed is based on the instability of people when hit by flood flows, either due to sprawling moments or shear forces. The hazard level classification is adjusted to the characteristics (height and weight) of Indonesian people in general.

Based on the analysis results based on the depth of the flood (D), a high level of flood hazard has a larger area than the level of flood hazard based on DV. So that in determining the level of flood hazard in the study area, it is determined based on the most critical level of the area, which is based on the depth of the flood (D).

Figure 4. Flood hazard map to people as a function of depth (D) for TR 2 years.

3.3. *Flood vulnerability analysis*

Based on the spatial map of the inundation depth resulting from the flood simulation and the vulnerability value of each element at risk, the vulnerability level can then be made based on the classification, as shown in Table 3. In the study area, the flood-exposed areas are settlements and activity sites grouped into the categories of settlements and paddy fields and fields/fields. Grouped in the agricultural category. The level vulnerability of elements at risk is used to calculate the potential for financial flood losses.

3.4. *Flood risk analysis*

3.4.1. Economic losses. The value of economic losses at the study site is calculated based on the level of risk elements exposed to flooding. The accuracy of the loss value is influenced by the division of the height of the inundation depth and the area of the affected area. The calculation of losses in this study is divided into height and depth for each 0.1 m interval.

In the study location, the majority of flood-affected areas are classified as residential and agricultural. In calculating this loss, no field survey was carried out for the level of losses in the field so that the unit value was used as an assumption. The value of the housing unit is assumed after depreciation of IDR 2,040,000 - per m^2, while the value of agricultural production refers to data based on the Central Statistics Agency in 2019, which is IDR 17,200,000 - per ha per growing season.

Table 3. Category vulnerability classification based on water depth (m).

Category	Residential	Industry	Agriculture	Factor Damage (%)
Very Low	<2	<2	<0.5	< 20
Low	2 - 3.5	2 - 3.5	0.5 - 1	20 - < 40
Medium	3.5 - 4.5	3.5 - 4	2-Jan	40 - < 60

Category	Residential	Industry	Agriculture	Factor Damage (%)
High	4.5 - 5.5	4 - 4.5	3-Feb	60 - < 80
Very High	> 5.5	> 4.5	> 3	80 - 100

3.4.2. Social losses. Social losses are losses to humans caused by floods. Calculation of social losses based on the components of depth and velocity of flooding by modifying the Wallingford and Cox equations [9,10].

Social loss refers to the number of people exposed to flooding. In this social loss analysis, the population is assessed by sub-district. The results of the flood inundation analysis show that there are at least ten flood-affected districts spread over three districts. Population data used data from the Central Statistics Agency in 2019.

In calculating social losses, it is also necessary to calculate regional vulnerability. The regional vulnerability parameter refers to DEFRA, as shown in Table 3 [11]. The vulnerability level for the Dengkeng watershed based on the speed of flooding has a value of 1 (low risk), based on land use, it has a value of 2 (medium risk), while based on the early warning system using existing standards which is two so that the level of regional vulnerability in the Dengkeng watershed has a value of 5.
The analysis of flood losses on humans produces predictions of the number of people injured and dead, as shown in Table 4.

Table 4. Predicting social losses using DV map delineation.

Return Period (year)	Risk To Humans		Annual Risk
	Injuries	Fatalities	Economical (x Billion Rupiah)
2	68	0	204.97
5	118	1	109.38
10	142	1	63.92
20	170	1	37.11
25	183	1	31.31
50	238	2	19.23
100	263	2	9.93

3.5. Evaluation of return period of designed flood
Risk criteria that occur in the Dengkeng watershed are assessed based on economic losses and social losses. The economic risk criteria are based on the GRDP value in the flood-affected areas, while the social risk criteria are based on the number of people exposed to floods who were injured or died. Based on the risk level matrix, losses in the Dengkeng watershed are classified as very high risk. In ensuring the security of an area, the risk level of an area is at least at a moderate level or a risk that can be tolerated. Calculation of the evaluation of the amount of the planned return period is carried out by conducting an annual risk assessment. The value of the annual loss is compared with the standard category of consequences; in this case, the focus is on the economic value. Based on the analysis, results show that the risk value can be tolerated or allowed at a probability of 0.1% - 1%. In the Dengkeng watershed, to reduce regional risk, it is necessary to design a flood design intervention with a return period of at least 50-100 years. Evaluation of the Flood Return Period in the Dengkeng Watershed as shown in Figure 4.

4. Conclusion

4.1. Conclusion

The conclusions obtained from this research are:

In general, the Dengkeng River can no longer accommodate the planned 2-year return period flood discharge. At the initial point of the flood, the Dengkeng Lower River segment has a cross-sectional capacity of 122.6 m^3/s, a 1.4-year return period discharge.

Assessment of the level of flood hazard for humans in the Dengkeng watershed is carried out spatially based on inundation depth (D) and multiplication of depth and velocity (D x V), but analysis of flood hazard level based on inundation depth (D) gives safer results than DxV .

The level of economic loss in the Dengkeng watershed is at a very high level in the 2-year plan return period, while the level of human loss is only moderate, so it can be concluded that in the 2-year return period, the risk level of the Dengkeng watershed is at a very high level.

To reduce risk in the Dengkeng watershed, it is necessary to evaluate the return period of the flood management plan to an acceptable level, namely a minimum return period of 100 years.

VH
H
M
L
VL

		Range Prob.				Tinggi (H)	
probabilitas tahunan kurang dari 10%, Tr<10 tahun	Pasti (*Low*)	<0.1	M	M	H	VH	Risk of Dengkeng
probabilitas tahunan 4% - 10% Tr : 10-25 tahun.	Sering (*Frequent*)	0.04-0.1	L	M	H	H	VH
probabilitas tahunan 2% -5 % (Tr : 2 5 dan 50 tahun)	Mungkin (*Probable*)	0.02-0.04	L	M	M	H	H
Probabilitas tahunan sekitar 1 % - 2 % (Tr: 50-100 tahun)	Kadang-kadang (*Occasional*)	0.01-0.02	L	M	M	M	H
Probabilitas tahunan sekitar 2 % - 1 % (Tr: 100-1000 tahun)	Kemungkinannya sangat kecil (*Remote*)	0.01-0.001	VL	L	L	M	M
Probabilitas tahunan <0.1%. (TR>1000 tahun).	Tidak Mungkin (*Improbable*)	<0.001	VL	VL	L	L	M
		Indicator	Tidak berarti (*Insignificant*)	Rendah (*Minor or Low*)	Sedang (*Moderate*)	Tinggi (*High/Major*)	Sangat Tinggi (*Catastrophic or Very High*)
		Safety (injury/loss of life)	Tidak ada luka/meninggal	Luka ringan <10, sedang=1	Luka ringan 10-100, sedang=1-10, berat=1 org	Luka ringan 100-1000, sedang=10-100, berat=1-10, meninggal=1 org	Luka ringan >1000, sedang=>100, berat=>10, meninggal=>1 org
		Economi (monetery losses)	Kerugian Ekonomi < 0.1 % dari PDRB	masih bisa diatasi pememerintah daerah Kab./Kota, 0.1% - 2 % dari PDRB	2% - 5 % dari PDRB	5% - 10 % dari PDRB	lebih 10 % dari PDRB
		Social&Cultural	tidak terlalu mengganggu, tidak ada pengungsian	Gangguan sistem< 24 jam	Gangguan sistem sedang, 1-7 hari	Gangguan sistem sedang, 7-14 hari, terjadi pengungsian	Gangguan sistem berdampak nasional, >30 hari, terjadi pengungsian >14 hari
		Ekologi (flora&fauna)	Kerugiankecil, lokasi terbatas	perbaikan kerusakan dalam waktu pendek	Kerusakan cukup luas	Kerusakan tersebar sangat luas, perlu waktu lama	Kerusakan tersebar sangat luas, perlu waktu lama bahkan tidak bisa diperbaiki

Figure 5. Evaluation of the flood return period in the Dengkeng Watershed.

4.2. Suggestion

Suggestions from the author for the next research are:

In this study, the study area covers the entire watershed, so for non-structural analysis of flood reduction that directly touches the community, such as determining evacuation areas and preparedness, it is necessary to study in more detailed areas, such as at the kelurahan or RW level.

In the analysis of structural flood risk reduction, such as the construction of embankments, it is necessary to take into account the maximum discharge of the return period that may occur, as an uncertainty value based on the discharge.

In a more detailed analysis of risk reduction in the Dengkeng watershed, a return period evaluation matrix needs to be developed based on more detailed conditions in the Dengkeng watershed.

IOP Conf. Series: Earth and Environmental Science **894** (2021) 012045 doi:10.1088/1755-1315/894/1/012045

References

[1] Kusuma M S B, Rahayu H P, Farid M, Adityawan M B, Setiawati T, Silasari R 2010 *J Tek Sipil* **17(2)**

[2] Westen C J V, Alkema D, Kerle N, Demen M, Kingma N C, Lubszynska M, Woldai T, Parodi G, Frigoerio S, Rusmini M 2011 *Multi-hazard risk assessment. Distance education course. Guide book* (United Nations University – ITC School on Disaster Geoinformation Management (UNU-ITC DGIM))

[3] Merz B, Thieken Ah and Gocht M 2007 *Advances in natural and technological hazards research* p 231-251 ed Begum S, Stive M and Hall J (Dordrecht, Netherlands: Springer)

[4] Sagala S A H 2006 *Analysis of Flood Physical Vulnerability in Residential Areas, Case Study: Naga City, The Phillipines. MSc Thesis* (Enschede Netherlands: ITC, International Institute for Geo-Information Science and Earth Observations)

[5] Munich Re 2005b *Natural catastrophes in 2004* p 60 (Munich Re, Munich)

[6] Langhammer J 2006 *Geografie – Sbornfk ČGS* **111(1)** 33–50

[7] Merz B, Kreibick H, Thieken A and Schmidtke R 2004 *Nat Hazardsand Earth Syst Sci* **4** 153-163

[8] Sarminingsih A, Soekarno I, Hadihardaja, Iwan K, Kusuma B, M Syahri 2014 *Int J Appl Eng Res* **9(23)** p 22921-22940

[9] Wallingford H R 2005 *Flood Risks to People, Phase 2, R&D Outputs, FD2321/TR1 The Flood Risk to People Methodology* (Defra/ Environment Agency Flood and Coastal Defence R&D Programme)

[10] Cox R J, T D Shand and M J Blacka 2010 *Australian Rainfall & Runoff, Appropriate Safety Criteria for People The Institution of Engineers* (Australia)

[11] Sayers P B, Gouldby B P, Simm J D, Meadowcroft I, Hall J 2003 *Risk, performance and uncertainty in flood and coastal defence—a review. R&D Technical Report FD2302/TR1* (Defra/ Environment Agency Flood and Coastal Defence R&D Programme)

AUTHOR INDEX

Institute of Physics
Dirac House, Temple Back
Bristol BS1 6BE UK

ISSN: 1755-1307
ISBN 978-1-7138-4876-9